MOLECULES IN LASER FIELDS

MOLECULES IN LASER FIELDS

EDITED BY

ANDRÉ D. BANDRAUK
Université de Sherbrooke
Sherbrooke, Quebec, Canada

Marcel Dekker, Inc. New York • Basel • Hong Kong

Library of Congress Cataloging-in-Publication Data

Molecules in laser fields / edited by André D. Bandrauk.
p. cm.
Includes bibliographical references and index.
ISBN 0-8247-9175-4 (alk. paper)
1. Laser Photochemistry. I. Bandrauk, André D.
QD716.L37M65 1993
539'.6–dc20 93-40776
CIP

The publisher offers discounts on this book when ordered in bulk quantities. For more information, write to Special Sales/Professional Marketing at the address below.

This book is printed on acid-free paper.

Marcel Dekker, Inc.
270 Madison Avenue, New York, New York 10016

Current printing (last digit):
10 9 8 7 6 5 4 3 2 1

PRINTED IN THE UNITED STATES OF AMERICA

Preface

Current laser technology has made such significant progress that it is now possible to tailor laser pulses to specific experimental demands. More precisely, it is possible to shape and modify pulses in a controlled manner. Furthermore, new compression techniques allow experimentalists to concentrate radiation energy to very intense levels on short time scales, approaching the time of nuclear motion (1 picosecond, 10^{-12} sec) and electronic motion (1 femtosecond, 10^{-15} sec). This new technology makes it possible therefore to envisage the study of the effect of intense laser fields on molecules, that is, one can now investigate highly nonlinear (multiphoton), nonperturbative, laser-induced optical effects in molecular systems. High intensities imply highly coherent effects since radiative transition rates can exceed intra- and intermolecular relaxation rates. Such coherence effects can also be achieved at low intensities, in the perturbative regime of laser-molecule interactions, through the coherence properties of laser light and the quantum nature of molecular processes themselves. This last observation suggests the enticing possibility of coherent control of molecular photochemical and photophysical processes by using more than one laser source to induce interfering multiphoton transitions in molecules. Experimentalists and theorists alike have at their disposal three parameters to be explored in laser-matter interaction problems: intensity, pulse shape and phase. Thus in principle, photochemical and photophysical processes need to be mapped out completely as a function of these three parameters.

In fact it is only in the last two to three years that important advances have been made in the experimental and theoretical investigation of laser radiation interaction with molecules. The advances have been made on two fronts: high intensity effects and coherent control. First, by high intensity, one means the nonlinear response of molecules to intense laser pulses such that radiative perturbations considerably modify the nuclear and electronic states of molecules embedded in such pulses. New nonlinear optical or multiphoton processes such as above-threshold ionizations (ATI) (or more appropriately, nonlinear photoelectron spectroscopy), above-threshold photodissociation (ATD), and Coulomb explosions have been discovered and characterized for simple diatomic molecules. Such highly nonperturbative effects open new challenges for the emerging new sciences of nonlinear photochemistry and photophysics. It is to be expected that this area of frontier research will expand into the realm of more complex systems such as polyatomic and condensed matter systems. Second, the subject of coherent control of molecular multiphoton photochemical and photophysical processes is arousing the interest of more and more experimentalists in view of the accessibility of an ever-improving laser technology. Thus, coherent excitations on the nuclear and electronic time scales mentioned above can now be envisaged. The idea of using pulse shaping and more than one laser to control photoexcitation pathways also shows promise as being the ultimate means of controlling photochemical reactions at the microscopic (i.e., molecular) level.

The aim of this book is therefore to bring to the photochemical, photophysical and chemical physics community new developments in multiphoton or equivalently in nonlinear molecular photochemistry and photophysics. The emphasis is on the investigation of small molecules as it is for these cases that detailed experiments are being carried out and accurate theoretical modeling is possible. The contributors to this volume are all experts in their respective areas. Each has contributed his or her state-of-the-art knowledge in areas related to molecules in laser fields.

The first chapter presents a concise, self-contained summary of quantum electrodynamics, with the use of coherent states of the electromagnetic field to bridge the quantum picture of dressed states and traditional semiclassical electrodynamics. The second, third and fourth chapters deal with simple molecules in intense electromagnetic fields, for which highly nonlinear, nonperturbative field-induced effects occur. The fifth chapter examines the properties of highly excited states of H_2, since these serve as the intermediate (doorway) states to the subsequent ionization induced by intense fields. The final three chapters have been included to herald the emergence of a new discipline, the control of molecular photochemistry and photophysics by coherent laser excitations.

This book was prepared mostly while I was a visiting professor at the Institute for Molecular Science (IMS), Okazaki, Japan. It is thus with great pleasure that I acknowledge the generous hospitality of my host, Professor Hiroki Nakamura, director of the Theoretical Department at IMS. Spontaneous discussions with Professors Keiji Morokuma and Iwao Ohmine of IMS have on many an occasion shed new light on seemingly well-known concepts. Finally, I would like to acknowledge a new Canadian venture, the Federal Centre of Excellence in Molecular and Interfacial Dynamics (CEMAID). My enthusiastic participation in CEMAID helped motivate me to undertake the editing of this book.

André D. Bandrauk

Contents

Contributors

Eric E. Aubanel, Ph.D. Postdoctoral Fellow, Faculty of Sciences, Laboratory of Theoretical Chemistry, Université de Sherbrooke, Sherbrooke, Quebec, Canada

Mark A. Banash, Ph.D.[1] Department of Chemistry and the Princeton Center for Photonics and Opto-Electronic Materials, Princeton University, Princeton, New Jersey

André D. Bandrauk, Ph.D., FRSC Professor, Faculty of Sciences, Laboratory of Theoretical Chemistry, Université de Sherbrooke, Sherbrooke, Quebec, Canada

---

[1] *Current affiliation*: Grintel Optics, Newtown, Pennsylvania

Paul Brumer, Ph.D. Professor, Chemical Physics Theory Group, Department of Chemistry, University of Toronto, Toronto, Ontario, Canada

Philip H. Bucksbaum, Ph.D. Professor, Department of Physics, University of Michigan, Ann Arbor, Michigan

P. B. Corkum, Ph.D. Steacie Institute for Molecular Sciences, National Research Council of Canada, Ottawa, Ontario, Canada

Peter Dietrich, Ph.D. Steacie Institute for Molecular Sciences, National Research Council of Canada, Ottawa, Ontario, Canada

Jean-Marc Gauthier, M.Sc.[2] Université de Sherbrooke, Sherbrooke, Quebec, Canada

Masahiro Iwai, Ph.D.[3] Research Associate, Department of Theoretical Studies, Institute for Molecular Science, Myodaiji, Okazaki, Japan

Michel Laberge, Ph.D. Physicist, CREO Products Inc., Burnaby, British Columbia, Canada

Sungyul Lee, Ph.D. Associate Professor, Department of Chemistry, College of Natural Sciences, Kyunghee University, Yongin-kun, Kyungki-Do, Korea

Hiroki Nakamura, Ph.D. Professor, Department of Theoretical Studies, Institute for Molecular Science, Myodaiji, Okazaki, Japan

Moshe Shapiro, Ph.D. Chemical Physics Department, The Weizmann Institute of Science, Rehovot, Israel

[2]*Current affiliation*: Department of Chemistry, Faculty of Sciences, Université de Montréal, Montreal, Quebec, Canada
[3] deceased

Donna T. Strickland, Ph.D.[4] Department of Chemistry, Princeton University, Princeton, New Jersey

David J. Tannor, Ph.D. Associate Professor, Department of Chemistry and Biochemistry, University of Notre Dame, South Bend, Indiana

Warren S. Warren, Ph.D. Professor, Department of Chemistry and the Princeton Center for Photonics and Opto-Electronic Materials, Princeton University, Princeton, New Jersey

Anton Zavriyev, Ph.D.[5] Graduate Student, Department of Physics, University of Michigan, Ann Arbor, Michigan

---

[4] *Current affiliation*: Steacie Institute for Molecular Sciences, National Research Council of Canada, Ottawa, Ontario, Canada

[5] *Current affiliation*: Steacie Institute for Molecular Sciences, National Research Council of Canada, Ottawa, Ontario, Canada

MOLECULES IN LASER FIELDS

1

Quantum and Semiclassical Electrodynamics

André D. Bandrauk

Université de Sherbrooke, Sherbrooke, Quebec, Canada

1.1 INTRODUCTION

The advent of new laser technology allows one to prepare intense electromagnetic pulses which when applied to atomic and molecular systems produce highly nonperturbative nonlinear effects [1]. Thus nonperturbative theoretical treatments are required to describe adequately the interaction of intense electromagnetic fields with atoms and molecules. A complete, consistent treatment requires the simultaneous description of the electromagnetic field (i.e., Maxwell's equations) coupled to the microscopic atomic and molecular systems which are described by the time-dependent Schrödinger equation. Such an approach might more appropriately be called the Maxwell-Schrödinger equation. This should allow one to treat consistently and on an equal basis the quantum behavior of atoms and molecules via their proper Schrödinger equation coupled to the electromagnetic fields, either classical or quantum [2–7]. The effect of the microscopic molecular systems on the propagation of these electromagnetic fields, is then described by Maxwell's equations [8–12].

This first chapter will summarize the relevant concepts and equations of classical electromagnetic theory in order to arrive at a convenient quantization scheme to generate quantum electrodynamics. In turn, in the high intensity limit where wave interference effects are predominant, one must recover the classical electrodynamic limit, as opposed to the limit of pure quantum (i.e., *photon* states) where the wave nature of the electromagnetic field is suppressed and the photon is a relevant concept. The conceptual link between the two regimes, for the electromagnetic field is the *coherent* state first introduced by Glauber [13]. This link as we will show in this chapter will allow us to justify the use of a semiclassical quantum picture of molecule-electromagnetic field interactions, the *dressed state* approach.

An important concept in electromagnetic theory is that of *gauge transformations*, which allows one by a judicious choice of coordinates, to simplify the matter-field equations, which in the quantum case are the Maxwell-Schrödinger equations. In the classical theory these are appropriately obtained from the total Lagrangean of the system [4, 7], whereas in the quantum theory these transformations become unitary transformations of the operators (Heisenberg representation) or of the states or functions (Schrödinger representation). Although the quantum transition amplitudes or probabilities between the field-matter states must be independent of gauge (i.e., gauge invariance is a necessary condition for a consistent theory), different gauges present different physical representations. In this chapter we will focus on three different gauges or representations, the Coulomb or radiation gauge (minimal coupling), the electric field gauge (Goppert-Mayer transformation), and the Bloch-Nordsieck representation (space translation) [14]. Each representation has been used with varying degrees of success in atoms [5, 6, 15], and the last has been examined recently for the molecular case [16]. We will focus on the problem of gauge invariance of transition amplitudes and on exact model solutions demonstrating the important effect of ponderomotive potentials at high intensities. Finally we will address the question of separability of equations of motion in an intense electromagnetic field. These and other relevant aspects of field-particle interaction will occur throughout this volume.

1.2 MAXWELL'S EQUATIONS

These fundamental equations of classical electromagnetic theory describe the motion of charged particles under the influence of the electric field $\mathbf{E}(\mathbf{r},t)$ and the magnetic field $\mathbf{B}(\mathbf{r},t)$. The particle motion is described through their density $\rho(\mathbf{r},t)$ and their currents $\mathbf{j}(\mathbf{r},t)$

$$\rho(\mathbf{r},T) = \sum_i q_i\, \delta(\mathbf{r} - \mathbf{r}_i(T)) \;;\mathbf{j}(\mathbf{r},t) = \sum_i q_i\, \dot{\mathbf{r}}_i(t)\, \delta(\mathbf{r} - \mathbf{r}_i(t)) \tag{1.1}$$

for particles of charge q_i, position $\mathbf{r}_i(t)$, and velocity $\dot{\mathbf{r}}_i(t)$. The basic field equations form coupled equations relating the fields $\mathbf{E}$ and $\mathbf{B}$ to ρ and $\mathbf{j}$, (c = velocity of light)

$$\text{a): } \nabla \cdot \mathbf{E}(\mathbf{r},T) = \rho(\mathbf{r},t); \qquad \text{b) } \nabla \cdot \mathbf{B}(\mathbf{r},t) = 0 \tag{1.2}$$

$$\text{a): } \nabla \times \mathbf{E}(\mathbf{r},t) = -\frac{1}{c}\frac{\partial}{\partial t}\mathbf{B}(\mathbf{r},t); \quad \nabla \times \mathbf{B}(\mathbf{r},t) = \frac{1}{c}\frac{\partial \mathbf{E}(\mathbf{r},t)}{\partial t} + \frac{\dot{\mathbf{j}}}{c}(\mathbf{r},t) \tag{1.3}$$

These equations are supplemented by the particle equations of motion (Lorentz equations) describing the particle dynamics

$$m_i \frac{d^2\mathbf{r}_i(t)}{dt^2} = q_i[\mathbf{E}_i(\mathbf{r}_i(t),t) + \frac{\dot{\mathbf{r}}_i(t)}{c} \times \mathbf{B}(\mathbf{r}_i(t),t)] \tag{1.4}$$

Starting from a Lagrangean field theoretic description, one can define rigorously a total Hamiltonian for the field-particle system [4, 7]

$$\mathbf{H} = \frac{1}{2}\sum_1 m_i \dot{\mathbf{r}}_i^2(t) + \frac{1}{2}\int d^3r[\mathbf{E}^2(\mathbf{r},t) + \mathbf{B}^2(\mathbf{r},t)] \tag{1.5}$$

$\mathbf{H}$ is the total energy of the global system, field + particle, and can be shown to be a constant of the motion (i.e., $\mathbf{H}$ is independent of t for the total system).

Equations 1.2 and 1.3 can be written in terms of potentials, which is the starting point of Lagrangean theories

$$\mathbf{E}(\mathbf{r},t) = -\frac{1}{c}\frac{\partial \mathbf{A}(\mathbf{r},t)}{\partial t} - \nabla U(\mathbf{r},t); \quad \mathbf{B}(\mathbf{r},t) = \nabla \times \mathbf{A}(\mathbf{r},t) \tag{1.6}$$

where **A**(**r**,t) is a vector field called the *vector* potential, and U is a scalar potential. Thus since $\nabla \cdot \nabla \times \mathbf{A} = \nabla \times \nabla U = 0$ (because ∇ is $\perp$ to $\nabla \times \mathbf{A}$; $\nabla \parallel \nabla$), then Equations 1.2b and 1.3a are automatically satisfied. Substituting Equation 1.6 into Equations 1.2a and 1.3b gives equations for **A** and U (using $\nabla \times \nabla \times \mathbf{A} = \nabla(\nabla \cdot \mathbf{A}) - \nabla^2 \mathbf{A}$)

$$\nabla U(\mathbf{r},t) = -\rho(\mathbf{r},t) - \frac{\nabla}{c} \cdot \frac{\partial \mathbf{A}}{\partial t}(\mathbf{r},t) \tag{1.7}$$

$$\Box\, \mathbf{A}(\mathbf{r},t) = \frac{\mathbf{j}}{c}(\mathbf{r},t) - \nabla \left[\nabla \cdot \mathbf{A}(\mathbf{r},t) + \frac{1}{c}\frac{\partial U}{\partial t}(\mathbf{r},t)\right] \tag{1.8}$$

where Δ is the Laplacian:

$$\sum_{i=1}^{3} \frac{\partial^2}{\partial x_i^2}$$

$\Box$ is the d'Alembertian

$$\frac{1}{c^2}\frac{\partial^2}{\partial t^2} - \Delta$$

Upon examining Equation 1.6, one observes readily that **E** and **B** are invariant under the gauge transformation

$$\mathbf{A}'(\mathbf{r},t) = \mathbf{A}(\mathbf{r},t) + \nabla F(\mathbf{r},t); \quad U'(\mathbf{r}) = U(\mathbf{r},t) - \frac{1}{c}\frac{\partial}{\partial t} F(\mathbf{r},t) \tag{1.9}$$

where F(**r**,t) is a scalar field and is a function of **r** and t only. Thus the potentials are not unique since the same physical fields **E** and **B** can be obtained with different potentials **A** and U (this is reminiscent of classical mechanics where adding an arbitrary constant to any potential does not change the force, and hence the equations of motion). This nonuniqueness in the electromagnetic potentials can be removed by the choice of a gauge condition which fixes $\nabla \cdot \mathbf{A}$ (note that $\nabla \times \mathbf{A}$ already determines **B**; Equation 1.6). The two most commonly used gauges are the Lorentz gauge which gives rise to relativistically covariant equations and the Coulomb gauge which results in noncovariant equations but is most convenient for nonrelativistic electrodynamics. The Lorentz gauge condition corresponds to setting the divergence of the 4-vector (**A**,U) to zero (i.e., the parenthesis in the right hand side of Equation 1.8).

The Coulomb (or radiation) gauge is simply defined by the gauge condition $\nabla \cdot \mathbf{A}(\mathbf{r},t) = 0$, so that Equations 1.7–1.8 become

$$\Delta U(\mathbf{r},t) = -\rho(\mathbf{r},t); \quad \Box \mathbf{A}(\mathbf{r},t) = \frac{\mathbf{j}}{c}(\mathbf{r},t) - \frac{\nabla}{c}\frac{\partial}{\partial t} U(\mathbf{r},t) \tag{1.10}$$

These are second-order partial differential equations which are coupled together, the solutions of which then furnish the physical fields **E** and **B** defined in Equation 1.6.

The Maxwell Equations 1.2–1.3 which embody the known laws of electrodynamics (Coulomb, Faraday, Biot-Savard, and Ampere laws) imply certain directional properties such as longitudinal and transversal. These are best seen by following the prescription of Cohen-Tannoudji et al. [7], namely by examining these equations in momentum (reciprocal) space. This is advantageous since the field-free solution of (Equation 1.10) is $\mathbf{A}_0(\mathbf{r},t) = \mathbf{A}_0 \exp i\,(\mathbf{k} \cdot \mathbf{r} = \omega t)$, with $\omega = ck$. Thus let $E(\mathbf{k},t)$ be the Fourier spatial transform of $\mathbf{E}(\mathbf{r},t)$ as defined by the usual relations

$$\mathbf{E}(\mathbf{k},t) = (2\pi)^{-3/2} \int d^3r\, \mathbf{E}(\mathbf{r},t)\, e^{-i\mathbf{k}\cdot\mathbf{r}}; \quad \mathbf{E}(\mathbf{r},t) = (2\pi)^{-3/2} \int d^3k\, \mathbf{E}(\mathbf{k},t) e^{i\mathbf{k}\cdot\mathbf{r}} \tag{1.11}$$

Since $\mathbf{E}(r,t)$ is real, therefore $\mathbf{E}^*(\mathbf{k},t) = \mathbf{E}(-\mathbf{k},\mathbf{t})$ where $\mathbf{E}^*$ is the complex conjugate. From Equation 1.11, two essential identities follow

$$\int d^3r F^*(\mathbf{r}) G(\mathbf{r}) = \int d^3k F^*(\mathbf{k}) G(\mathbf{k});$$

$$\int d^3r' F(\mathbf{r}') G(\mathbf{r} - \mathbf{r}') = \int d^3k F(\mathbf{k}) G(\mathbf{k})\, e^{+i\mathbf{k}\cdot\mathbf{r}} \tag{1.12}$$

where $F(\mathbf{k})$ and $G(\mathbf{k})$ are Fourier transforms of $F(\mathbf{r})$ and $G(\mathbf{r})$. $\mathbf{k}$ is the propagation vector direction for the free waves $e^{i\mathbf{k}\cdot\mathbf{r}}$.

Returning now to the Maxwell Equations 1.2–1.3, and invoking the canonical equivalence of ∇_r to **ik**, one obtains the equivalent momentum space equations

$$\text{a): } \mathbf{ik}\cdot\mathbf{E}(\mathbf{k},t)=\rho(\mathbf{k},t); \qquad \text{b): } \mathbf{ik}\cdot\mathbf{B}(\mathbf{k},t)=0 \tag{1.13}$$

$$\text{a): } \mathbf{ik}\times\mathbf{E}(\mathbf{k},t)=\frac{\dot{\mathbf{B}}(\mathbf{k},t)}{c}; \qquad \text{b): } \mathbf{ik}\times\mathbf{B}(\mathbf{k},t)=\frac{\dot{\mathbf{E}}(\mathbf{k},t)}{c}+\frac{\mathbf{j}(\mathbf{k},t)}{c} \tag{1.14}$$

where $\dot{\mathbf{E}}, \dot{\mathbf{B}}$ are time derivatives.

Maxwell's equations in **k** space are now ordinary differential equations as opposed to partial differential equations in **r** space. The relationship between the fields and potentials, Equation 1.6, now become simpler

$$\mathbf{E}(\mathbf{k},t)=-\frac{\dot{\mathbf{A}}}{c}(\mathbf{k},t)-\mathbf{ik}\mathrm{U}(\mathbf{k},t); \qquad \mathbf{B}(k,t)=\mathbf{ik}\times\mathbf{A}(\mathbf{k},t) \tag{1.15}$$

Keeping in mind that the Coulomb gauge condition in both respective spaces is

$$\nabla\cdot\mathbf{A}(\mathbf{r},t)=\mathbf{ik}\cdot\mathbf{A}(\mathbf{k},t)=0 \tag{1.16}$$

then one sees that **A** is purely *transversal* (i.e., $\mathbf{a}_{\parallel}=0$), with respect to the propagation direction **k**. Thus from Equations 1.13–1.15 one obtains the longitudinal (||) and transversal ($\perp$) components of the field with respect to **k** in **r** space

$$\mathbf{E}_{\perp}=-\frac{1}{c}\dot{\mathbf{A}}_{\perp}, \quad \mathbf{E}_{\parallel}=-\nabla\mathrm{U} \tag{1.17}$$

$$\mathbf{B}_{\perp}=\nabla\times\mathbf{a}_{\perp}, \quad \mathbf{B}_{\parallel}=0 \tag{1.18}$$

It follows that the longitudinal and transverse parts of **E** are associated in the Coulomb gauge with the scalar field U and the vector field **A** respectively. Since ∇F (Equation 1.9) is longitudinal, then by virtue of Equation 1.16, $\mathbf{A}_{\perp}$ is gauge *invariant* in the Coulomb gauge but $\mathbf{A}_{\parallel}$ is not.

1.3 HAMILTONIAN OF FIELD-PARTICLE SYSTEMS

From Equation 1.5 one now calculates the total energy of the electromagnetic field in terms of longitudinal (||) and transversal components ($\perp$). Thus using the momentum space representation and the orthogonality relation $\mathbf{E}_{\|} \cdot \mathbf{E}_{\perp} = 0$, one obtains

$$\int d^3r\, \mathbf{E}^2 = \int d^3k\, [|\mathbf{E}_{\|}(\mathbf{k},t)|^2 + |\mathbf{E}_{\perp}(\mathbf{k},t)|^2] \tag{1.19}$$

so that the total transversal energy is (since $\mathbf{B} = \mathbf{B}_{\perp}$)

$$\mathbf{H}_{\perp} = \frac{1}{2}\int d^3r\, [\mathbf{E}_{\perp}^2(\mathbf{r},t) + \mathbf{B}^2(\mathbf{r},t)] \tag{1.20}$$

and the longitudinal energy is

$$\mathbf{H}_{\|} = \frac{1}{2}\int d^3k\, |\mathbf{E}_{\|}(\mathbf{k},t)|^2 = \frac{1}{2}\int d^3k \left[\rho^*(\mathbf{k},t)\frac{\rho(\mathbf{k},t)}{k^2}\right]$$

$$= \frac{1}{8\pi}\int d^3r d^3r' \frac{\rho(\mathbf{r},t)\,\rho(\mathbf{r}',t)}{|\mathbf{r}-\mathbf{r}'|} = V_C \tag{1.21}$$

For this last calculation we have used the relation between the longitudinal electric field $\mathbf{E}(\mathbf{k},t)$ and the charge distribution $\rho(\mathbf{k},t)$ (Equation 1.13a)

$$\mathbf{E}_{\|}(\mathbf{k},t) = -i\rho(\mathbf{k},t)\frac{\mathbf{k}}{k^2} \tag{1.22}$$

and also the Coulomb Fourier transform $1/(4\pi\mathbf{r}) \to 1/(2\pi^{3/2}\,k^2)$ (Equation 1.12). Thus $\mathbf{H}_{\|}$ is the electrostatic energy of the system of charges, V_C, so that finally one has the total energy of the field-particle system in the Coulomb gauge

$$\mathbf{H}_C = \frac{1}{2}\sum m_i\dot{\mathbf{r}}_i^2(t) + V_C + \mathbf{H}_{\perp} \tag{1.23}$$

The total momentum of the particle-field system can be obtained from the rigorous Lagrangean formation as the sum of the particle mechanical momentum (mv) and the field Poynting vector [4, 7]

$$\mathbf{P} = \sum_i m_i \dot{\mathbf{r}}_i(t) + \frac{1}{c} \int d^3r\, [\mathbf{E}(\mathbf{r},t) \times \mathbf{B}(\mathbf{r},t)] \tag{1.24}$$

This can also be shown to be a constant of motion of the total system [7]. Thus the longitudinal field momentum is defined as

$$\mathbf{P}_{\|} = \frac{1}{c} \int d^3r\, [\mathbf{E}_{\|}(\mathbf{r},t) \times \mathbf{B}(\mathbf{r},t)] = \frac{1}{c} \int d^3k\, [\mathbf{E}_{\|}^*(\mathbf{k},t) \times \mathbf{B}(\mathbf{k},t)] \tag{1.25}$$

Using Equation 1.22 for $\mathbf{E}_{\|}$, Equation 1.15 for $\mathbf{B}$ and $\mathbf{A}$, and the vector identity $\mathbf{a} \times (\mathbf{b} \times \mathbf{c}) = (\mathbf{a} \cdot \mathbf{c})\mathbf{b} - (\mathbf{a} \cdot \mathbf{b})\mathbf{c}$, one can rewrite Equation 1.25 in terms of ρ and $\mathbf{A}$ as

$$\mathbf{P}_{\|} = \frac{i}{c} \int d^3k\, \frac{\rho^*}{k^2}\, \mathbf{k} \times (i\mathbf{k} \times \mathbf{A}) = \frac{1}{c} \int d^3k\, \rho^*\, [\mathbf{A} - \hat{\mathbf{k}}(\hat{\mathbf{k}} \cdot \mathbf{A})] \tag{1.26}$$

where k is the unit vector in the direction of propagation. Since $\mathbf{A}$ is transversal, $k \cdot \mathbf{A} = 0$, by transforming back to coordinate space, one obtains

$$\mathbf{P}_{\|} = \frac{1}{c} \int d^3r\, \rho\, \mathbf{a}_{\perp} = \sum_i q_i \cdot \mathbf{A}_{\perp}(\mathbf{r},t) \tag{1.27}$$

where Equation 1.1 has been used for ρ to get the final answer.

Defining now a new momentum $\mathbf{p}_i$ for each particle as

$$\mathbf{p}_i = m_i \dot{\mathbf{r}}_i(t) + \frac{q_i}{c} \mathbf{A}_{\perp}(\mathbf{r}_i,t) \tag{1.28}$$

then one can describe the total momentum of the field-particle system as

$$\mathbf{P} = \sum_i \mathbf{p}_i(\mathbf{r}_i,t) + \mathbf{P}_{\perp} \tag{1.29}$$

where $\mathbf{P}_\perp$ is a transversal component

$$\mathbf{P}_\perp = \frac{1}{c}\int d^3r\,(\mathbf{E}_\perp \times \mathbf{B}) \tag{1.30}$$

to be defined later more explicitly (see Equation 1.42).

Rigorous derivation from the Lagrangean formulation (field theoretic) [4, 7], of the field-particle system identifies $\mathbf{p}_i(\mathbf{r}_i,t)$ (Equation 1.28) as the canonical momentum conjugate to the particle coordinate $\mathbf{r}_i$ (i.e., $\mathbf{p}_i$ is the generalized momentum). This is why one associates $q_i\mathbf{A}_\perp$ as the longitudinal momentum although it depends on a transverse field $\mathbf{A} = \mathbf{A}_\perp$. Thus in the Coulomb gauge, the difference between the *conjugate* momentum $\mathbf{p}_i$ and the *mechanical* momentum $m_i\dot{\mathbf{r}}_i$ is the longitudinal momentum of the field for each particle, $(q_i/c)\,\mathbf{A}(\mathbf{r}_i,t)$. We conclude therefore by rewriting the total Hamiltonian of the field-particle system in the Coulomb-gauge for which $\mathbf{A}_\perp = \mathbf{A}$, $\nabla \cdot \mathbf{A} = 0$, as

$$\mathbf{H}_C = \frac{1}{2}\sum m_i\left[\mathbf{p}_i(\mathbf{r}_i,t) - \frac{q_i}{c}\,\mathbf{A}(\mathbf{r}_i,t)\right]^2 + V_C + \mathbf{H}_\perp \tag{1.31}$$

The first term in brackets is the kinetic energy of the particle $(m_i\dot{\mathbf{r}}_i^2/2)$, the second the static Coulomb interaction, and the third term is the energy of the transverse electromagnetic field. Equation 1.31 is the fundamental input for the time-dependent Schrödinger equation describing the total system (field-particle) as a quantum system. The Hamiltonian (Equation 1.31) is called the *minimal* coupling Hamiltonian as it can be derived consistently from a Lagrangean involving only the dynamical variables $\mathbf{r}_i$, $\mathbf{A}$ and their velocities $\dot{\mathbf{r}}_i$, $\dot{\mathbf{A}}$ [4, 7], where $-\dot{\mathbf{A}}/c$ is the transverse electric field $\mathbf{E}_\perp$ (see Equation 1.17) in the Coulomb gauge.

1.4 TRANSVERSE FIELDS

Equation 1.31 shows that the time-dependent electromagnetic fields $\mathbf{E}$ and $\mathbf{B}$ enter the Coulomb gauge Hamiltonian as transverse fields (see Equations 1.17–1.18) and that the energy of the longitudinal field $\mathbf{E}_\parallel$ has been transformed into the static Coulomb potential V_C. We will now focus on

Equation 1.14 that gives the time-dependence of the transverse fields, which we rewrite (using the relation $\mathbf{k} \cdot \mathbf{E} \equiv 0$) as

$$\dot{\mathbf{E}} = \mathrm{i}c\mathbf{k} \times \mathbf{B} - \mathbf{j}; \quad \frac{\mathbf{k}}{\mathbf{c}} \times \dot{\mathbf{B}} = \mathrm{i}k^2\mathbf{E} \tag{1.32}$$

In Equation 1.32 all variables are now *transverse* (i.e., $\mathbf{E} = \mathbf{E}_\perp$, $\mathbf{B} = \mathbf{B}_\perp$, $\mathbf{j} = \mathbf{j}_\perp$. This means that the longitudinal part of Equation 1.14b, $\mathbf{E}_\| + \mathbf{j}_\| = 0$, is equivalent to the continuity equation $\rho + \mathbf{ik} \cdot \mathbf{j} = 0$, since $\mathbf{k} \cdot \mathbf{j}_\| = \mathbf{k} \cdot \mathbf{j}$ and we have used Equation 1.22 for $\rho(\mathbf{k},t)$). For zero currents (static charges), $\mathbf{j} = \mathbf{j}_\perp = 0$, one can reexpress Equation 1.32 as

$$\frac{\partial}{\partial t}(\mathbf{E} \pm \hat{\mathbf{k}} \times \mathbf{B}) = \pm\, \mathrm{i}\omega\, (\mathbf{E} \pm \hat{\mathbf{k}} \times \mathbf{B}) \tag{1.33}$$

where

$$\omega = ck, \quad \hat{\mathbf{k}} = \frac{\mathbf{k}}{k} \tag{1.34}$$

Let us now define a new composite field variable, as suggested by Equation 1.33

$$\mathbf{a}(\mathbf{k},t) = -\frac{\mathrm{i}}{2}\,[\mathbf{E}(\mathbf{k},t) - \hat{\mathbf{k}} \times \mathbf{B}(\mathbf{k},t)] \tag{1.35}$$

From the reality of $\mathbf{E}$ and $\mathbf{B}$ and Equation 1.12 relating the Fourier transforms, one derives

$$\mathbf{a}^*(-\mathbf{k},t) = +\frac{\mathrm{i}}{2}\,[\mathbf{E}(\mathbf{k},t) + \hat{\mathbf{k}} \times \mathbf{B}(\mathbf{k},t)] \tag{1.36}$$

from which follows new expressions for the transverse fields as linear combinations of the new variables $\mathbf{a}$ and $\mathbf{a}^*$

$$\mathbf{E}(\mathbf{k},t) = \mathrm{i}[\mathbf{a}(\mathbf{k},t) - \mathbf{a}^*(-\mathbf{k},t)]; \quad \mathbf{B}(\mathbf{k},t) = \mathrm{i}\hat{\mathbf{k}} \times [\mathbf{a}(k,t) + \mathbf{a}^*(-k,t)] \tag{1.37}$$

where we have used $\hat{\mathbf{k}} \times \hat{\mathbf{k}} \times \mathbf{B} = -\mathbf{B}$.

The Maxwell equations 1.32–1.33 now reduce simply to

$$\dot{\mathbf{a}}(\mathbf{k},t) + i\omega\mathbf{a}(\mathbf{k},t) = i\frac{\mathbf{j}}{2}(\mathbf{k},t) \tag{1.38}$$

For the case of a free field, $\mathbf{j} = 0$, the solution (Equation 1.38) is a pure harmonic oscillation (i.e., a normal mode of the free field). Since the current depends on the velocity (Equation 1.1), adding the particle equation of motion (Equation 1.4) results in coupled sets of Equations to be solved for the field variables $\mathbf{a}$ and the particle variables. $\mathbf{a}$ is a transverse vector field since $\mathbf{E}$ and $\mathbf{B}$ are both transverse also (i.e., it is orthogonal to the propagation vector $\mathbf{k}$ and is thus two-dimensional). This means that there are two orthogonal directions $\lambda = 1,2$ such that one can define two unit vectors $\hat{\varepsilon}_{k\lambda}$, where $\mathbf{k} \cdot \hat{\varepsilon}_{k\lambda} = 0$, in a plane normal to $\mathbf{k}$. One can thus write

$$\mathbf{a}(\mathbf{k},t) = \sum_{\lambda=1}^{2} \hat{\varepsilon}_{k\lambda}\, \mathbf{a}_\lambda(\mathbf{k},t) \tag{1.39}$$

We will now proceed to express all transverse field properties in terms of these $\mathbf{a}_\lambda$'s. In view of their analogy to normal modes this will lead us then to direct quantization of the electromagnetic field.

The total transverse field energy is obtained from Equations 1.19–1.20. Thus substituting the values from Equation 1.37 and using for $\mathbf{B}$ the relation $(\hat{\mathbf{k}} \times \hat{\varepsilon}_{k\lambda}) \cdot (\hat{\mathbf{k}} \times \hat{\varepsilon}_{k\lambda'}) = \delta_{\lambda\lambda'}$, gives

$$\mathbf{H}_\perp = \int d^3k \sum_{\lambda=1} [\mathbf{a}_\lambda^*(\mathbf{k},t)\, \mathbf{a}_\lambda(\mathbf{k},t) + \mathbf{a}_\lambda(-\mathbf{k},t)\, \mathbf{a}_\lambda^*(-\mathbf{k},t)] \tag{1.40}$$

or using the relation $\mathbf{a}(-\mathbf{k},t) = \mathbf{a}^*(\mathbf{k},t)$, $\mathbf{a}^*(-\mathbf{k},t) = \mathbf{a}(\mathbf{k},t)$ (see Equation 1.11), one obtains a sum of energies for the normal modes of the field.

$$\mathbf{H}_\perp = \int d^3\mathbf{k} \sum_{\lambda} [\mathbf{a}_\lambda^*(\mathbf{k},t)\, \mathbf{a}_\lambda(\mathbf{k},t) + \mathbf{a}_\lambda(\mathbf{k},t)\, \mathbf{a}_\lambda^*(\mathbf{k},t)] \tag{1.41}$$

One can also use the normal modes $\mathbf{a}$ to calculate the transverse momentum $\mathbf{P}$ from Equation (1.30), which gives (using $\hat{\varepsilon}_{k\lambda} \cdot \hat{\mathbf{k}} = 0$ from the transversality condition)

$$\mathbf{P}_\perp = \int d^3k \frac{\hat{k}}{c} \sum_{\lambda=1} [\mathbf{a}^*_\lambda(\mathbf{k},t)\, \mathbf{a}_\lambda(\mathbf{k},t) + \mathbf{a}_\lambda(\mathbf{k},t)\, \mathbf{a}^*_\lambda(\mathbf{k},t)] \tag{1.42}$$

This momentum is clearly to be identified with the momentum of the field and therefore is called the transverse momentum (Equation 1.30).

Using the Fourier transforming Equation (1.37) and using the fact that the free field solution (1.38) can be written as $\mathbf{a}(\mathbf{k},t) = \mathbf{a}(\mathbf{k})e^{(-i\omega t)}$, one can obtain an expansion for $\mathbf{E}(\mathbf{r},t)$ in travelling plane waves

$$\mathbf{E}(\mathbf{r},t) = i \int \frac{d^3k}{(2\pi)^{3/2}} \sum_{\lambda=1}^{2} \hat{\varepsilon}_{k\lambda}[\mathbf{a}_\lambda(\mathbf{k},t)e^{i(\mathbf{k}\cdot\mathbf{r}-\omega t)} - \text{c.c.}] \tag{1.43}$$

One can proceed further and express the transverse vector potential $\mathbf{A}(\mathbf{r},t)$ also in terms of the normal modes $\mathbf{a}(\mathbf{k},t)$. Thus using Equation 1.17 one obtains

$$\mathbf{A}(\mathbf{r},t) = \int \frac{d^3k}{(2\pi)^{3/2}} \sum_{\lambda} \hat{\varepsilon}_{k\lambda} \frac{c}{\omega} [\mathbf{a}_\lambda(\mathbf{k})e^{i(\mathbf{k}\cdot\mathbf{r}-\omega)} + \text{c.c.}] \tag{1.44}$$

This is transverse since $\hat{\varepsilon} \perp \mathbf{k}$. Fourier transforming back to k space, one gets

$$\mathbf{A}(\mathbf{k},t) = \frac{c}{\omega} [\mathbf{a}(\mathbf{k},t) + \mathbf{a}^*(-\mathbf{k},t)] \tag{1.45}$$

Using the normal mode Equation 1.37 for $\mathbf{E}(\mathbf{k},t)$ combined with Equation 1.45 one obtains new expressions for the normal modes $\mathbf{a}$ in terms of the transverse vector potential $\mathbf{A}(\mathbf{k},t)$ and the transverse electric field $\mathbf{E}(\mathbf{k},t)$

$$\mathbf{a}(\mathbf{k},t) = \frac{1}{2}\left[\frac{\omega}{c}\mathbf{A}(\mathbf{k},t) - i\mathbf{E}(\mathbf{k},t)\right] \tag{1.46}$$

Since $\mathbf{E} = -(1/c)\,\dot{\mathbf{A}}$ (Equation 1.17), the normal mode (Equation 1.46) is reminiscent of a harmonic oscillator normal coordinate ($\omega x - ip$), so that $\mathbf{A}/c$ and $\mathbf{E}$ can be considered as canonical conjugate variables. This is in fact a rigorous result obtained from a full Lagrangean approach [4, 7].

It is now convenient to pass to a discrete representation of the momentum variables k and replace Fourier integrals (Equation 1.11) by Fourier series, corresponding to a finite volume L^3 for the electromagnetic field. The correspondence between the continuous and discrete representation is obtained by the following transformation

$$\int d^3k\, f(k) \leftrightarrow \sum_{i=1}^{\infty} \left(\frac{2\pi}{L}\right)^3 f(k_i) \tag{1.47}$$

This discrete representation also allows us to fix normalizations for all the fields encountered so far in order to correspond to the well-known quantum results for the harmonic oscillator. We therefore use the following expansions in terms of the new discrete classical variables

$$\mathbf{a}_j(t) = \frac{(2\pi/L)^{3/2}}{(\hbar\omega_j)^{1/2}}\, \mathbf{a}_{\lambda j}(\mathbf{k}_j, t) \tag{1.48}$$

$$\mathbf{H}_\perp = \sum_j \frac{\hbar\omega_j}{2} (\mathbf{a}_j^* \mathbf{a}_j + \mathbf{a}_j \mathbf{a}_j^*) \tag{1.49}$$

$$\mathbf{P}_\perp = \sum_j \frac{\hbar\mathbf{k}_j}{2} (\mathbf{a}_j^* \mathbf{a}_j + \mathbf{a}_j \mathbf{a}_j^*) \tag{1.50}$$

$$\mathbf{A} = \sum_j \mathbf{A}_j\, \hat{\varepsilon}_j\, (\mathbf{a}_j\, e^{i\mathbf{k}_j \cdot \mathbf{r}} + c \cdot c \cdot) \tag{1.51}$$

$$\mathbf{E} = i \sum_j \mathbf{E}_j\, \hat{\varepsilon}_j\, (\mathbf{a}_j\, e^{i\mathbf{k}_j \cdot \mathbf{r}} - c \cdot c \cdot) \tag{1.52}$$

$$\mathbf{B} = i \sum_j \mathbf{B}_j\, \hat{\mathbf{k}}_j \times \hat{\varepsilon}_j\, (\mathbf{a}_j\, e^{i\mathbf{k}_j \cdot \mathbf{r}} - c \cdot c \cdot) \tag{1.53}$$

The mode coefficients $\mathbf{A}_j$, $\mathbf{E}_j$, $\mathbf{B}_j$ are related in the following way

$$\frac{\mathbf{A}_j}{c} = \left[\frac{\hbar}{2\omega_j L^3}\right]^{1/2}; \quad \mathbf{E}_j = \frac{\omega \mathbf{A}_j}{c} = \mathbf{B}_j \tag{1.54}$$

which is consistent with the Coulomb gauge definitions of the transverse fields (Equations 1.17–1.18).

1.5 QUANTUM FIELDS

The fundamental dynamical variables in the classical formalism developed in the previous section are the particle positions $\mathbf{r}_i$ and momenta $\mathbf{p}_i$, and the field's normal mode variables $\mathbf{a}_j$ and $\mathbf{a}_j^*$. From the discussion following Equation 1.46 the modes are analogous to the normal mode representation of a harmonic oscillator, $\mathbf{a} = (\omega x \pm ip)/(2\hbar\omega)^{1/2}$, with $x = \mathbf{A}/c$ and $p = \mathbf{E}$ (or vice versa). We can therefore quantize the field as an ensemble of harmonic oscillators. Thus in a full quantum description we impose the commutation relations

$$\text{particles: } [\hat{\mathbf{r}}_{i\alpha}, \hat{\mathbf{r}}_{i\beta}] = [\hat{\mathbf{p}}_{i\alpha}, \hat{\mathbf{p}}_{i\beta}] = 0, \quad [\hat{\mathbf{r}}_{i\alpha}, \hat{\mathbf{p}}_{j\beta}] = i\hbar\delta_{ij}\delta_{\alpha\beta} \tag{1.55}$$

$$\text{field: } [\hat{\mathbf{a}}_k, \hat{\mathbf{a}}_j] = [\hat{\mathbf{a}}_k^+, \hat{\mathbf{a}}_j^+] = 0; \quad [\hat{\mathbf{a}}_k, \hat{\mathbf{a}}_j^+] = \delta_{jk} \tag{1.56}$$

The first (Equation 1.55) is equivalent to the Schrödinger Equation in terms of coordinates $\mathbf{r}$ or $\mathbf{p}$. The second (Equation 1.56) also gives a Schrödinger Equation in terms of the variables $\mathbf{E}$ or $\mathbf{A}$ [14, 16]. Since the $\hat{\mathbf{a}}$'s are in principle time-dependent, the commutation relations (Equation 1.56) must be calculated at the same time so that the time-dependence made explicit in Equation (1.43) vanishes and the commutation relations become time-independent.

The physical variables (Equations 1.49–1.53) describing the electromagnetic field now become *operators* since the variable $\mathbf{a}$ is now an operator $\hat{\mathbf{a}}$. Thus we have the quantum representation for the field observables

$$\mathbf{A}_\perp = \hat{\mathbf{A}}(r,t) = \sum_j A_j\hat{\varepsilon}_j\, [\hat{\mathbf{a}}_j(t)\, e^{i\mathbf{k}_j \cdot \mathbf{r}} + \hat{\mathbf{a}}_j^+(t)\, e^{i\mathbf{k}_j \cdot \mathbf{r}}] \tag{1.57}$$

$$\hat{\mathbf{E}}_\perp = \hat{\mathbf{E}}(\mathbf{r},t) = i\sum_j \mathbf{E}_j\hat{\varepsilon}_j\,[\hat{\mathbf{a}}_j(t)\,e^{i\mathbf{k}_j\cdot\mathbf{r}} - \hat{\mathbf{a}}_j^+(t)\,e^{i\mathbf{k}_j\cdot\mathbf{r}}] \tag{1.58}$$

$$\hat{\mathbf{B}}(\mathbf{r},t) = i\sum_j \mathbf{B}_j\hat{\mathbf{k}}_j \times \hat{\varepsilon}_j\,[\hat{\mathbf{a}}_j(t)\,e^{i\mathbf{k}_j\cdot\mathbf{r}} - \hat{\mathbf{a}}_j^+(t)\,e^{i\mathbf{k}_j\cdot\mathbf{r}}] \tag{1.59}$$

$$\hat{\mathbf{H}}_\perp = \sum_j \frac{\hbar\omega_j}{2}\,[\hat{\mathbf{a}}_j^+\hat{\mathbf{a}}_j + \hat{\mathbf{a}}_j\hat{\mathbf{a}}_j^+] = \sum_j \hbar\omega_j\,[\hat{\mathbf{a}}_j^+\hat{\mathbf{a}}_j + 1/2] \tag{1.60}$$

where $\mathbf{A}_j$, $\mathbf{E}_j$, and $\mathbf{B}_j$ are amplitudes defined in the previous section. (Note that $\mathbf{H}_\perp$ is time-independent for free fields).

$$\hat{\mathbf{P}}_\perp = \sum \frac{\hbar\mathbf{k}_j}{2}\,[\hat{\mathbf{a}}_j^+\hat{\mathbf{a}}_j + \hat{\mathbf{a}}_j\hat{\mathbf{a}}_j^+] = \sum \hbar\mathbf{k}_j\,\hat{\mathbf{a}}_j^+\hat{\mathbf{a}}_j \tag{1.61}$$

where for the last equation we have used the result

$$\sum_j \mathbf{k}_j = 0$$

for finite volume (L^3) normalization.

We recapitulate by defining the total quantum Hamiltonian of the field-particle system. Thus in the Coulomb gauge we have from Equation 1.31

$$\hat{\mathbf{H}}_C = \sum_i \frac{1}{2m_i}\left[\hat{\mathbf{p}}_i - \frac{q_i}{c}\,\hat{\mathbf{A}}(\mathbf{r}_i)\right]^2 + \frac{1}{2}\sum_{i\neq j}\frac{q_i q_j}{|\mathbf{r}_j - \mathbf{r}_j|} + \sum_j \hbar\omega_j\left(\hat{\mathbf{a}}_j^+\mathbf{a}_j + \frac{1}{2}\right) \tag{1.62}$$

where $\hat{\mathbf{A}}$ is now a *field operator* defined in Equation 1.56. Furthermore, it follows from the definition of the longitudinal momentum $\mathbf{P}_{\|}$ (Equation 1.27) and the transverse momentum $\mathbf{P}_\perp$ (Equations 1.30 and 1.61) that the total momentum (field + particle) is also an operator

$$\hat{\mathbf{P}} = \sum_i \hat{\mathbf{p}}_i + \sum_j \hbar\mathbf{k}_j\hat{\mathbf{a}}_j^+\hat{\mathbf{a}}_j \tag{1.63}$$

and must be a constant of motion as in the classical case (Equation 1.63). In Equations 1.62 and 1.63 we have suppressed the time-dependence which occurs in the definition of the operator equations (1.57–1.61) (i.e., we are adopting the Schrödinger representation). The time evolution of the system is obtained using the *Heisenberg* Equations of motion for operators $\hat{\mathbf{O}}$

$$i\hbar \frac{d\,\hat{\mathbf{O}}(t)}{dt} = [\hat{\mathbf{O}}(t), \hat{\mathbf{H}}(t)] \tag{1.64}$$

Then one can show by explicit calculation

$$\dot{\hat{\mathbf{r}}}_i = (i\hbar)^{-1}\,[\hat{\mathbf{r}}_i, \hat{\mathbf{H}}] = \frac{1}{m_i}\left(\hat{\mathbf{p}}_i - \frac{q_i}{c}\,\hat{\mathbf{A}}(r_i)\right) \tag{1.65}$$

$$m_j\ddot{\hat{\mathbf{r}}}_j = \frac{m_j}{i\hbar}\,[\dot{\hat{\mathbf{r}}}_j, \hat{\mathbf{H}}] = q_j\hat{\mathbf{E}}(r_j) + \frac{q_j}{2c}\,[\dot{\hat{\mathbf{r}}}_j \times \hat{\mathbf{B}}(r_j) - \hat{\mathbf{B}}(r_j) \times \dot{\hat{\mathbf{r}}}_j] \tag{1.66}$$

$$\dot{\hat{\mathbf{a}}}_j = \frac{1}{i\hbar}\,[\hat{\mathbf{a}}_j, \hat{\mathbf{H}}] = -\,i\left[\omega_j\hat{\mathbf{a}}_j - \frac{\hat{\mathbf{j}}_j}{2\hbar\omega_j^{1/2}}\right] \tag{1.67}$$

where in Equation 1.66, $\mathbf{E} = \mathbf{E}_{\parallel} + \mathbf{E}_{\perp}$ is the total electric field. Thus the quantum equations of motion parallel in all aspects the classical equations of motion for particle and field variables. In particular Equation 1.67 can be called the *quantum Maxwell* equation as it reflects the fact that the classical Maxwell Equations (1.38) apply also to the quantum field operators.

It is possible to obtain stationary states of the system by looking for the eigenstates of the Hamiltonian (Equation 1.62) at a fixed time t and then using these eigenstates to study the time-evolution of the system following a prescription called the *Schrödinger representation.* In this picture the various observables of the system field-particles are fixed and it is the state vector (function) $|\varphi(t)\rangle$ which evolves according to the time-dependent Schrödinger Equation,

$$i\hbar \frac{\partial}{\partial t}|\varphi(t)\rangle = \hat{\mathbf{H}}_C|\varphi(t)\rangle \tag{1.68}$$

where $\hat{\mathbf{H}}_C$ is considered to be time-independent. Then one can separate the total Hamiltonian (Equation 1.62) into a particle Hamiltonian

$$\hat{\mathbf{H}}_p = \sum_i \frac{1}{2m_i}\hat{\mathbf{p}}_i^2 + \frac{1}{2}\sum_{i\neq j}\frac{q_i q_j}{|\mathbf{r}_i - \mathbf{r}_j|} \tag{1.69}$$

a radiation part

$$\hat{\mathbf{H}}_R = \sum_j \hbar\omega_j\left(\hat{\mathbf{a}}_j^+\hat{\mathbf{a}}_j + \frac{1}{2}\right) \tag{1.70}$$

and an interaction part

$$\hat{\mathbf{H}}_I = \sum_i \left[\frac{-q_i}{m_i c}\hat{\mathbf{p}}_i \cdot \hat{\mathbf{A}}(\mathbf{r}_i) + \frac{q_i^2}{2m_i c^2}\hat{\mathbf{A}}(\mathbf{r}_i)^2\right] \tag{1.71}$$

One can therefore use the free (nonperturbed) eigenstate of $\hat{\mathbf{H}}_o = \hat{\mathbf{H}}_p + \hat{\mathbf{H}}_R$ as basis functions for expanding the exact state function $|\varphi(t)>$. The Hilbert space Ω of the system is the tensor product $\Omega = \Omega_p \otimes \Omega_R$ where $\Omega_p$is spanned by the eigenvectors $|p>$ of the particle Hamiltonian $\hat{\mathbf{H}}_p$ and Ω_R is defined by the radiation harmonic oscillator eigenstates $|n>$. The basis states corresponding to the Hamiltonian decomposition (Equations 1.69–1.71) become

$$|p>|n_1>|n_2, \ldots, |n_j> = |p;n_1, \ldots, n_j> \tag{1.72}$$

The true state $|\varphi(t)>$ will be a linear combination of the states (Equation 1.72) with time-dependent coefficients. $|p>$ is an orthonormal eigenstate basis of the particles and $|n_j>$ are the quantum radiation field states of quantum number n_j. We turn now to the description of these field states.

The eigenstates of the operator $\hat{\mathbf{a}}_j^+\hat{\mathbf{a}}_j$ are well known. These are harmonic oscillator eigenstates $|n_j>$ with eigenvalues n_j given by

$$\hat{\mathbf{a}}_j^+\hat{\mathbf{a}}_j|n_j> = n_j|n_j>, \quad n_j = 0, 1, 2, \ldots \tag{1.73}$$

In coordinate space, the corresponding eigenfunctions $<x|n_j>$ are the Hermite polynomials so that in fact one can use wave functions to describe these photon states [14, 16] provided we set x = **A** or **E**

$$<x|n> = \left(\frac{\omega}{\pi\hbar}\right)^{1/4} 2^{-n/2} (n!)^{-1/2} e^{-\omega x^2/2\hbar} \mathbf{H}_n\left(x\frac{\omega}{\hbar}\right)^{1/2} \tag{1.74}$$

The eigenstates $|n_j>$ obey the well-known operator relations since in coordinate space

$$\hat{\mathbf{a}} = \frac{\omega x + ip}{\hbar\omega^{1/2}} = \frac{\omega x + \hbar d/dx}{\hbar\omega^{1/2}}$$

whereas in state space

$$\hat{\mathbf{a}}_j^+|n_j> = (n_j + 1)^{1/2}|n_j + 1>, \quad \hat{\mathbf{a}}_j|n_j> = n_j^{1/2}|n_j - 1>$$

$$|n_j> = (n_j!)^{-1/2}(\hat{\mathbf{a}}_j^+)^{n_j}|0> \tag{1.75}$$

Since the oscillators are independent when free, $\hat{\mathbf{a}}_j^+$ and $\hat{\mathbf{a}}_i$ commute; the tensor products of $|n_j>$ satisfy the following eigenequations

$$\hat{\mathbf{H}}_R|n_i, ..., n_j> = \sum_j (n_j + 1/2)\, \hbar\omega_j\, |n_i, ..., n_j>$$

$$\hat{\mathbf{P}}_R|n_i, ..., n_j> = \sum_j n_j\, \hbar\mathbf{k}_j\, |n_i, ..., n_j> \tag{1.76}$$

The state $|n_i, ..., n_j>$ is clearly an excited state with respect to the ground state $|0>$, with relative energy

$$\mathbf{E}_R = \sum_j n_j\hbar\omega_j$$

and total momentum

$$\mathbf{P}_R = \sum_j n_j\hbar\mathbf{k}_j$$

This can be interpreted as an ensemble of particles each of energy $\hbar\omega_j$ and momentum $\hbar\mathbf{k}_j$. Thus quantization of the electromagnetic fields described by Maxwell's Equation leads to the concept of a field particle called the *photon.* The ground state with zero photons is designated as the vacuum although there is a large zero point energy

$$E_o = \sum_j \frac{\hbar\omega_j}{2}$$

The total number of photons is the eigenvalue of the total number operator

$$\hat{N} = \sum_j \hat{\mathbf{a}}_j^+ \hat{\mathbf{a}}_j \tag{1.77}$$

Returning to the normal mode expansion (Equation 1.58) for the electric field, one sees that the vacuum state corresponds in fact to zero average field but nonzero variance

$$<0|\hat{\mathbf{E}}_\perp(\mathbf{r})|0> = 0, \quad <0|[\hat{\mathbf{E}}_\perp(\mathbf{r})]^2|0> = \sum_i \frac{\hbar\omega_i}{2L^3} \tag{1.78}$$

Thus in the quantum theory, zero fields undergo fluctuations due to the noncommutativity of the canonical variables, the electric field $\hat{\mathbf{E}}$ and the vector potential $\hat{\mathbf{A}}$ which are now operators.

1.6 SEMICLASSICAL FIELD STATES

Since we will be describing the interaction of lasers with molecules, the high intensity and coherence inherent in these sources begs the question as to what is the limit of the quantum radiation fields described in the previous section in that situation. We may ask first what is the number of photons present in a certain laser beam per unit volume. Thus since the flux (or intensity) I is the energy per unit volume carried per unit time, one deduces the following relation between intensity I and number n of photons of frequency ω

$$I = \frac{N\hbar\omega}{V} c \tag{1.79}$$

Let us assume an intensity I = 1 W/cm^2 with a photon energy of 1 ev (ω = 8000 cm^{-1}) the number of photons present per cm^3 is N/cm^3 $\cong$ 2×10^8/cm^3. Clearly this is a very large number of photons implying very high quantum numbers of the radiation field. Since large quantum numbers usually imply that a semiclassical description should be more than adequate, we now undertake such a derivation.

In our classical description of the transverse radiation fields, $\mathbf{E}_\perp$ and **B** (Section 1.4), we introduced classical normal modes **a**(k,t), Equation (1.35) which were then subsequently quantized in Section 1.5. Thus comparing the classical field Hamiltonian $\mathbf{H}_\perp$ (Equation 1.49) and the quantum equivalent Hamiltonian $\mathbf{H}_R$ (Equation 1.70) we have

$$\mathbf{H}_{\text{classical}} = \sum_j \hbar\omega_j \mathbf{a}_j^* \mathbf{a}_j; \quad \mathbf{H}_{\text{quantum}} = \sum_j \hbar\omega_j \left(\hat{\mathbf{a}}_j^+ \hat{\mathbf{a}}_j + \frac{1}{2} \right) \tag{1.80}$$

Thus, since for quantum states $<n_j|\hat{\mathbf{a}}_j^+\hat{\mathbf{a}}_j|n_j> = n_j$ (the number of photons in mode j), then for large n's one seems to have a correspondence between the classical field variable $\mathbf{a}_j$ and the equivalent quantum operator $\hat{\mathbf{a}}_j$. This correspondence would be exact if one could find eigenstates $|\alpha_j>$ such that

$$\text{a)}\ <\alpha_j|\hat{\mathbf{a}}_j|\alpha_j> = \alpha_j; \quad \text{b)}\ <\alpha_j|\hat{\mathbf{a}}_j^+\hat{\mathbf{a}}_j|\alpha_j> = \alpha_j^*\alpha_j = n_j \tag{1.81}$$

Equation 1.81 would then establish the exact equivalence between classical and quantum field operators (Equations 1.52 and 1.58) and concurrently equivalence between the classical and quantum energies (Equations 1.49 and 1.70).

Such states can be simply constructed if we impose the eigenvalue equation conditions

$$\hat{\mathbf{a}}_j|\alpha_j> = \alpha_j|\alpha_j>; \quad <\alpha_j|\hat{\mathbf{a}}_j^+ = \alpha_j^*<\alpha_j| \tag{1.82}$$

where the $\hat{\mathbf{a}}$'s are quantum operators and the α's are the normal modes of the classical radiation field defined previously. These new states originally derived by Schrödinger [17] have been introduced by Glauber [13] to describe the electromagnetic field in terms of coherence and correlation of photons. We rewrite the transverse classical electromagnetic field (Equation 1.52) in a volume L^3 in terms of positive and negative propagation components for one mode

$$\mathbf{E}_j(\mathbf{r},t) = \mathbf{E}_j^+ + \mathbf{E}_j^- = i\left(\frac{\hbar\omega_j}{2L^3}\right)^{1/2}\left[\mathbf{a}_j(t)\, e^{+i\mathbf{k}_j\cdot\mathbf{r}} - \mathbf{a}_j^*(t)\, e^{-i\mathbf{k}_j\cdot\mathbf{r}}\right] \tag{1.83}$$

The corresponding quantum field operators become

$$\hat{\mathbf{E}}_j^+ = +i\left(\frac{\hbar\omega_j}{2L^3}\right)^{1/2}\hat{\mathbf{a}}_j(t)\, e^{+i\mathbf{k}_j\cdot\mathbf{r}}; \quad \hat{\mathbf{E}}_j^- = -i\left(\frac{\hbar\omega_j}{2L^3}\right)^{1/2}\hat{\mathbf{a}}_j^+(t)\, e^{-i\mathbf{k}_j\cdot\mathbf{r}} \tag{1.84}$$

By the quantum postulate we seek a solution for the fields in the form

$$\hat{\mathbf{E}}|E\rangle = E|E\rangle \text{ or the adjoint } \langle E|\hat{\mathbf{E}}^+ = \langle E|E^* \tag{1.85}$$

By choosing the eigenstates of $\hat{\mathbf{a}}_j$ and $\hat{\mathbf{a}}_j^+$ (as defined in Equation 1.82), then the eigenvalue of the field operators $\hat{\mathbf{E}}_j^+$ and $\hat{\mathbf{E}}_j^-$ are the classical field components

$$\mathbf{E}_j^+(\mathbf{r},t) = +i\left(\frac{\hbar\omega_j}{2L^3}\right)^{1/2}\alpha_j\, e^{i\mathbf{k}_j\cdot\mathbf{r}}; \quad \mathbf{E}_j(\mathbf{r},t) = -i\left(\frac{\hbar\omega_j}{2L^3}\right)^{1/2}\alpha_j^*\, e^{-i\mathbf{k}_j\cdot\mathbf{r}} \tag{1.86}$$

The properties of the *coherent* states have been described by Goldin [18] and Perelomov [19] and we summarize some of their salient features. Expanding these states in a complete set of photon number states $|n_j\rangle$ for mode j of frequency ω_j

$$|\alpha_j\rangle = \sum_{n_j} |n_j\rangle\langle n_j|\alpha_j\rangle \tag{1.87}$$

and using the relations in Equation 1.75

$$<n_j|\hat{\mathbf{a}}_j = (n_j + 1)^{1/2} <n_j + 1|, \quad <n_j|\hat{\mathbf{a}}|\alpha_j> = \alpha_j <n_j|\alpha_j> \tag{1.88}$$

one obtains

$$<n_j|\alpha_j> = \frac{(n_j + 1)^{1/2}}{\alpha_j} <n_j + 1|\alpha_j> \tag{1.89}$$

Introducing the definitions for $<n + 1| = <0|\hat{\mathbf{a}}^{n+1} \; [(n + 1)!]^{-1/2}$, Equation 1.75 becomes

$$<n_j|\alpha_j> = \alpha_j^{n_j}(n_j!)^{-1/2} <0|\alpha_j> \tag{1.90}$$

The expansion for the coherent state is therefore

$$|\alpha_j> = <0|\alpha_j> \sum_{n_j} \frac{\alpha_j^{n_j}}{(n_j!)^{+1/2}} |n_j> \tag{1.91}$$

From the normalization condition $<\alpha_j|\alpha_j> = 1$, one obtains by further manipulation

$$|<0|\alpha_j>|^2 = e^{-|\alpha_j|^2}$$

so that finally

$$|\alpha_j> = e^{(-|\alpha_j|^2)/2} \sum_{n_j} \frac{\alpha_j^{n_j}}{(n_j!)^{1/2}} |n_j> \tag{1.92}$$

We see therefore that the coherent state $|\alpha_j>$, which is the eigenvector of the quantum field operators $\hat{\mathbf{E}}_j^{\pm}$ (Equation 1.84) with eigenvalues the classical fields $\mathbf{E}_j^{\pm}$ (Equation 1.83), is the linear superposition of an infinite number of photon states. These states which are the quantum equivalent of the classical field states are called *semiclassical* states. The probability P_{n_j} of finding n_j photons in such a state is a Poisson distribution

$$P_{n_j} = e^{-|\alpha_j|^2} |\alpha_j|^{2n} \frac{j}{n_j}! \tag{1.93}$$

with the average $<n_j>^2$ and variance Δn_j^2 being equal:

$$<E_j> = \left(\alpha_j^2 + \frac{1}{2}\right)\hbar\omega_j; \quad <E_j^2> = \left(\alpha_j^4 + 2\alpha_j^2 + \frac{1}{4}\right)\hbar^2\omega_j^2$$

so that

$$\Delta n_j^2 = <n_j^2> - <n_j>^2 = |\alpha_j|^2 \tag{1.94}$$

This gives for large n_j negligible error in the average photon number

$$\frac{\Delta n_j}{n_j} \simeq \frac{1}{n_j} \to 0$$

We arrive therefore at the conclusion that the coherent state $|\alpha_j>$, which is a quantum state of the radiation field in the limit of large number of photons n_j, has fluctuations Δn_j much smaller than the number of photons n_j ($\Delta n_j \simeq n_j^{1/2}$) and has as eigenvalue the classical field E. In fact the coherent state $|\alpha_j>$ is a minimum uncertainty state for which the product of the number fluctuation and the phase is a minimum since in general phase and number also satisfy an uncertainty principle [19, 20]. Therefore, we remind the reader that states of exact photon number (i.e., the eigenstates $|n_j>$) have indeterminate (random) phases, whereas the opposite extreme, precise phase implies indefinite photon number as in the classical wave limit (see Equation 1.104). Thus a state $|n_j>$ is the quantum analog of a single mode classical field of well-defined energy $n_j\hbar\omega_j$ but with a random phase equally distributed from 0 to 2π. A detailed analysis of interference between two independent laser beams shows no interference if each mode has a well defined photon number value whereas for semiclassical states, the two beams interfere as two classical Maxwell waves [7].

This can be more readily understood in terms of the electric field variable **E**. Thus from the definitions of the field quantum oscillator operator $\hat{\mathbf{a}}$ (Equations 1.46, 1.57, and 1.58) one can associate the electric field E as the coordinate x of the corresponding photon eigenfunctions $<x|n>$ (Equation

1.74) (i.e., there is a corresponding photon Schrödinger Equation for the photon states with **E** as the coordinate [16]). The average value of the electric field, <n|E|n>, is clearly null for any pure quantum state of photon number n. This arises from the fact that for large n, the oscillations of the photon wavefunction become more random as n increases, with corresponding uncertain values of $\overline{E}$ (i.e., the phase of the field is undetermined). Coherent quantum states satisfy the eigenvalue condition (Equation 1.82) so that using Equation 1.46 for $\hat{\mathbf{a}}$, one can rewrite the coherent state eigenvalue Equation as a function of the field variable E (we use the canonical relation

$$<E|\frac{\hat{\mathbf{A}}}{c} = i\hbar \frac{\partial}{\partial E} <E|$$

since **E** and $\dot{\mathbf{A}}$/c are canonical variables in the classical case)

$$\frac{-i}{(2\hbar\omega)^{1/2}}\left(\hbar\omega \frac{\partial}{\partial E} + E\right)<E|\alpha> = \alpha<E|\alpha> \tag{1.95}$$

which gives as normalized coherent state solution

$$\varphi_\alpha(E) = <E|\alpha> = (\pi\hbar)^{-1/4}\, e^{-[(2\hbar\omega)^{1/2} E - i\alpha]^2} \tag{1.96}$$

We note that the coherent states are *not* orthogonal thus admitting complex eigenvalues since they are nothing but the ground state of the quantum field (vacuum state) displaced in the coordinate E by the amount iα. Thus the average value of the electric field $<\varphi_\alpha(E)|E|\varphi_\alpha(E)> = (2\hbar\omega)^{1/2}\, i\alpha$ is no longer null, but equals the classical value (Equation 1.86), after appropriate normalization. One can further show (using the coherent state solution; Equation 1.96) that the state $\varphi_\alpha(E)$ is one of minimum uncertainty. There is no spreading of the time-dependent wavepacket constructed from $\varphi_\alpha(E)$ so that fluctuations in values of E are minimum [18, 19].

Returning to the quantum Hamiltonian (Equation 1.62), which is time-independent in the Schrödinger Equation (1.68), one would like to replace the quantum field operators in the interaction (Equation 1.71) by the corresponding semiclassical field states described previously in order to handle current laser problems where the number of photons is usually quite large. However, since as we have repeatedly emphasized the quantum states are time-independent in the Schrödinger representation, and classical states are in principle time-dependent, we perform first a unitary transformation on the Schrödinger Hamiltonian in order to introduce a time-dependence via the fields. This can be achieved by transforming to an interaction representation which removes the field Hamiltonian $\mathbf{H}_R$ (Equation 1.70) by the unitary transformation

$$|\varphi(t)\rangle = \exp(-i\,\hat{\mathbf{H}}_R\, t/\hbar)\,|\varphi'(t)\rangle \tag{1.97}$$

This transformation gives a new time-dependent Hamiltonian

$$\hat{\mathbf{H}}_C(t) = \sum_i \left(\hat{\mathbf{p}}_i - \frac{q_i}{c}\,\hat{\mathbf{A}}(\mathbf{r},t)\right)^2 + V_C \tag{1.98}$$

where a time-dependence now enters into the quantum vector potential

$$\begin{aligned}\hat{\mathbf{A}}(\mathbf{r}_i,t) &= \exp\left(i\hat{\mathbf{H}}_R \frac{t}{\hbar}\right)\hat{\mathbf{A}}(\mathbf{r}_i)\exp\left(-i\hat{\mathbf{H}}_R \frac{t}{\hbar}\right)\\ &= \sum_j \mathbf{A}_j\,\hat{\varepsilon}_j\,[\hat{\mathbf{a}}_j\, e^{i(\mathbf{k}_j\cdot\mathbf{r}-\omega_j t)} + \hat{\mathbf{a}}_j^{+}\, e^{-i(\mathbf{k}_j\cdot\mathbf{r}-\omega_j t)}]\end{aligned} \tag{1.99}$$

$\mathbf{A}_j$ is defined in Equation 1.54. To obtain Equation 1.99 we have used the commutation relations $[\hat{\mathbf{N}},\hat{\mathbf{a}}] = -\hat{\mathbf{a}}$, $[\hat{\mathbf{N}},\hat{\mathbf{a}}^{+}] = \hat{\mathbf{a}}^{+}$ from Equations 1.56 and 1.78.

The expectation value of $\hat{\mathbf{A}}$ with respect to photon states $|n_j\rangle$ being zero, then semiclassical states are obtained by assuming that the electromagnetic field is in a coherent state. One can therefore average the new Hamiltonian (Equation 1.98) over this state and obtain the quantum vector potential operator expressed as a classical field. In view of the preceding results on coherent states (Equations 1.81–1.96)

$$\mathbf{A}(\mathbf{r},t) = \langle\alpha_1 \cdot\cdot\, \alpha_j, \ldots |\hat{\mathbf{A}}|\alpha_1, \cdot\cdot\, \alpha_j, \ldots\rangle = \sum_j n_j^{1/2}\,\mathbf{A}_j \left\{\hat{\varepsilon}_j \exp[i(\mathbf{k}_j\cdot\mathbf{r} - \omega_j t) + \text{c.c.}]\right\} \tag{1.100}$$

where n_j is the photon number in mode j, which is related to the intensity by Equation 1.79. The quantum operator (Equation 1.99) is now a time-dependent function (c-number) of the variables n_j, $\mathbf{r}$, and t.

The final semiclassical expression (Equation 1.100) can be generalized to include the relative *phase* of modes (i.e., the correct initial conditions). Thus since in general one can write a travelling wave as $\exp i(\mathbf{k} \cdot \mathbf{r} - \omega t - \varphi)$, then in analogy with the Dirac representation $\langle x|p\rangle = \exp(ikx)/(2\pi)^{1/2}$ for $-\infty \leq x \leq \infty$, $-\infty \leq k \leq \infty$, then one can define a number-phase conjugate representation [5, 20]

$$\langle \varphi_j | n_j \rangle = \exp \frac{(in_j\varphi_j)}{(2\pi)^{1/2}} \tag{1.101}$$

where n_j is the photon number eigenvalue for the number operator $\hat{n}_j$ (defined in Equation 1.97) for a particular mode j, and $0 \leq \varphi_j \leq 2\pi$. Clearly one has the conjugate relation

$$\hat{n}_j = \frac{1}{i}\frac{\partial}{\partial \varphi_j} = \hat{\mathbf{a}}_j^+ \hat{\mathbf{a}}_j \tag{1.102}$$

This leads to the convenient definition of the creation and annihilation operators

$$\hat{\mathbf{a}}_j^+ = \left(\frac{1}{i}\frac{\partial}{\partial \varphi_j}\right)^{1/2} e^{i\varphi_j}; \quad \hat{\mathbf{a}}_j = e^{-i\varphi_j}\left(\frac{1}{i}\frac{\partial}{\partial \varphi_j}\right)^{1/2} \tag{1.103}$$

satisfying the commutation relations

$$[\hat{\mathbf{a}}_j, \hat{\mathbf{a}}_{j'}^+] = \delta_{jj'}; \quad [\hat{\varphi}_j, \hat{n}_j] = i \tag{1.104}$$

The expressions in Equation 1.103 will be shown to lead to the expected semiclassical results. Thus in the limit of large photon number N_j, we approach the semiclassical limit discussed previously and so we can write

$$n_j = N_j + n_j', \quad |n_j\rangle = \exp(iN_j\varphi_j)\,|n_j'\rangle \tag{1.105}$$

which gives finally new expressions for the photon operators (we note that Equation 1.105 corresponds to a translation in the space of photon quantum numbers n_j)

$$\hat{\mathbf{a}}_j^+ = \left(N_j + \frac{1}{i}\frac{\partial}{\partial\varphi_j}\right)^{1/2} e^{i\varphi_j}; \quad \hat{\mathbf{a}}_j = e^{-i\varphi_j}\left(N_j + \frac{1}{i}\frac{\partial}{\partial\varphi_j}\right)^{1/2} \tag{1.106}$$

For large semiclassical expectation values N_j, one can expand to first order (this implies phase fluctuations much less than the order of photon numbers)

$$\hat{\mathbf{a}}_j^+ = N_j^{1/2}\left[1 - \frac{i}{2N_j}\frac{\partial}{\partial\varphi_j}\right] e^{i\varphi_j}; \quad \hat{\mathbf{a}}_j = e^{-i\varphi_j} N_j^{1/2}\left[1 - \frac{i}{2N_j}\frac{\partial}{\partial\varphi_j}\right] \tag{1.107}$$

Introducing these final expressions into Equation 1.100 transforms **A** and each exponential into the desired form $\exp[i(\mathbf{k}_j \cdot \mathbf{r} - \omega_j t - \varphi_j)] + \text{c.c.}$ Quantum fluctuations $\delta\mathbf{A}$ due to phase fluctuations will arise from the corrections (Equation 1.106)

$$-\frac{i}{2} N_j^{1/2} \frac{\partial}{\partial\varphi_j}$$

These are small in the semiclassical limit, $N_j >> 1$ (e.g., for a CO_2 laser at intensity $I = 10^{12}$ W/cm^2, $N = 10^{27}$/m^3). Furthermore, since the exact relative phases of a multimode laser are difficult to define, one usually averages out these parameters.

1.7 DRESSED, FLOQUET STATES AND TRANSITION AMPLITUDES

The new field-particle Hamiltonian (1.98) will now become *time-dependent* if one inserts (1.100) for **A** (i.e., explicit quantum features of the electromagnetic field have been removed). Only the quantization of the particles remains in the presence of a time-dependent perturbation expressed through the semiclassical radiation vector potential **A**(**r**,t), Equation (1.100).

One can therefore proceed in two fashions. First one can consider the interaction Hamiltonian $\mathbf{H}_I$ (Equation 1.71) as being time-dependent with the semiclassical fields incorporated in the description. Such a description is most appropriate for intense fields (i.e., large number of photons) where in addition the field amplitude is a time-dependent pulse envelope corresponding to a complicated photon wave packet. In such a case solving the time-dependent Schrödinger Equation (1.68) with now a time-dependent Hamiltonian **H**(t) is the most appropriate way of solving laser-matter interaction problems. Examples of this approach will appear in later chapters.

It is also of course possible to solve for the time-independent, stationary eigenstates of **H** (Equations 1.68–1.71) and thus retain the quantized nature of the field. Thus the true eigenfunctions can be expressed as linear combinations of the time-independent quantized field-particle states (Equation 1.72). Such an approach is called a *dressed* state approach [5, 6], and can be used to interpret both perturbative and nonperturbative nonlinear phenomena induced by laser fields [21–23]. Such a description is most useful when only a few essential states are coupled by the matter-field interaction and when the field is constant over a period longer than the lifetime of the essential states.

The relationship between the time-dependent and time-independent formalism can be demonstrated by the use of Green's function methods. For a time-independent Hamiltonian, $\hat{\mathbf{H}}$, the formal time-dependent solution of Equation 1.68 is

$$|\varphi(t)\rangle = \exp(-i\hat{\mathbf{H}}t/\hbar)\, |\varphi_i\rangle = \hat{\mathbf{U}}(t)\, |\varphi_i\rangle \tag{1.108}$$

where $\hat{\mathbf{U}}$ is called the *evolution* operator and $\varphi_i = \varphi(0)$ is the initial eigenstate of $\hat{\mathbf{H}}_o = \hat{\mathbf{H}}_p + \hat{\mathbf{H}}_R$. Taking the Laplacean transformation of Equation 1.108 one obtains

$$L|\varphi(t)\rangle = \int_o^\infty e^{-st}\, |\varphi(t)\rangle\, dt = \frac{1}{s + i\hat{\mathbf{H}}/\hbar}\, |\varphi_i\rangle \tag{1.109}$$

where the integral is assumed to be convergent for a given s [24]. The inverse transform relates $|\varphi(t)\rangle$ to an inverse function of the Hamiltonian $\hat{\mathbf{H}}$

$$|\varphi(t)\rangle = \frac{1}{2\pi i} \int_{\eta - i\infty}^{\eta + i\infty} ds\, e^{st} \frac{1}{s + i\hat{\mathbf{H}}/\hbar}\, |\varphi_i\rangle \tag{1.110}$$

where $\eta > 0$. We can obtain a Fourier transform by defining $s = -iE/\hbar$, $\varepsilon = \eta\hbar$

$$|\varphi(t)\rangle = \frac{1}{2\pi i} \int_{i\varepsilon+\infty}^{i\varepsilon-\infty} dE\, e^{-iEt/\hbar} \frac{1}{E - \hat{\mathbf{H}}} |\varphi_i\rangle \tag{1.111}$$

This integral will be convergent provided the singularities of the resolvent $(E - \hat{\mathbf{H}})^{-1}$ are below the imaginary line $E = i\varepsilon$. Thus real eigenvalues E_r and eigenvalues with negative imaginary parts, such as resonances [21-23] (i.e., $E = E_R - i\Gamma_r$), where Γ_r is the width of such a resonance, are therefore admissible. (This implies for the real part of s, Res > o in Equation 1.110). One can therefore define a new operator, the *Green's function* or resolvent operator

$$\hat{\mathbf{G}}^+(E) = \lim_{\varepsilon \to o} \frac{1}{E + i\varepsilon - \hat{\mathbf{H}}} \tag{1.112}$$

so that now one can define a real energy integral relating the initial state φ_i and the total (field + particle) function at any time t

$$|\varphi(t)\rangle = -\frac{1}{2\pi i} \int dE\, \hat{\mathbf{G}}^+(E)\, e^{-iEt/\hbar} |\varphi_i\rangle \tag{1.113}$$

Therefore, we see that the system wave function is the Fourier transform of the Green's function operator $\hat{\mathbf{G}}^+(E)$. Thus for a conservative system, the energies are real because $\hat{\mathbf{H}}$ is Hermitian, all relevant information is contained in $\hat{\mathbf{G}}(E)$ which is stationary. In particular, the transition amplitude between an initial, φ_i and final φ_f, field-particle state, is expressed in terms of $\hat{\mathbf{G}}(E)$

$$T_{fi}(t) = \langle\varphi_r(t)|\varphi(t)\rangle = -\frac{1}{2\pi i} \int_{-\infty}^{\infty} dE \langle\varphi_f|\hat{\mathbf{G}}^+(E)|\varphi_i\rangle \exp\left[-i(E - E_f)t/\hbar\right] \tag{1.114}$$

where we use the fact that at t, $\varphi_f(t) = \varphi_f\, e^{-iE_f t/\hbar}$.

The initial φ_i and final φ_f states are stationary time-independent eigenstates of the time-independent zeroth-order (initial and final) Hamiltonian $\hat{\mathbf{H}}_o = \hat{\mathbf{H}}_p + \hat{\mathbf{H}}_R$. These zeroth-order (uncoupled) eigenstates are the field-particle states defined in Equation 1.72. $\hat{\mathbf{G}}(E)$ which involves $\hat{\mathbf{H}}_o$ and the time-independent interaction $\hat{\mathbf{H}}_I$ (Equation 1.71), can be calculated by expressing the intermediate states as linear combinations of these uncoupled eigenstates. These time-independent, linear combinations of the states (Equation 1.72) representing the true eigenstates of the field-particle state are called the *dressed states* of the total system [5, 6, 21–23].

A. Dressed and Floquet States

Let us show the equivalence of the two methods, time-dependent and time-independent, in a single mode calculation. We start with an atom or molecule in the presence of a single mode semiclassical field $\mathbf{A}(t) = \hat{\varepsilon}A_\omega \cos\omega t$. We neglect spatial dependence. (This corresponds to the dipole approximation whereby the A^2 term in Equation 1.71 can be removed as a phase factor; see next section). The time-dependent Schrödinger Equation becomes

$$\text{a)}\quad i\hbar \frac{\partial|\varphi(t)\rangle}{\partial t} = [\hat{\mathbf{H}}_p + \hat{\mathbf{H}}_I(t)]\,|\varphi(t)\rangle$$

$$\text{b)}\quad \hat{\mathbf{H}}_I(t) = V[e^{i\omega t} + e^{-i\omega t}], \qquad V = A_\omega \hat{\varepsilon}\cdot\frac{\mathbf{p}}{2} \tag{1.115}$$

and $\hat{\mathbf{H}}_p\,|\varphi_i\rangle = E_i\,|\varphi_i\rangle$ defines the particle eigenstates in absence of the field. We now develop the general solution (Equation 1.113) in a Fourier series

$$|\varphi(t)\rangle = \frac{1}{2\pi i}\sum_{n=-\infty}^{\infty}\int_{-\infty}^{\infty} dE\, \mathbf{G}_n(E)\,|\varphi_i\rangle\, e^{-i(E-n\hbar\omega)t/\hbar} \tag{1.116}$$

This is consistent with the periodicity condition $|\varphi(t+\tau) = |\varphi(t)\rangle$ since H_I is periodic with period $\tau = 2\pi/\omega$. Substituting Equation 1.116 into Equation 1.115 gives

$$\sum_{n=-\infty}^{\infty}\int_{-\infty}^{\infty} dE\,(E - \hat{\mathbf{H}}_p - n\hbar\omega)\,\mathbf{G}_n(E)\, e^{-i(E-n\hbar\omega)/t\hbar} =$$

$$\sum_{n=-\infty}^{\infty} \int_{-\infty}^{\infty} dE\, V\mathbf{G}_{n+1}\, e^{-i(E-(n+1)\hbar\omega)/t\hbar} + \mathbf{G}_{n-1}\, e^{-i(E-(n-1)\hbar\omega)/t\hbar} \tag{1.117}$$

Equating term by term in Equation 1.117 for any time t yields

$$E - \hat{\mathbf{H}}_p - n\hbar\omega\mathbf{G}_n(E) = V[\mathbf{G}_{n-1}(E) + \mathbf{G}_{n+1}(E)] \tag{1.118}$$

The expansion of Equation 1.116 in terms of a Fourier series of harmonic frequencies $n\omega$ is a standard method used in periodic time-dependent problems and is called a Floquet expansion [25]. From Equation 1.118 one can further identify a Hamiltonian with matrix elements

$$H_{nn} = \hat{\mathbf{H}}_p + n\hbar\omega; \quad H_{n,n\pm 1} = V \tag{1.119}$$

This is called a *Floquet* Hamiltonian [6, 25]. We thus see that the Fourier coefficients $\mathbf{G}_n(E)$ of the Green's function operator are coupled with their neighbors $\mathbf{G}_{n\pm 1}$ in the Floquet expansion.

We now turn to the quantum description (Equation 1.68) where we write for a single mode

$$i\hbar\frac{\partial|\varphi(t)\rangle}{\partial t} = \left[\hat{\mathbf{H}}_p + \hbar\omega(\hat{\mathbf{a}}^{+}\hat{\mathbf{a}}) + \left(\frac{\hbar\omega}{2L^3}\right)^{1/2}\hat{\varepsilon}\cdot\mathbf{p}(\hat{\mathbf{a}}^{+} + \hat{\mathbf{a}})\right]|\varphi(t)\rangle \tag{1.120}$$

$|\varphi(o)\rangle = |n_j\rangle\, |\varphi_i\rangle$, where $n_i\rangle$ and $|\varphi_i\rangle$ are the initial photon and particle quantum states. We expand $|\varphi(t)\rangle$ in terms of the photon state $|n' + n_i\rangle$, that is, we write

$$|\varphi(t)\rangle = \sum_{n'} \varphi_{n'}(t)\, |n' + n_i\rangle \tag{1.121}$$

Substituting in Equation 1.120 and projecting onto some final state $|n + n_i\rangle$, one finds

$$i\hbar\frac{\partial\varphi_n(t)}{\partial t} = [\hat{\mathbf{H}}_p + (n + n_i)\hbar\omega]\varphi_n(t) + \frac{\hat{\varepsilon}\cdot\mathbf{p}}{2}[A_{n+1}\varphi_{n+1}(t) + A_{n-1}\varphi_{n-1}(t)] \tag{1.122}$$

where

$$A_{n\pm1} \simeq A_n = A_{n_i+n} = \left(\frac{n_i\hbar\omega_i}{2L^3}\right)^{1/2}$$

for $n_i >> n$, is related to the incident field intensity I by Equations 1.79 and 1.100.

Expanding now the coefficient functions $\varphi_n(t)$ as in Equation 1.113

$$\varphi_n(t) = -\frac{1}{2\pi i}\int_{-\infty}^{\infty} dE\, e^{-iEt/\hbar}\, \mathbf{G}_n(E) \tag{1.123}$$

one obtains

$$[E - \hat{\mathbf{H}}_p - (n + n_i)\hbar\omega]\mathbf{G}_n(E) = \frac{\hat{\varepsilon}\cdot\mathbf{p}}{2} A_{n_i} [\mathbf{G}_{n+1}(E) + \mathbf{G}_{n-1}(E)] \tag{1.124}$$

This is identical to the semiclassical Floquet Equation 1.118 where

$$V = \frac{\hat{\varepsilon}\cdot\mathbf{p}}{2} A_\omega$$

and therefore in the semiclassical limit one has the equivalence between Equations 1.124 and 1.118 since $A_w \simeq A_{n_i}$. Thus comparing the semiclassical Floquet Equations (1.118) to the quantum Equations (1.124), one sees that in the large photon number limit, one can retain photon numbers changes n with respect to the initial total large photon number n_i in order to characterize the couplings between different Fourier components with frequency $n\omega$. Thus couplings between these different components are equivalent to quantum transitions between photon states. Finally, neglecting $\mathbf{G}_{n-1}$ or $\mathbf{G}_{n+1}$ terms in Equations 1.118–1.119 corresponds to the rotating wave approximation (RWA).

B. Transition Amplitudes

We now turn to the problem of evaluating transition amplitudes. It is possible to expand the total **G** (Equation 1.112) in terms of the original unperturbed

Hamiltonian $\hat{\mathbf{H}}_o = \hat{\mathbf{H}}_p + \hat{\mathbf{H}}_R$ and the interaction $\hat{\mathbf{H}}_I = V$ to generate a perturbation theory expansion of multiphoton transitions which is related to Born expansions of scattering theory [21]. We thus define first the unperturbed resolvent

$$\hat{\mathbf{G}}^o(E) \underset{\varepsilon\to o}{=} \frac{1}{E - \hat{\mathbf{H}}_o + i\varepsilon} \tag{1.125}$$

which by inserting into and then expanding Equation 1.112 gives a power series in V and $\hat{\mathbf{G}}^o$ (or using $A^{-1} = B^{-1} + B^{-1}(B - A)A^{-1}$)

$$\begin{aligned}\hat{\mathbf{G}}(E) &= (1 - \hat{\mathbf{G}}^o V)^{-1}\hat{\mathbf{G}}^o = \hat{\mathbf{G}} + \hat{\mathbf{G}}^o V\hat{\mathbf{G}}^o + \hat{\mathbf{G}}^{o+} V\hat{\mathbf{G}}^o V\hat{\mathbf{G}}^o + \ldots \\ &= \hat{\mathbf{G}}^o + \hat{\mathbf{G}}^o V\hat{\mathbf{G}}^o \end{aligned} \tag{1.126}$$

A transition matrix (T-matrix) may now be defined by the infinite perturbation series (Born expansion [26])

$$\begin{aligned}\hat{\mathbf{T}}(E) &= V + V\hat{\mathbf{G}}^o V + V\hat{\mathbf{G}}V\hat{\mathbf{G}}^o V + \ldots \\ &= V + V\hat{\mathbf{G}}^o\hat{\mathbf{T}} = V + V\hat{\mathbf{G}}V \end{aligned} \tag{1.127}$$

which yields a new expression for **G** [21, 26] from Equations 1.126 and 1.127

$$\hat{\mathbf{G}}(E) = \hat{\mathbf{G}}^o + \hat{\mathbf{G}}^o\hat{\mathbf{T}}(E)\hat{\mathbf{G}}^o \tag{1.128}$$

In obtaining the transition amplitude between an initial state $|\varphi_i\rangle$ and a final state $|\varphi_f\rangle$ (Equation 1.113), it is convenient to separate from the total Hilbert space of unperturbed eigenstates of $\hat{\mathbf{H}}_o$, the two states $|\varphi_i\rangle$ and $|\varphi_f\rangle$ by a projection operator $\hat{\mathbf{Q}}$ from all other states, $\hat{\mathbf{P}}$

$$\hat{\mathbf{P}} = 1 - \hat{\mathbf{Q}}; \quad \hat{\mathbf{Q}} = |\varphi_i\rangle\langle\varphi_i| + |\varphi_f\rangle\langle\varphi_f| \tag{1.129}$$

We can now define a new transition matrix **t** which excludes the initial and final states

$$\begin{aligned}\hat{\mathbf{t}} &= V + V\hat{\mathbf{G}}^o\hat{\mathbf{P}}V + V\hat{\mathbf{G}}^o\hat{\mathbf{P}}V\hat{\mathbf{G}}^o\hat{\mathbf{P}}V + \ldots \\ &= (1 - V\hat{\mathbf{G}}^o\hat{\mathbf{P}})^{-1}V \end{aligned} \tag{1.130}$$

Such a transition matrix takes into account interaction to all orders of the field-particle interaction V with intermediate states only (i.e., excluding completely the initial and final state).

Thus the total transition matrix becomes

$$\hat{\mathbf{T}} = V(1 + \hat{\mathbf{G}}^o\hat{\mathbf{Q}}\hat{\mathbf{T}}) + V\hat{\mathbf{G}}^o\hat{\mathbf{P}}\hat{\mathbf{T}} \tag{1.131}$$

Solving for **T** in Equation 1.131 by transposing the last term to the left gives

$$\hat{\mathbf{T}} = (1 - V\hat{\mathbf{G}}^o\hat{\mathbf{P}})^{-1}\, V(1 + \hat{\mathbf{G}}^o\hat{\mathbf{Q}}\hat{\mathbf{T}}) = \hat{\mathbf{t}}(1 + \mathbf{G}^o\hat{\mathbf{Q}}\hat{\mathbf{T}}) \tag{1.132}$$

Taking matrix elements of Equation 1.132 between the initial and final states $|\varphi_i>$ and $|\varphi_f>$ we find the inelastic transition amplitude T_{fi}

$$T_{fi} = t_{fi} + t_{fi}G^o_{ii}T_{ii} + t_{ff}G^o_{ff}T_{fi} \tag{1.133}$$

and the total elastic scattering amplitude for state i [21],

$$T_{ii} = t_{ii} + t_{ii}G^o_{ii}T_{ii} + t_{if}G^o_{ff}T_{fi} \tag{1.134}$$

The two equations (1.133 and 1.134) can be used to eliminate T_{ii} so that finally we have an expression for the inelastic transition amplitude

$$T_{fi} = \frac{t_{fi}}{[1 - G^o_{ii}t_{ii}]\,[1 - G^o_{ff}t_{ff}] - t_{fi}t_{if}G^o_{ii}G^o_{ff}} \tag{1.135}$$

where

$$G^o_{ff} = <\varphi_f|\hat{G}^o|\varphi_f> = \lim_{\varepsilon \to o} (E - E^o_f + i\varepsilon)^{-1}$$

This inelastic transition amplitude T_{fi} between the initial state $|\varphi_i>$ and $|\varphi_f>$ is expressed as a transition amplitude t_{fi} in which perturbations to the states $|\varphi_i>$ and $|\varphi_f>$ by intermediate states are excluded. All perturbations to the initial and final state occur only in the denominator as a diagonal matrix element of t, since as an example, $1 - G^o_{ff}t_{ff} = G^o_{ff}(E - E^o_f - t_{ff})$. We thus see that $t_{ii(ff)}$ corresponds to the energy shift of the initial (final) states $\varphi_{i(f)}$ and in the case of the presence of continua in the intermediate states, $t_{ii(ff)}$ will contain an imaginary part $-i\Gamma_{i(f)}$ where Γ corresponds to the linewidth of the initial (final) state (i.e., $t_{ff} = \Delta E_f - i\Gamma_f$) [21, 27].

In order to obtain the transition amplitude $T_{fi}(t)$ from the time-dependent theory (Equation 1.114), one needs to express the total time- independent resolvent $\hat{G}(E)$ in terms of T_{fi} by using Equations 1.128 and 1.135, which gives

$$G_{fi}(E) = \frac{t_{fi}(E)}{[E - E_i^o - t_{ii}(E)]\,[E - E_f^o - t_{ff}(E)] - t_{fi}(E)\,t_{if}(E)} \tag{1.136}$$

The poles of the Green's function matrix element (Equation 1.136) give the true energies of the initial states $|\varphi_i>$ and $|\varphi_f>$ due to the particle-field interactions. These are clearly separated as diagonal (elastic) terms, $t_{ii(ff)}$, corresponding to the intermediate states only. The (inelastic) nondiagonal term, $t_{fi}t_{if} = |t_{if}|^2$, is due to the coupling between states φ_i and φ_f to the intermediate states only. We can therefore factorize the denominator in terms of the eigenenergies,

$$E'_i = E_i^o + \Delta_i - i\Gamma_i \;; \quad E'_f = E_f^o + \Delta_f - i\Gamma_f \tag{1.137}$$

where now an imaginary part can appear representing decay into contiua in the intermediate states which gives rise to photodissociation or photoionization or both [21, 27]. The time-dependent transition amplitude (Equation 1.114) now becomes

$$T_{fi} = -\frac{1}{2\pi i}\int_{-\infty}^{\infty} dE \exp[-i(E - E_f)t/\hbar]\,\frac{t_{fi}(E)}{(E - E_i')\,(E - E_f')} \tag{1.138}$$

which can be rewritten as

$$T_{fi}(t) = -\frac{1}{2\pi i}\int_{-\infty}^{\infty} dE\, t_{fi}(E)\, g(E - E_i')\, g(E - E_f') \exp[-i(E - E_f)t/\hbar] \tag{1.139}$$

where

$$\underset{\varepsilon\to o}{g(E)} = -\frac{i}{\hbar}\int_{o}^{\infty} e^{i(E + i\varepsilon)t/\hbar}\, dt = \frac{1}{E + i\varepsilon} = \frac{PP}{E} - i\pi\delta(E) \tag{1.140}$$

Since g(E) can be expressed as $(E - i\varepsilon)/(E^2 + \varepsilon^2)$, one sees that the real part PP is the principal part of the function 1/E (i.e., E = 0 is to be excluded) and δ(E) is the Dirac delta function. The principal part PP of Equation 1.140 takes into account *nonresonant* processes whereas the imaginary part corresponds to *resonant* processes through the delta function δ(E).The product of the two g functions can be separated in to partial fractions as follows

$$g(E - E_i')g(E - E_f') = g(E_i' - E_f')\,[g(E - E_i') - g(E - E_f')] \tag{1.141}$$

These functions are clearly related to the Green's functions introduced earlier, $\mathbf{G}^o(E)$ (Equation 1.125) and $\mathbf{G}(E)$ (Equation 1.126). They have the following limiting property from their definition (Equation 1.139) [28]

$$g(E)e^{-iEt/\hbar} = -\frac{i}{\hbar}\int_{-t}^{\infty} e^{iE\tau/\hbar}\,d\tau = \begin{cases} -2\pi i\delta(E) & t \to +\infty \\ 0 & t \to -\infty \end{cases} \tag{1.142}$$

Carrying out the integration (Equation 1.140) and substituting $E_f' - E_f = (E_i' - E_f) - (E_i' - E_f')$, one gets

$$\underset{t\to\infty}{T_{fi}(t)} = t_{fi}(E_i')g(E_i' - E_f')e^{+i(E_i' - E_f)t/\hbar} - t_{fi}(E_f')e^{-i(E_i' - E_f)t/\hbar}\,g(E_i' - E_f')e^{+i(E_i' - E_f')t/\hbar} \tag{1.143}$$

In the limit $t \to \infty$ the second term in Equation 1.143 is zero as a result of Equation 1.142. We now introduce the approximation that the interaction time is small so that the perturbative effects on the energy levels are negligible, (i.e., we assume $E_i' = E_i$, $E_f' = E_f$) which implies from Equations 1.137 and 1.43, that

$$\frac{(\Delta - i\Gamma)t}{\hbar} << 1 \tag{1.144}$$

for any initial and final state $|\varphi_i>$ and $|\varphi_f>$, respectively.

Thus finally

$$\underset{t\to\infty}{T_{fi}(t)} = t_{fi}(E_i)\;g(E_i - E_f)\;e^{-i(E_i - E_f)t/\hbar} \tag{1.145}$$

The rate of transition W_{fi} per unit time then follows from the definition

$$W_{fi} = \frac{d}{dt} |T_{fi}(t \to \infty)|^2 = \frac{2\pi}{\hbar} |t_{fi}(E_i)|^2 \, \delta(E_i - E_f) \tag{1.146}$$

where we have used the relations (see Equation 1.140)

$$Eg(E) = 1; \quad g^*(E) = -2\pi i \delta(E) \tag{1.147}$$

Thus one obtains a constant rate of transition under the assumption that perturbations to initial and final states are negligible. For Equation 1.146 to be meaningful, interaction times must be larger than typical atomic periods since one needs to calculate a matrix element t_{fi} with respect to stationary states $|\varphi_i>$ and $|\varphi_f>$ but short enough that these states can still be considered as stationary. For weak fields, these conditions can be satisfied and the expansion of t in terms of the radiative interaction, V, and Equation 1.129 yields the usual perturbative multiphoton transition amplitudes. In the non-perturbative limit (i.e., strong field limit) both initial and final states are nonstationary. Energy shifts and widths can become large and energy dependent. The transition amplitudes (probabilities) can become time-dependent and cross-sections are more difficult to define [21, 27, 29]. As an example in Figure 1.1, we illustrate the population of the 1*s* level of the H atom obtained by finite element calculation of the ionization probabilities at an intensity of $I = 1.75 \times 10^{14}$ W/cm^2 [30]. One clearly sees different rates at short, intermediate, and long times due to the fact that the interaction is nonperturbative (i.e., the initial and final states cannot be considered as stationary states). In the case of Figure 1.1, calculating the rate at long times gives an ionization rate $W_{oi} = 10^{14}$/s (i.e., a lifetime of 10 femtoseconds).

In general, the exact eigenstate $|\varphi(t)>$ can be expanded in terms of complete set of the field-particle states

$$|\varphi(t)> = \sum_{p,n_1,n_2,n_j} C_{p,n_1,n_2n_j}(t) \, |p; n_1,n_2, ..., n_j> \tag{1.148}$$

Equation (1.114) allows one to calculate the transition probabilities

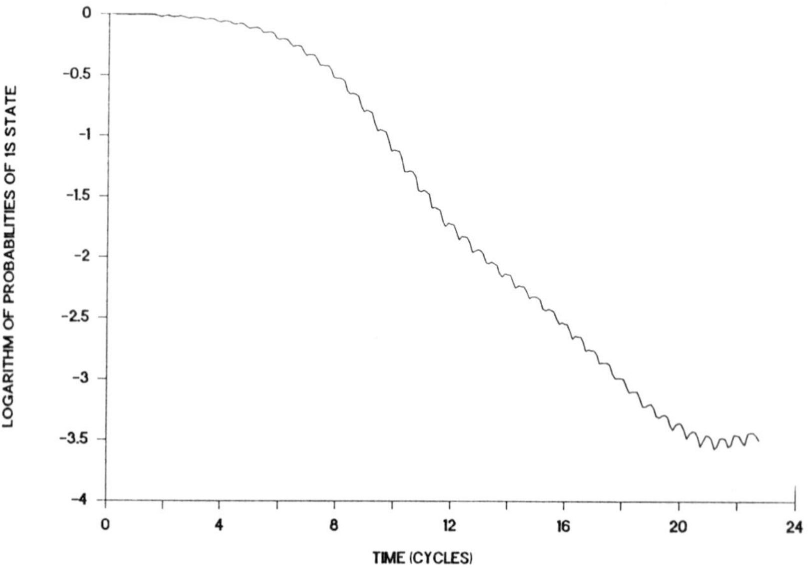

Figure 1.1 H Atom in a laser field of intensity $I = 1.75 \times 10^{14}$ W/cm^2 at frequency $\omega = 0.2$ au. Log of occupation probability of initial 1s state as a function of time in cycles (t(cycle) = 8×10^{-16} sec).

$$|T_{fi}|^2 = |C^{(t \to \infty)}_{p,n_1,n_2,\ldots,n_j}|^2$$

using the field-particle states as basis sets. Alternatively, one can solve a linear system of differential equations for the amplitudes $C_{p,n_1,n_2,n_j}(t)$ occurring in Equation 1.148 by putting the expansion into the time-dependent Schrödinger Equation with the quantized time-dependent vector potential (Equation 1.99). In the case of short pulses, this necessitates expanding the pulse in terms of the coherent states which are themselves linear combinations of the photon states (Equation 1.92). Examples of these approaches will appear in various chapters of this book.

In conclusion of this section, we emphasize that classical or semiclassical electromagnetic fields can always be expanded in terms of photon states so that a dressed state representation can always be used to interpret the results as transitions between these states. On the other hand we have alluded previously to the fact that there exist gauge transformations which can change the equations of motion. Thus, in the Coulomb gauge, it was pointed out that the vector potential $\mathbf{A}(\mathbf{r},t)$ is not unique (see Equation 1.9). In quantum mechanics, one can usually obtain general unitary transformations of the time-dependent Schrödinger Equation (1.68) for both time-dependent and time-independent Hamiltonians alike. These transformations change the physical interpretations since the Equations of motion in the classical and therefore also in the quantum theory are changed. We turn next to this general problem in particle-radiation interactions.

1.8 GAUGE AND UNITARY TRANSFORMATIONS

In Equation 1.9 we pointed out that the potentials **A** and U of electromagnetic theory are not unique so that gauge transformations can be introduced via the transformation function $\mathbf{F}(\mathbf{r},t)$. Such a function defines gauge transformations in the Lagrangean formulation of field-particle interactions and must not depend on the velocities $\dot{\mathbf{r}}_j$, so that the new Lagrangean obtained from the transformation must like the original one depend only on the dynamical variables $\mathbf{r}_j$ and $\dot{\mathbf{r}}_j$, **A** and $\dot{\mathbf{A}}$ and *not* on the acceleration [4, 7]. Thus proper gauge transformations are defined for the field-particle system by the gauge functions $\mathbf{F}(\mathbf{r}_j;\mathbf{A}(\mathbf{r});t)$ where now **A** is to be considered as the field dynamical variable and

$$\mathbf{E} = -\frac{1}{c}\frac{\partial \mathbf{A}}{\partial t}$$

is the conjugate momentum (see Equations 1.17 and 1.46). Each change in gauge will in principle introduce now longitudinal components to **A** (i.e., $\mathbf{A}_{\parallel} \neq 0$) and will also introduce a change in the quantum representation. With each of these transformations one can show one has associated with it a unitary transformation operator $\hat{\mathbf{T}}$ of the form [4, 7],

$$\hat{\mathbf{T}} = \exp \frac{1}{\hbar}\,\mathbf{F}(\mathbf{r}_j;\,\mathbf{p}_j;\,\mathbf{A}(\mathbf{r}_j);\,\mathbf{E}(\mathbf{r}_j);\,t) \tag{1.149}$$

where **F** is a function of the generalized coordinates of the system, $\mathbf{r}_j$ and $\mathbf{A}(\mathbf{r}_j)$. The unitary transformation defined above depends only on the generalized coordinates. However one can consider more general unitary transformations depending also on the conjugate momenta

$$\hat{\mathbf{T}} = \exp \frac{i}{\hbar} \mathbf{F}(\mathbf{r}_j;\mathbf{p}_j;\mathbf{A}(\mathbf{r}_j);\mathbf{E}(\mathbf{r}_j);\, t) \tag{1.150}$$

The presence of the conjugate momenta do not permit one to call these gauge transformations in the strict Lagrangean formalism but rather these are now general unitary quantum mechanical transformations which lead to changes in quantum representation (i.e., unitary transformations are more general than gauge transformations). Because of the noncommutativity of coordinates and momenta, the presence of these in **F** (Equation 1.150) implies that the transformation will change the coordinates and momenta in general in going from one representation to another. Thus states $|\varphi(t)\rangle$ and operators $\hat{\mathbf{O}}(t)$ of some original system will change as

$$|\varphi'(t)\rangle = \hat{\mathbf{T}}|\varphi(t)\rangle \tag{1.151}$$

$$\hat{\mathbf{O}}'(t) = \hat{\mathbf{T}}\hat{\mathbf{O}}\hat{\mathbf{T}}^{+} \tag{1.152}$$

where $\hat{\mathbf{T}}^{+}\hat{\mathbf{T}} = 1$ so that

$$\langle\varphi(t)|\hat{\mathbf{O}}|\varphi(t)\rangle = \langle\varphi'(t)|\hat{\mathbf{O}}'(t)|\varphi'(t)\rangle \tag{1.153}$$

remains invariant. The transformation (Equation 1.151) will induce a change in the Hamiltonian, which implies a change in physical interpretation

$$i\hbar\frac{d|\varphi'\rangle}{\partial t} = \left[i\hbar\frac{d\hat{\mathbf{T}}}{dt} + \hat{\mathbf{T}}\hat{\mathbf{H}}(t)\right]|\varphi(t)\rangle = \left[i\hbar\frac{d\hat{\mathbf{T}}}{dt}\,\hat{\mathbf{T}} + \hat{\mathbf{T}}\hat{\mathbf{H}}(t)\hat{\mathbf{T}}^{+}\right]|\varphi'(t)\rangle \tag{1.154}$$

The Hamiltonian in the new primed representation with new basis functions $|\varphi'(t)\rangle$ is now

$$\hat{\mathbf{H}}'(t) = \hat{\mathbf{T}}\hat{\mathbf{H}}(t)\hat{\mathbf{T}}^{+} + i\hbar\,\frac{d\hat{\mathbf{T}}}{dt}\,\hat{\mathbf{T}}^{+} \tag{1.155}$$

Thus any new operator $\hat{\mathbf{O}}'$ in the new representation is related to the original operator $\hat{\mathbf{O}}$ by Equation 1.152 and so that by Equation 1.153 all physical predictions such as evolution operators are identical. It is to be noted that Equation 1.155 implies the existence of unitary equivalent Hamiltonians $\hat{\mathbf{H}}_T = \hat{\mathbf{T}}\hat{\mathbf{H}}\hat{\mathbf{T}}^+$. In fact by choosing the gauge transformation function $\mathbf{F}$ to be

$$\frac{q}{c}\chi(\mathbf{r},t)$$

and the relation

$$\hat{\mathbf{T}}\hat{\mathbf{p}}\hat{\mathbf{T}}^+ = \hat{\mathbf{p}} - \frac{q}{c}\nabla\chi$$

then one can rewrite the time-dependent single particle Schrödinger Equation (1.154) as

$$i\hbar\frac{\partial|\varphi'\rangle}{\partial t} = \left[\frac{1}{2m}\left(\mathbf{p} - \frac{q}{c}\mathbf{A}'\right)^2 + \mathbf{U}'\right]|\varphi'\rangle \tag{1.159}$$

where $\mathbf{A}'$ and $\mathbf{U}'$ have been defined previously in Equation 1.9. Clearly, the new state $|\varphi'\rangle$, which differs from the previous state $|\varphi\rangle$, only by the phase factor $\exp(iq\chi/\hbar c)$ is the eigenstate of a new Schrödinger Equation which has the same form as the original equation but with new gauge transformed potentials $\mathbf{A}'(\mathbf{r},t)$ and $\mathbf{U}'(\mathbf{r},t)$. We note that this *gauge invariance* is only valid for $\mathbf{F}$ (or χ) being scalar functions of $\mathbf{r}$ only (i.e., they must commute with the vector potential $\mathbf{A}$). Later, we will encounter further on more general transformation functions (Equation 1.150) which will not commute with $\mathbf{A}$ and therefore cannot be considered as gauge functions in the Lagrangean sense. Finally we note that due to the time-dependence of $\hat{\mathbf{T}}$, $\hat{\mathbf{H}}'(t) \neq \hat{\mathbf{T}}\hat{\mathbf{H}}(t)\hat{\mathbf{T}}^+$ (i.e., $\hat{\mathbf{H}}'(t)$ and $\hat{\mathbf{H}}(t)$ are no longer unitary equivalent and they no longer possess the same spectrum).

We now examine the invariance of transition amplitudes to gauge transformations. We return to the time-independent quantum field description from Section 1.5, for which the evolution operator $\hat{\mathrm{U}}(t)$ (Equation 1.108) involving the time-independent quantum Hamiltonian $\hat{\mathbf{H}}$ was shown to give time-independent transition amplitudes T_{fi} (see Equation 1.161), expressible in terms of the time-independent radiative interaction $V = \mathbf{H}_I$ (Equation 1.71). It was shown further in Section 1.7 that the time-dependent theory based on the time-dependent semiclassical field (Equation 1.100) gave identical equations for the Green's functions and thus implicitly for the evolution operator also as in the time-independent formulation based on quantum fields (Equations 1.118 versus 1.124). In the present discussion, we examine the effect of time-dependent gauge transformations on the transition amplitudes defined according to the evolution Equation 1.108 where $\hat{\mathbf{H}}$ is time-independent.

Thus for any time t, we have from Equations 1.154 and 1.155

$$|\varphi'(t)\rangle = \hat{\mathbf{T}}(t)\hat{\mathbf{U}}(t)\,|\varphi(0)\rangle = \hat{\mathbf{T}}(t)\hat{\mathbf{U}}(t)\hat{\mathbf{T}}^{+}(0)|\varphi'(0)\rangle$$

or

$$\hat{\mathbf{U}}'(t) = \hat{\mathbf{T}}(t)\hat{\mathbf{U}}(t)\hat{\mathbf{T}}^{+}(0) \tag{1.160}$$

By definition (see Equation 1.115), we have the identity

$$T_{fi} = \lim_{t\to\infty} \langle\varphi_f'(t)|\hat{\mathbf{U}}'(t)|\varphi_i'(0)\rangle = \lim_{t\to\infty}\langle\varphi_f(t)|\hat{\mathbf{U}}(t)|\varphi_i(0)\rangle \tag{1.161}$$

that is, the transition amplitude is gauge invariant because of the unitarity of the transformation operator $\hat{\mathbf{T}}(t)$. We note that if $\hat{\mathbf{T}}(t)$ and $\hat{\mathbf{T}}(0)$ are $\neq 1$, then $\varphi_f'(t) \neq \varphi_f(t)$, $\varphi_i'(0) \neq \varphi_i(0)$.

We now consider the problem of interaction of the particle system with a pulse; therefore, one has to consider the possibility of a time-dependent interaction $\mathbf{H}_I(t)$ due to a rise and fall of the field amplitudes. We thus choose our time-dependent Hamiltonian as

$$\hat{\mathbf{H}}(t) = \hat{\mathbf{H}}_o + \gamma(t)\,\hat{\mathbf{H}}_I \tag{1.162}$$

where $\hat{\mathbf{H}}_o$ is the uncoupled field-particle Hamiltonian (Equation 1.126) and $\hat{\mathbf{H}}_I$ is the time-independent quantum radiative interaction (Equation 1.125). Anticipating the fact that both $\hat{\mathbf{H}}_I$ and the gauge transformation functions $\hat{\mathbf{F}}$ will have the same field dependence through the field operators $\hat{\mathbf{A}}(\mathbf{r})$ and $\hat{\mathbf{E}}(\mathbf{r})$, we choose the gauge transformation operators as

$$\hat{\mathbf{T}}[\gamma(t)] = e^{i\gamma(t)\hat{\mathbf{F}}/\hbar} \tag{1.163}$$

Since $\hat{\mathbf{T}}(t)$ now depends on time via the rise and fall (turning on and off) parameter $\gamma(t)$, the evolution in the new representation $|\varphi'(t)>$ is defined by the time-dependent Hamiltonian (Equation 1.158)

$$\hat{\mathbf{H}}'[\gamma(t)] = \hat{\mathbf{H}}_o + \hat{\mathbf{H}}_I'[\gamma(t)] - \dot{\gamma}(t)\,\hat{\mathbf{F}} \tag{1.164}$$

where $\hat{\mathbf{H}}_1'[\gamma(t)]$ contains all terms involving $\gamma(t)$ after the gauge transformation.

In view of the preceding discussion, transition amplitudes calculated using Equations 1.162 and 1.164 must be identical. Furthermore, imposing the condition that $\gamma(t \to \infty) = \gamma(0) = 0$, defines proper unperturbed initial $|\varphi_i>$ and final $|\varphi_f>$ eigenstates of $\hat{\mathbf{H}}_o$. Hence transition amplitudes remain gauge invariant providing the full-gauge transformed Hamiltonian $\hat{\mathbf{H}}'[\gamma(t)]$ is used. For slow varying pulses (i.e., both slow envelope and phase variation) one is in the *adiabatic* limit: $\dot{\gamma}(t) << 1$, so that amplitudes calculated using $\hat{\mathbf{H}}_I$ and $\hat{\mathbf{H}}_I'$ will give equivalent results. Rapidly varying pulse envelopes (through the Fourier components, $\mathbf{A}_j$ (Equation 1.100) and phases φ_j (Equation 1.107) clearly give rise to time-dependent nonadiabatic transitions, which must be included in order to preserve gauge invariance of the transition amplitudes T_{fi}).

1.9 ELECTRIC FIELD GAUGE, BLOCH-NORDSIECK REPRESENTATION, AND DIPOLE APPROXIMATION

The Coulomb or radiation gauge was defined in Section 1.2 as that gauge where the vector potential satisfies the transversality condition $\nabla \cdot \mathbf{A} \equiv 0$. It gives rise to a Hamiltonian (Equations 1.69-1.71) where the radiation potential $\mathbf{A}$ is coupled to the particle canonical momentum operators $\hat{\mathbf{p}}$, which are only equal to $m\dot{\mathbf{r}}$ when $\mathbf{A} \equiv 0$. In the case where the field wavelength λ is large with respect to the atomic or molecular size which is usually of the order of 1 Å (10^{-8} cm), then the spatial variation of the electromagnetic field can be neglected. This approximation is called the long-wavelength approximation and corresponds to setting $e^{i\mathbf{k}\cdot\mathbf{r}} \simeq 1$ in the field Equations (1.57–1.59). Thus the fields become purely time-dependent. If the long wavelength approximation is made before applying any unitary transformation to the total Hamiltonian (Equation 1.68), one obtains the usual dipole-approximation.

The long wavelength approximation in general does not commute with the unitary transformations [31, 32]. Using this dipole approximation, a unitary transformation was introduced by Goppert-Mayer in 1931 which leads to a description of field-particle interactions in terms of the classical scalar dipole-electric field potential $\mathbf{E} \cdot \mathbf{r}$ [33]. This we call the electric field gauge (EF). Such a gauge is based on physically measurable quantities such as the electric field and the particle dipole moment. This gauge has been generalized beyond the dipole approximation by Powers and Zienau [34, 35] so that the total Hamiltonian $\hat{\mathbf{H}}$ (Equations 1.69–1.71) is expressible in terms of the physical fields $\mathbf{E}(\mathbf{r},t)$ and $\mathbf{B}(\mathbf{r},t)$ and electric $\mathbf{P}$ and magnetic $\mathbf{M}$ polarizations, in close analogy to classical electricity and magnetism. Finally, one should add that absorption lineshapes are easily calculated in the EF gauge (as shown by Lamb [36]) whereas in the Coulomb gauge a much more elaborate calculation involving nonresonant transitions is required [37, 38], thus suggesting that the EF gauge is usually more convenient.

A. EF Gauge

We choose as origin the coordinate $\mathbf{r} = 0$, so that in the dipole approximation ($r/\lambda < 1$), the total semiclassical Hamiltonian (Equation 1.98) can be written as

$$\hat{\mathbf{H}}_C(r,t) = \sum_j \frac{1}{2m_j}\left[\hat{\mathbf{p}}_j - \frac{q_j}{c}\mathbf{A}(t)\right]^2 + V_C \tag{1.165}$$

where V_C is the coulomb energy. The form of $\hat{\mathbf{H}}_C(t)$ suggests the introduction of a unitary operator $\hat{\mathbf{T}}$ which translates each momentum operator $\hat{\mathbf{p}}_j$ by $q_i/c\, A(t)$ in order to eliminate the vector potential. This is achieved by choosing

$$\hat{\mathbf{T}} = \exp\left[\frac{-i}{\hbar} c \sum_j q_j \mathbf{r}_j \cdot \mathbf{A}(t)\right] = \exp\left[\frac{-i}{\hbar}\mathbf{d} \cdot \frac{\mathbf{A}(t)}{c}\right] \tag{1.166}$$

where

$$\mathbf{d} = \sum q_j \mathbf{r}_j$$

is the total electric dipole moment of the charge distribution with respect to the origin. From the relation $e^{-i\alpha x}\,\hat{\mathbf{p}}e^{+i\alpha x} = \hat{\mathbf{p}} + \alpha$, one readily obtains the translation

$$\hat{\mathbf{T}}\mathbf{p}_j\hat{\mathbf{T}}^+ = \mathbf{p}_j + \frac{q_i}{c}\,\mathbf{A}(t) \tag{1.167}$$

The transformed Hamiltonian is obtained from Equation 1.158, where now

$$\hat{\mathbf{T}}\hat{\mathbf{H}}_C(t)\hat{\mathbf{T}}^+ = \sum_j \frac{\mathbf{p}_j^2}{2m_j} + V_C \tag{1.168}$$

$$i\hbar\left(\frac{d\hat{\mathbf{T}}}{dt}\right)\hat{\mathbf{T}}^+ = \mathbf{d}\cdot\frac{\dot{\mathbf{A}}(t)}{c} = -\,\mathbf{d}\cdot\mathbf{E}(t) \tag{1.169}$$

The total new EF gauge Hamiltonian now reads

$$\hat{\mathbf{H}}_{EF}(t) = \sum_j \frac{\hat{\mathbf{p}}_j^2}{2m_j} + V_C - \mathbf{d}\cdot\mathbf{E}(t) \tag{1.170}$$

Both interaction terms ($q\mathbf{A}\cdot\mathbf{p}/mc$) and ($q^2\,\mathbf{A}^2/2mc^2$) in Equation 1.165 have been replaced by the dipole interaction. Two comments are now in order. First, since the unitary transformation (Equation 1.166) involves only particle and field coordinates and no momenta, it corresponds to a proper gauge transformation as it also can be shown to give rise to a new Lagrangean [4, 7]. Secondly, in Equation 1.169 we have used the equivalence (Equation 1.17)

$$\mathbf{E}_\perp = -\frac{1}{c}\frac{\partial \mathbf{A}_\perp}{\partial t} = -\frac{1}{c}\,\dot{\mathbf{A}}(t)$$

so that **E** is in principle transversal. Thus in comparing calculations in the Coulomb gauge and the EF gauge, one must respect this relation between **E** and **A**. As an example, if one chooses $\mathbf{A}(t) = A(t)\hat{\varepsilon}\cos\omega t$, then in the EF gauge one must use

$$\mathbf{E}(t) = \frac{\hat{\varepsilon}}{c}\,[A(t)\,\omega \sin \omega t - \dot{A}(t) \cos \omega t]$$

in order to have perturbative transition matrix elements (Equation 1.162) remain invariant.

We now consider the same EF gauge transformation in a *quantized* representation, appropriate to the description of dressed states discussed in Section 1.7. The vector field operator becomes in the dipole approximation ($\mathbf{k} \cdot \mathbf{r}_j \ll 1$)

$$\hat{\mathbf{A}} = \sum_j \mathbf{A}_{\omega_j} \hat{\varepsilon}_j\,[\hat{\mathbf{a}}_j + \hat{\mathbf{a}}_j^+] \tag{1.171}$$

We now apply the equivalent unitary transformation (Equation 1.165) in quantized form to $\hat{\mathbf{H}}_C$

$$\hat{\mathbf{T}} = \exp\left[-\frac{i}{\hbar}\sum q_j\,\hat{\mathbf{r}}_j \cdot \frac{\hat{\mathbf{A}}}{c}\right] = \exp\left[-\frac{i}{\hbar}\mathbf{d} \cdot \hat{\mathbf{A}}\right] \tag{1.172}$$

In Equation 1.172 $\hat{\mathbf{A}}$ is now a time-independent field operator whereas in Equation 1.167 $\mathbf{A}(t)$ is the equivalent semiclassical time-dependent field. We can therefore rewrite the quantum unitary operator (Equation 1.172) as

$$\hat{\mathbf{T}} = \exp\left[-i\sum_j \gamma_j\,(\hat{\mathbf{a}}_j + \hat{\mathbf{a}}_j^+)\right]; \quad \gamma_j = \frac{\hat{\varepsilon}_j \cdot \mathbf{d}}{(2\hbar\omega_j L^3)^{1/2}} \tag{1.173}$$

where the γ's are c-numbers. From these last two equations it follows that

$$\hat{\mathbf{r}}_j' = \hat{\mathbf{T}}\hat{\mathbf{r}}_j\hat{\mathbf{T}}^+ = \frac{\hat{\mathbf{p}}_j}{m_j} \tag{1.174}$$

$$\hat{\mathbf{T}}\hat{\mathbf{a}}_k\hat{\mathbf{T}}^+ = \hat{\mathbf{a}}_k + i\gamma_k; \quad \hat{\mathbf{T}}\hat{\mathbf{a}}_k^+\hat{\mathbf{T}}^+ = \hat{\mathbf{a}}_k^+ - i\gamma_k \tag{1.175}$$

We see that in the EF representation, both semiclassical and quantum, the particle momentum $\mathbf{p}_j$ which was defined in the Coulomb gauge as

$$\mathbf{p}_j' = m_j\hat{\dot{\mathbf{r}}}_j + \frac{q_j}{c}\mathbf{A}(t) \tag{1.176}$$

and included the longitudinal field momentum now becomes exactly the mechanical momentum

$$\mathbf{p}_j' = m_j\hat{\dot{\mathbf{r}}}_j \tag{1.177}$$

It is the velocity $\hat{\dot{\mathbf{r}}}_j$ which is truly gauge invariant but not the momentum $\mathbf{p}_j$. Thus in the EF gauge one is dealing with physical variables only, i.e., $\hat{\dot{\mathbf{r}}}_j$ and $\mathbf{E}$. Finally, in the quantum representation, one sees that the quantum field operators $\hat{\mathbf{a}}(\hat{\mathbf{a}}^+)$ are displaced by the quantities γ_k which depend on the particle dipole moment only. This gives a small correction to the transverse zero point field energy [7, 14, 16].

B. BN Representation

One of the advantages of the EF gauge is that it depends only on physical variables (i.e., $\mathbf{p}_j = m\hat{\dot{\mathbf{r}}}_j$) and $\mathbf{E}$. This was achieved by a unitary transformation (Equation 1.172) which translated the momentum operator in the Coulomb gauge. Clearly another possibility would be to translate the particle coordinates, thus suggesting the transformation function (in the dipole approximation)

$$\mathbf{F} = \frac{q_j}{m_j c}\mathbf{p}_j \cdot \int \mathbf{A}(t)\,dt + \frac{q_j^2}{2m_j c^2}\int \mathbf{A}^2(t)\,dt \tag{1.178}$$

where in the dipole approximation, the

$$\frac{\mathbf{A}^2}{2m_j c^2}$$

term is but a phase factor which can be easily removed by adding it to $\mathbf{F}$. The effect of this transformation is to translate the particle coordinates $\mathbf{r}_j$

$$\hat{\mathbf{T}}\mathbf{r}_j\hat{\mathbf{T}}^+ = \mathbf{r}_j + \alpha_j(t) \tag{1.179}$$

where

$$\alpha_j(t) = \frac{q_j}{m_j c} \int \mathbf{A}(t)\, dt \tag{1.180}$$

or

$$\ddot{\alpha}_j = \frac{q_j}{m_j c} \dot{\mathbf{A}}(t) = \frac{q_j}{m_j} \mathbf{E}(t) \tag{1.181}$$

Thus the particle j is displaced along the classical trajectory that it would follow in the electric field $\mathbf{E}(t)$.

This transformation was first used by Bloch and Nordsieck [14] and then by Pauli and Fierz [39] in order to remove the infrared catastrophe in early quantum electrodynamics. It has been cited as the Bloch and Nordsieck transformation [40, 41], the Pauli-Fierz-Kramers transformation [42, 43], and the space translation representation [15, 44]. We refer to it as the BN (Bloch and Nordsieck) transformation [16, 32]. In this new representation the total field-particle Hamiltonian now becomes

$$\hat{\mathbf{H}}_{BN} = \frac{\hat{\mathbf{p}}^2}{2m} + V(\mathbf{r}_j + \alpha_j(t)) \tag{1.182}$$

In the quantum BN representation, one obtains the following relation between the Coulomb gauge dynamical operator, $\hat{\mathbf{O}}$, and the new BN operators, $\hat{\mathbf{O}}'$, following the recipe given in Equation 1.155 in the dipole approximation [31, 41, 42]

$$\hat{\mathbf{p}}_j' = \hat{\mathbf{p}}_j; \quad \mathbf{r}_j' = \mathbf{r}_j - \frac{q_j}{m_j \omega_k^2} \hat{\mathbf{E}}_k \tag{1.183}$$

$$\hat{\mathbf{a}}_k' = \hat{\mathbf{a}}_k - \frac{i\hat{\varepsilon}_k}{(2\hbar\omega_k^3 L^3)^{1/2}} \cdot \sum_j \frac{q_j \hat{\mathbf{p}}_j}{m_j c} \tag{1.184}$$

We see that the new momentum $\mathbf{p}'$ is unchanged, the particle coordinates are now translated by the electric field, whereas the quantized field operators are translated by particle momenta instead of the particle coordinates. Thus the BN representation is in some sense canonical to the EF gauge, since in the latter field variables are translated by particle coordinates (Equation 1.175), and in the former by particle momenta (Equation 1.184). In fact these particle momenta displacements of the radiation field operators in the BN representation can be shown to generate interparticle magnetic interactions when one goes beyond the dipole approximation [32, 41].

We conclude that the BN transformation cannot be considered as a gauge transformation since the appearance of the particle momentum $\mathbf{p}_j$ in the transformation function $\mathbf{F}$ (Equation 1.178) would generate accelerations in the equations of motion and hence one cannot consider $\mathbf{F}$ as a generating function for a Lagrange transformation. Rather, the BN transformation corresponds to a unitary transformation which is more general than gauge transformations. Of note is the fact that the BN representation in the dipole approximation describes the field-particle interactions in terms of a particle coordinate displacement $\alpha(t)$ which satisfies the classical Equations of motion of the particle in an electric field $\mathbf{E}(t)$.

We now compare the radiative matrix elements in the three representations—Coulomb, EF, and BN—in order to assess the convergence properties of any perturbation expansion for these representations. For a classical oscillator, one has the relation $\mathbf{p}/m = \mathbf{v} = \omega\mathbf{r}$ between the physical variables of the system. In the quantum picture this becomes an Equation for matrix elements between states $|\alpha\rangle$ and $|\beta\rangle$

$$i\hbar = \frac{\mathbf{p}_{\alpha\beta}}{m} = \langle\alpha[\hat{\mathbf{r}},\hat{\mathbf{H}}_o]|\beta\rangle = \hbar\omega_{\alpha\beta}\mathbf{r}_{\alpha\beta} \tag{1.185}$$

where $\hbar\omega_{\alpha\beta} + E_\alpha - E_\beta$ and $\mathbf{H}_o$ is the field-free Hamiltonian. Using the equivalence $\mathbf{A} = \mathbf{E}c/\omega$ gives the following relation between the momentum radiative matrix elements in the Coulomb gauge and the dipole radiative matrix elements in the EF gauge as

$$\frac{\mathbf{A}\cdot\mathbf{p}_{\alpha\beta}}{mc} = i\,\frac{\omega_{\beta\alpha}}{\omega}\,\mathbf{r}_{\alpha\beta}\cdot\mathbf{E} \tag{1.186}$$

Thus the radiative matrix elements are only equal in resonance (i.e., $\omega = \omega_{\beta\alpha}$). For nonresonant and virtual transitions, then large matrix elements are obtained in the Coulomb gauge for $\mathbf{p}_{\alpha\beta}$ when $\omega_{\beta\alpha}/\omega >> 1$ (i.e., high energy virtual nonresonant excitations). In the case of multiphoton transitions, such nonresonant transitions will therefore be important in calculating transition amplitudes in the Coulomb gauge.

A comparison of the Coulomb and BN radiative transition matrix elements is obtained by expanding the radiatively shifted potential (Equation 1.182)

$$V[\mathbf{r} + \alpha(t)] = V(\mathbf{r}) + \alpha(t) \cdot \nabla V + \ldots \tag{1.187}$$

and using the commutation relation

$$i\hbar = \nabla V = [\hat{\mathbf{H}}_0, \hat{\mathbf{p}}] \tag{1.188}$$

one obtains

$$(\nabla V)_{\alpha\beta} = i\omega_{\beta\alpha}\, \mathbf{p}_{\alpha\beta} \tag{1.189}$$

from which follows (assuming $A(t) = A_0 \cos \omega t$ in Equation 1.180)

$$\alpha(t) \cdot (\nabla V)_{\alpha\beta} = \frac{iA_0}{mc}\left(\frac{\omega_{\beta\alpha}}{\omega}\, \mathbf{p}_{\alpha\beta}\right) \tag{1.190}$$

This agrees with the Coulomb gauge radiative coupling only in the resonant case. Thus in the BN representation, nonresonant and virtual transitions are even more dominant when compared to Coulomb gauge calculations. Clearly, BN representation perturbation expansion will converge more slowly than Coulomb gauge expansions, the latter converging also more slowly than EF expansions. We note however that at very high frequencies (i.e., $\omega >> \omega_{\beta\alpha}$), then the BN matrix elements will be smaller than the corresponding elements in the other two representation. This has been therefore called a high frequency approximation [15] (i.e., one expects more rapid convergence of perturbative and nonperturbative calculations in the BN representation as the particle follows less and less high frequency fields).

In the molecular case, a comparison between all three representations is further complicated by the Born-Oppenheimer approximation which is always invoked in order to separate electronic and nuclear motion. Thus in the case of symmetric molecules (i.e., with inversion symmetry as in the atomic case), the previous comparison remains valid for molecules also. In the case of nonsymmetric molecules, both Coulomb and BN representation give zero radiative transition moments for intrastate (e.g., infrared) transitions. This is due to the fact that permanent dipole moments are zero in both Coulomb and BN representations. These are restored and become equal to the EF permanent moments only after introducing corrections to the Born-Oppenheimer approximation [16, 32].

In conclusion of this section, any representation can be used for multiphoton radiative calculations provided one uses complete basis sets and the complete Hamiltonian. In the molecular case, this latter requirement becomes imperative, since the Born-Oppenheimer approximation introduces an additional approximation beyond the atomic case, the adiabatic approximation. This approximation violates conservation of total momentum, nuclear and electronic, by setting the nuclear momentum to zero. Nonadiabatic corrections restore the total momentum conservation law [16].

1.10 THE FREE PARTICLE AND PONDEROMOTIVE ENERGIES

The free particle (i.e., $V_C = 0$) offers the possibility of obtaining exact solutions of the time-dependent Schrödinger Equation in presence of a semiclassical time-dependent electromagnetic field [43]. This will enable us to compare different gauges or representations and derive the important ponderomotive potentials which are present when particles interact with oscillatory fields [45]. At high intensities, these potentials can determine the dynamics of ionization [1, 2].

We will use a general unitary transformation, following the original method of Husimi [46], to enable us to go from one representation to another in a unified way. Thus we introduce the ansatz (we will set $\hbar = 1$) for the evolution operator $\hat{U}(t)$ (Equation 1.108)

$$\hat{U}(t) = e^{-if(t)}\, e^{-i\mathbf{g}(t)\cdot\mathbf{r}}\, e^{-i\mathbf{h}(t)\cdot\mathbf{p}/m}\, e^{-itp^2/2m} \tag{1.191}$$

The second and third exponential are essentially the EF and BN transformation operators discussed in the previous section. The first exponential is an arbitrary function of t and the last exponential corresponds to the free particle wave function. Using the relation $e^{-i\mathbf{g}\cdot\mathbf{r}}\,\hat{\mathbf{p}}\,e^{+i\mathbf{g}\cdot\mathbf{r}} = \mathbf{p} + \mathbf{g}$, one obtains an Equation of motion for $\hat{\mathbf{U}}(t)$,

$$i\frac{d\hat{\mathbf{U}}(t)}{dt} = \left[\left(\dot{\mathbf{f}} - \frac{\dot{\mathbf{h}}^2}{2m}\right) + \mathbf{r}\cdot\dot{\mathbf{g}} + \frac{1}{2m}(\mathbf{p} + \mathbf{g} + \dot{\mathbf{h}})^2\right]\hat{\mathbf{U}}(t) \tag{1.192}$$

where $\dot{\mathbf{f}} = df/dt$, etc. Since the evolution operator also satisfies the time-dependent Schrödinger Equation, $id\hat{\mathbf{U}}/dt = \hat{\mathbf{H}}(t)\hat{\mathbf{U}}(t)$, this allows us to specify the functions **f**, **g**, **h** for each gauge or representation

i) $$\mathbf{H}_C = \frac{(\mathbf{p} + q\mathbf{A}/c)^2}{2m} = \frac{1}{2}m\dot{\mathbf{r}}^2 \tag{1.193}$$

$$\mathbf{g} = 0, \quad \dot{\mathbf{h}} = \frac{q\mathbf{A}(t)}{c}, \quad \dot{\mathbf{f}} = \frac{\dot{\mathbf{h}}^2}{2m} = \frac{q^2\mathbf{A}(t)^2}{2mc^2} \tag{1.194}$$

ii) $$\mathbf{H}_{EF} = \frac{\mathbf{p}^2}{2m} + q\mathbf{r}\cdot\mathbf{E}(t) \tag{1.195}$$

$$\dot{\mathbf{g}} = \mathbf{E}(t), \quad \mathbf{g} = -\dot{\mathbf{h}}, \quad \dot{\mathbf{f}} = \frac{\dot{\mathbf{h}}^2}{2M} = \frac{\mathbf{g}^2}{2m} \tag{1.196}$$

iii) $$\mathbf{H}_{BN} = \frac{\mathbf{p}^2}{2m} + \frac{q^2\mathbf{A}^2(t)}{2mc^2} \tag{1.197}$$

$$\mathbf{g} = \mathbf{h} = 0, \quad \dot{\mathbf{f}} = \frac{q^2\mathbf{A}^2(t)}{2mc^2} \tag{1.198}$$

Introducing the field

$$\mathbf{E} = \mathbf{E}_o \cos \omega t = -\frac{1}{c}\frac{\partial \mathbf{A}}{\partial t}$$

so that $\mathbf{A} = -\mathbf{A}_o \sin \omega t$ where $A_o = E_o\, c/\omega$, one obtains the solutions for each gauge above as

$$|\varphi(p,t)\rangle = e^{-i[\mathbf{p}^2 t/2m + \phi(\mathbf{p},\mathbf{r},t)/\hbar]}\, |\varphi(p,o)\rangle \tag{1.199}$$

where

$$\phi_X = \mathbf{f} + \frac{\mathbf{g}\cdot\mathbf{p}}{m} = \frac{\mathbf{E}_o^2}{4m\omega^2}\left[t - \frac{\sin 2\omega t}{2\omega}\right] + \frac{\mathbf{E}_o\cdot\mathbf{p}}{m\omega^2}\cos\omega t \tag{1.200}$$

$$\begin{aligned}\phi_{E\Phi} &= \left(\mathbf{f} + \frac{\mathbf{g}\cdot\mathbf{h}}{m}\right) + \frac{\mathbf{p}}{m}\cdot(\mathbf{h} + \mathbf{g}t) + \frac{\mathbf{g}^2 t}{2m} + \mathbf{r}\cdot\mathbf{g} \\ &= \left(\frac{\mathbf{E}_o^2 t}{4m\omega^2} + \frac{3\mathbf{E}_o^2}{8m\omega^3}\sin 2\omega t\right) + \frac{\mathbf{E}_o\cdot\mathbf{p}}{m\omega^2}(\cos\omega t + \omega t \sin\omega t) + \\ &\quad \frac{\mathbf{E}_o^2 t}{2m\omega^2}\sin^2\omega t + \frac{\mathbf{E}_o\cdot\mathbf{r}}{\omega}\sin\omega t\end{aligned} \tag{1.201}$$

$$\phi_{BN} = \mathbf{f} = \frac{\mathbf{E}_o^2}{4m\omega^2}\left[t - \frac{\sin 2\omega t}{2\omega}\right] \tag{1.202}$$

In obtaining Equation 1.201 we have used the relation $e^{-i\mathbf{g}\cdot\mathbf{r}}\, F(\mathbf{p})\, e^{+i\mathbf{g}\cdot\mathbf{r}} = \mathbf{F}(\mathbf{p} + \mathbf{g}(t))$, and $\mathbf{F}(\mathbf{p}) = e^{-ip^2t/2m}$

The three ϕ's are phase factors accumulated during propagation of a quantum free particle in a time-dependent classic electromagnetic field. Since for the free particle, $\phi = Et$, one can define an average energy change for the particle during its interaction with an electromagnetic field for the time τ corresponding to many cycles (i.e., $\omega\tau \gg 1$). We thus define the average energy shift as

$$\Delta E = \frac{\int_o^\tau \partial\phi/\partial t\, dt}{\tau} \tag{1.203}$$

For all three phases defined above, only the terms linear in t remain after cycle averaging (i.e., $\langle \sin \omega t \rangle = \langle \cos \omega t \rangle = 0$), the final result being the same for all three representations (with $A_o\, \omega/c = E_o$), one obtains

$$\Delta E = \frac{E_o^2}{4m\omega^2} \tag{1.204}$$

This is the oscillatory energy acquired by the particle during its propagation in the oscillatory electromagnetic field obtained by solving Equation 1.181. Thus in the BN representation, this is the principle change in the phase of the particle wavefunction since according to Equation 1.181, this oscillatory motion is given exactly by the classical equations of motion in presence of the field **E**(t). In the other gauges, extra phase factors appear in the particle wavefunction since, as we have emphasized previously, different representations give rise to different physical interpretations.

This can be seen more clearly if one uses the generating function for Bessel functions [24]

$$e^{iz \cos \omega t} = \sum_{n=-\infty}^{+\infty} (-i)^n e^{in\omega t} J_n(z) \tag{1.205}$$

so that the term

$$\frac{\mathbf{A}_o \cdot \mathbf{p}}{m\omega c} \cos \omega t = \alpha_o \cdot \mathbf{p} \cos \omega t$$

in Equation 1.200 can be now written as

$$e^{i\alpha_o \cdot \mathbf{p} \cos \omega t}\, \varphi(p,o) = \sum_{n=-\infty}^{+\infty} (-i)^n e^{-i(\varepsilon_p + n\hbar\omega)t/\hbar} J_n(\alpha_o \cdot \mathbf{p})\, \varphi(p,o) \tag{1.206}$$

where $\varepsilon_p = \mathbf{p}^2/2m$ is the free particle energy. (This is essentially a Floquet expansion; see Section 1.7). Equation 1.206 demonstrates that the free particle state of momentum **p** now acquire energy components with energy $(\varepsilon_p + n\hbar\omega)$ for any integer n. The amplitude of these components are

$J_n(\alpha_o \cdot \mathbf{p})$ which are field-induced virtual modifications of the wave function. Clearly field-induced virtual effects vary from one gauge or representation to another. However, the average field-induced energy shift ΔE remains gauge or representation invariant and is called the *ponderomotive energy* acquired by the particle in the field [45]. Thus for the free particle, both classical and quantum ponderomotive energies (Equation 1.204) are equal after averaging the latter over many radiation cycles.

1.11 SEPARABILITY OF PARTICLE MOTION

In the previous section we have seen that for a free particle in an electromagnetic field the exact wavefunctions were obtainable in closed form and that different representations gave rise to different solutions. In general, one has to consider a many particle system, and we wish to examine here the consequences of gauge or representation transformations on the separability of the Equations of motion. This question has been addressed by Reiss [31] for two interacting nonrelativistic charged particles.

A. Dipole Approximation

We thus consider two particles (m_1, q_1) and (m_2, q_2) in the presence of a radiation field treated classically. The total Schrödinger then becomes in the Coulomb gauge

$$i\hbar \frac{\partial|\varphi(1,2)>}{\partial t} = (\hat{\mathbf{H}}_1 + \hat{\mathbf{H}}_2 + V_C)|\varphi(1,2)>, \mathbf{H}_j = \frac{1}{2m_j}\left[\hat{\mathbf{p}}_j - \frac{q_j}{c}\mathbf{A}(\mathbf{r}_j,t)\right]^2 \tag{1.207}$$

and V_C is the Coulomb interaction depending only on the relative particle distance $(\mathbf{r}_1 - \mathbf{r}_2)$. We now transform to center of mass and relative coordinates

$$\mathbf{R} = \frac{m_1\mathbf{r}_1 + m_2\mathbf{r}_2}{m_1 + m_2}; \quad \mathbf{r} = \mathbf{r}_1 - \mathbf{r}_2, \quad \mathbf{p}_1 + \mathbf{p}_2 = \mathbf{P}_R \tag{1.208}$$

We then obtain a new equation in these more appropriate coordinates

$$i\hbar = \frac{\partial|\varphi(1,2)>}{\partial t} = (\hat{\mathbf{H}}_1' + \hat{\mathbf{H}}_2' + V_C')\ \varphi(\mathbf{R},\mathbf{r},t) \tag{1.209}$$

where

$$\mathbf{H}_1' = \frac{1}{2m_1}\left[\left(\mathbf{p}_r + \frac{m_1}{m_1+m_2}\mathbf{P}_R\right) - \frac{q_1}{c}\mathbf{A}\left(\mathbf{R} + \frac{m_2}{m_1+m_2}\mathbf{r},t\right)\right]^2 \tag{1.210}$$

$$\mathbf{H}_2' = \frac{1}{2m_2}\left[\left(-\mathbf{p}_r + \frac{m_2}{m_1+m_2}\mathbf{P}_R\right) - \frac{q_2}{c}\mathbf{A}\left(\mathbf{R} - \frac{m_1}{m_1+m_2}\mathbf{r},t\right)\right]^2 \tag{1.211}$$

$$V_C' = V_C(\mathbf{r}) \tag{1.212}$$

Clearly, the total Hamiltonian is not separable in terms of center of mass coordinates **R** and relative coordinates **r**. The electromagnetic potential through its spatial dependence couples these two coordinates. Taking the center of mass as the origin (**R** = 0) and assuming that the relative motion is limited to atomic or molecular dimensions smaller than the light wavelength (i.e., the dipole approximation, $r/\lambda < 1$), then the equations become completely separable. Equations 1.210-1.211 can be rearranged to give the physically transparent result, in the dipole approximation

$$i\hbar = \frac{\partial|\varphi(\mathbf{R},\mathbf{r},t)>}{\partial t} = [\hat{\mathbf{H}}_C(\mathbf{R},t) + \hat{\mathbf{H}}_C(\mathbf{r},t)]\ |\varphi(\mathbf{R},\mathbf{r},t)> \tag{1.213}$$

where

$$H_C(\mathbf{R},t) = \frac{1}{2m} = \left[\hat{\mathbf{P}}_R - \frac{Q}{c}\mathbf{A}(t)\right]^2 \tag{1.214}$$

$$H_C(\mathbf{r},t) = = \frac{1}{2M}\left[\mathbf{p}_r - \frac{q}{c}\mathbf{A}(t)\right]^2 + V_C(\mathbf{r}) \tag{1.215}$$

$$Q = q_1 + q_2; \quad q = \frac{m_2 q_1 - m_1 q_2}{M} \tag{1.216}$$

$$M = m_1 + m_2; \quad m = \frac{m_1 m_2}{M} \tag{1.217}$$

We see from the above equations that the same time-dependent electromagnetic potential **A**(t) acts in both center of mass and relative coordinate systems. One can now apply the dipole EF gauge transformation (Equation 1.166) separately to each independent coordinate **R** and **r**, to give the expected classical results, in the dipole approximation of course

$$\hat{\mathbf{H}}_{EF}(\mathbf{R},t) = \frac{1}{2M}\hat{\mathbf{P}}_R^2 - Q\mathbf{R}\cdot\mathbf{E}(t) \tag{1.218}$$

$$\hat{\mathbf{H}}_{EF}(\mathbf{r},t) = \frac{1}{2M}\hat{\mathbf{p}}_R^2 - q\mathbf{r}\cdot\mathbf{E}(t) + V_C(\mathbf{r}) \tag{1.219}$$

One could try to go beyond the dipole approximation by introducing the more general unitary transformations than Equation 1.166 by letting **A** have a spatial dependence

$$\hat{\mathbf{T}}' = \exp\left[-\frac{i}{\hbar}\sum_{j=1}^{2} q_j\mathbf{r}_j\cdot\mathbf{A}(\mathbf{r}_j,t)\right] \tag{1.220}$$

The new Hamiltonian (Equations 1.168 and 1.169) is derived after this new transformation and applying the long wavelength approximation **E**(**r**,t) = **E**(t) after the transformation

$$\hat{\mathbf{H}}'_{EF}(\mathbf{R},\mathbf{r},t) = \frac{1}{2M}\left[\hat{\mathbf{P}}_R + \frac{Q\mathbf{k}}{\omega}\mathbf{E}(t)\cdot\mathbf{R} + q\frac{\mathbf{k}}{\omega}\mathbf{E}(t)\cdot\mathbf{r}\right]^2 +$$

$$\frac{1}{2m}\left[\hat{\mathbf{p}}_r + \frac{q'\mathbf{k}}{\omega}\mathbf{E}(t)\cdot\mathbf{r} + \frac{q\mathbf{k}}{\omega}\mathbf{E}(t)\cdot\mathbf{R}\right]^2 -$$

$$Q\mathbf{R}\cdot\mathbf{E}(t) - q\mathbf{r}\cdot\mathbf{E}(t) + V_C(\mathbf{r}) \tag{1.221}$$

where

$$q' = \frac{q_1 m_2^2 + q_2 m_1^2}{M^2}$$

Since $\omega = ck$, then the corrections to the particle momenta are of order 1/c. Assuming that the relative motion in **r** is restricted to atomic or molecular dimensions, then one can introduce the dipole approximation for the relative coordinate **r**, so that we assume $\mathbf{k} \cdot \mathbf{r} < 1$ then Equation 1.221 reduces to

$$\hat{H}'_{EF}(\mathbf{R},\mathbf{r},t) = \frac{1}{2m}\left[\hat{\mathbf{P}}_R + \frac{Q\mathbf{k}}{\omega} \cdot \mathbf{E}(t)\mathbf{R}\right]^2 + \frac{1}{2m}\left[\hat{\mathbf{p}}_r + \frac{q\mathbf{k}}{\omega}\mathbf{E}(t) \cdot \mathbf{R}\right]^2 - \mathbf{E}(t) \cdot [Q\mathbf{R} + q\mathbf{r}] + V_C(\mathbf{r}) \tag{1.222}$$

Thus even in the limit of total neutral charge Q = 0, the relative charge motion can remain coupled to the center of mass motion especially at high intensities by terms of order E/c. The corrections to Equations 1.218–1.219 that are evident in Equation 1.221 and which will be important at high intensities, arise from multipole moments. A more general unitary transformation than Equation 1.220 for the EF gauge involving these moments in the definition of the transformation itself has been given by Power and Zienau; this generates a Hamiltonian containing in principle all the classical electric and magnetic moments of the particle system [34, 35].

In the case of the BN representation, the particle equations of motion have been examined by Bandrauk et al. [32]. The main effect of the BN transformation is to modify the particle-particle potentials via field-dependent displacements of the coordinates. Thus in the dipole approximation, electron-electron and nuclear-nuclear (i.e., like-particle) interactions remain invariant whereas oppositely charged particles will oscillate with opposite phases and the field effects on the corresponding potentials will occur, as in Equation 1.182. Corrections to like-particle interactions appear only when photon recoil terms are retained. These recoil terms are proportional to 1/c, (c = velocity of light), reflecting corrections to momentum conservation laws of the photon + particle system. These terms are expected to become important at high intensities. Thus for oppositely charged particles, field corrections to potentials depend on the electric field **E** via coordinate displacements, whereas for like particles, these corrections depend on the vector potential

A via momentum displacements [32]. Finally we remark, that a further complication arises for molecules, where one usually applies the Born-Oppenheimer (adiabatic) approximation to separate electronic (fast) and nuclear (slow) motion. Nonadiabatic couplings occur between excited electronic as corrections to the law of conservation of total momentum, both electronic and nuclear. The importance of these nonadiabatic couplings can vary from one gauge or representation to another [16, 32] and will be discussed in later chapters.

B. Permanent Dipole Moments

One final aspect of intense field particle motion separability which differs from the previous two-particle system treated previously, is the permanent dipole moment inherent in symmetrically charged molecules with differing isotope content. The simplest example is that of H_2^+ which has no total permanent dipole moment, whereas in HD^+ the nuclear center of mass and the electronic (or nuclear) center of charge do not coincide (as opposed to H_2^+). Thus HD^+ will exhibit a net total dipole moment which, under the presence of a static electric field, yields quite different dissociation dynamics. This problem was first treated by Hiskes [47] and we reconsider here as it is relevant to high intensity molecular photodissociation by infrared lasers (see also Chapter 4).

We consider two nuclei of charges q_1e and q_2e at positions $\mathbf{R}_1$ and $\mathbf{R}_2$, $\mathbf{R} = |\mathbf{R}_1 - \mathbf{R}_2|$, and electrons at position $\mathbf{r}_i$. The total dipole moment μ of such a diatomic system of nuclear masses, M_1 and M_2, is given with reference to the center of mass $\mathbf{R}_{cm}$ of the total system

$$\frac{\mu}{e} = q_1\mathbf{R}_1^* + q_2\mathbf{R}_2^* - \sum_{i=1}^{n} \mathbf{r}_i^* \tag{1.223}$$

where

$$\mathbf{R}_i^* + \mathbf{R}_{cm} = \mathbf{R}_i; \quad \mathbf{r}_i^* + \mathbf{R}_{cm} = \mathbf{r}_i \tag{1.224}$$

We next define electronic coordinates $\mathbf{r}_i'$ with respect to the nuclear center of mass

$$\mathbf{R}_o = \frac{M_1 \mathbf{r}_1 + M_2 \mathbf{R}_2}{M_1 + M_2}$$

With respect to these new coordinates, the nuclear dipole moment becomes after rearrangement

$$q_1 \mathbf{R}_1^* + q_2 \mathbf{R}_2^* = q_1 \mathbf{R}_1 + q_2 \mathbf{R}_2 - \frac{(q_1 + q_2)}{M} \left[M_1 \mathbf{R}_1 + M_2 \mathbf{R}_2 + m_e \sum_i (\mathbf{r}_i' - \mathbf{R}_o) \right] \tag{1.225}$$

$$= \frac{(q_1 M_2 - q_2 M_1)}{M_1 + M_2} \mathbf{R} - (q_1 + q_2) \frac{m_e}{M} \sum_i \mathbf{r}_i' \tag{1.226}$$

$$M = M_1 + M_2 + n\, m_e \tag{1.227}$$

In the coordinate system with respect to the nuclear center of mass $\mathbf{R}_o$, the nuclear dipole moment is separated into a nuclear part dependent on relative nuclear coordinates $\mathbf{R} = \mathbf{R}_1 - \mathbf{R}_2$ and an electronic component. We next define the distance $\mathbf{d}$ between total center of mass, $\mathbf{R}_{cm}$ and the nuclear center of mass $\mathbf{R}_o$

$$\mathbf{d} = \mathbf{R}_{cm} - \mathbf{R}_o = \frac{M_1 \mathbf{R}_1 + M_2 \mathbf{R}_2 + m_e \sum_i \mathbf{r}_i}{M} - \left(\frac{M_1 \mathbf{R}_1 + M_2 \mathbf{R}_2}{M_1 + M_2} \right)$$

$$= \frac{m_e}{M} \sum_i \mathbf{r}_i' \tag{1.228}$$

from which follows the relation between $\mathbf{r}_i^*$ and $\mathbf{r}_i'$

$$\sum_i \mathbf{r}_i^* = \sum_i (\mathbf{r}_i' - \mathbf{d}) = \left(1 - \frac{m_e}{M} \right) n \sum_i \mathbf{r}_i' \tag{1.229}$$

Adding Equations 1.226 and 1.229 gives the total dipole moment as a function of relative nuclear coordinates, **R**, and electronic coordinates $\mathbf{r}_i'$ with respect to the nuclear center of mass, $\mathbf{R}_o$

$$\frac{\mu}{e} = \frac{q_1 M_2 - q_2 M_1}{M_1 + M_2}\mathbf{R} - \left[1 + \frac{(q_1 + q_2 - n)m_e}{M}\right]\sum_i \mathbf{r}_i' \tag{1.230}$$

M is the total mass (Equation 1.227) and n is the number of electrons, each of mass m_e. The first term is the center of mass dipole moment of two classical particles [47].

The relative (**R**) and center of mass ($\mathbf{R}_o$) nuclear coordinate representation of the total dipole moment of a diatomic molecule has the convenient feature that it shows zero nuclear dipole moment for symmetric diatomics. Thus for H_2^+ and H_2 alike, all radiative coupling is purely electronic. For isotopes of symmetrically charged molecules, HD^+ or HD, two contributions now arise to the internal total dipole moment and thus to the radiative coupling: a permanent nuclear dipole moment and a purely electronic moment (Equation 1.230). For an electric field parallel to the z-axis, the internal motion radiative interaction $V^f = -\mathbf{E} \cdot \mu$ becomes

$$V^f = -eE\left(\frac{q_1 M_2 - q_2 M_1}{M_1 + M_2}\right)z + eE\left[1 + \frac{(q_1 + q_2 - n)m_e}{M}\right]\sum_i z_i' \tag{1.231}$$

(z = R for a molecule aligned with the electric field).

We now illustrate the effect of permanent dipoles by comparing the effect of an intense electrostatic field **E** on the electronic potentials of H_2^+ and HD^+. In the case of a time-dependent, electromagnetic, $E(t) = E_o \cos \omega t$, the field effect is called dressing of the potentials (see Chapter 3). The two most important electronic states are the $1\sigma_g$ and $1\sigma_u$ states or molecular orbitals, as the corresponding potentials become degenerate asymptotically ($R \to \infty$), dissociating into H^+ and H(1*s*). The electronic transition moment is $\langle 1\sigma g|z|1\sigma_u\rangle = R/2$, which diverges asymptotically [48].

The electronic Hamiltonian matrix of H_2^+ in the presence of the static electric field E coupling the two potentials $V_g(R)$ and $V_u(R)$, is simply

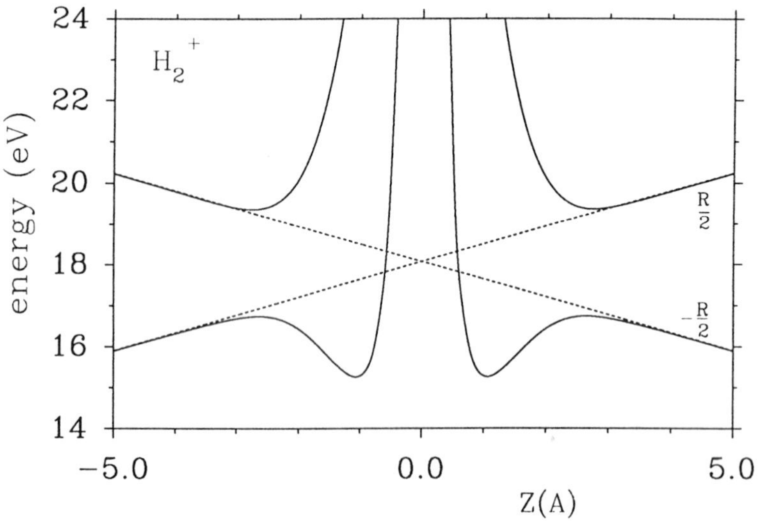

Figure 1.2 Ground ($W_+(R)$) and excited potentials ($W_-(R)$) of H_2^+ in a static electric field E of intensity $I = cE^2/4\pi$ corresponding to Equation 1.233.

$$H_{gu} = \begin{bmatrix} V_g(R) & eER/2 \\ eER/2 & V_u(R) \end{bmatrix} \tag{1.232}$$

Upon diagonalizing, this gives new *adiabatic*, static field-induced potentials for H_2^+

$$W_{\pm}(R) = \frac{V_g(R) + V_u(R)}{2} \pm \frac{1}{2}\left\{[V_g(R) - V_u(R)]^2 + (eER)^2\right\}^{1/2} \tag{1.233}$$

These new field-induced states of H_2^+ are shown in Figure 1.2, with the two possible separation of the fragments. This figure shows a symmetric potential system, with asymptotic $\pm$ ER/2 due to the electronic dipole moment coupling with the field. One sees further that dissociation of low vibrational levels of H_2^+ occurs by *tunnelling* through the laser-induced barrier in the lower potential $W_+(R)$ and that *trapping* of new field-induced vibrational states occurs in the upper laser-induced potential, $W_+(R)$. As shown in subsequent chapters, similar field-induced phenomena, tunnelling and trapping of molecular nuclear states in new field-induced potentials will be a common occurrence in the presence of time-dependent laser fields. Such states are then called dressed molecular states.

The molecular orbitals are defined in the Born-Oppenheimer approximation with respect to the charge center of mass, since the electronic Hamiltonian remains symmetric in both H_2^+ and HD^+. Thus the molecular orbitals of both molecules are identical with their center coinciding with the center of charge. The center of mass electronic dipole moment z' (Equation 1.231) can be expressed as

$$z' = z - z_0, \quad z_0 = \frac{R}{2}\left(\frac{M_1 - M_2}{M_1 + M_2}\right) \tag{1.234}$$

where z is defined with respect to the center of charge ($z = 0$), and z_0 is the separation between the center of mass and of charge. In HD^+, clearly z_0 is nonzero. The diagonal terms of the electronic Hamiltonian (Equation 1. 232) are now each augmented by $-eEz_0$ whereas the nondiagonal term remains the same: $\langle 1\sigma_g|z'|1\sigma_u\rangle = \langle 1\sigma_g|z|1\sigma_u\rangle = R/2$. Alternatively, the adiabatic field potentials $W_\pm(R)$ (Equation 1.233) are now modified by a constant term, a permanent dipole term, V_D

$$V_D = \frac{-eE}{M_1 + M_2}\left[(q_1M_2 - q_2M_1)R + (M_1 - M_2)\frac{R}{2}\right] \tag{1.235}$$

For HD^+, $V_D = -\ eER/6$, and since the asymptotic limits of $W_\pm(R)$ for H_2^+ (Equation 1.233) are $\pm\ eER/2$, we have the new asymptotic limits of $W_\pm(R)$ for HD^+

$$W_\pm(R)\ (\lim R \to \infty) = eE\begin{bmatrix} + & R/3 \\ - & 2R/3 \end{bmatrix} \tag{1.236}$$

This illustrated in Figure 1.3 where one sees that in the presence of the static field E the product $H^+ + D$ is preferred. This is consistent with the fact that for this product the electron has a lower electronic energy ($E_D^{1s} < E_H^{1s}$) due to reduced mass effects. Such permanent nuclear dipole moment effects

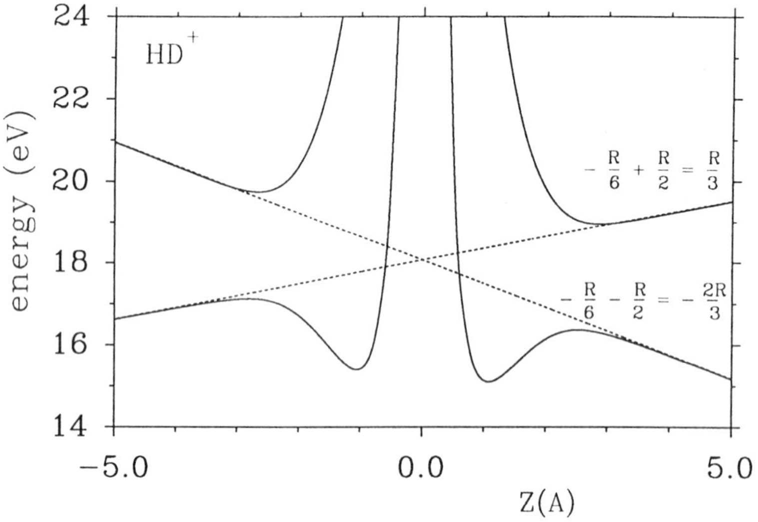

Figure 1.3 Ground ($W_+(R)$) and excited ($W_-(R)$) potentials of HD^+ showing the effect of the permanent dipole moments $V_D(R)$ (Equation 1.235).

are therefore always present in nonsymmetric molecules and influence the photodissociation dynamics at high field intensities (see Chapter 4). Finally we remark that a further complication arises for molecules where one usually applies the Born-Oppenheimer (adiabatic) approximation to separate electronic (fast) and nuclear (slow) motion. Nonadiabatic couplings occur between excited electronic as corrections to the law of conservation of total momentum, both electronic and nuclear. The importance of these nonadiabatic couplings can vary from one gauge or representation to another [16, 32] and will be discussed in later chapters.

1.12 INTERACTION REGIMES

Since present laser technologies allow experimentalists to prepare field intensities from weak ($I \sim 1$ W/cm^2) to intense ($I \sim 10^{18}$ W/cm^2), then consequently radiative interactions will induce perturbations or couplings ranging from the perturbative (Fermi Golden-rule) regime to nonperturbative (highly nonlinear) regimes. In this section we give estimates of these couplings and remark on their consequences for the appropriate interaction regimes.

Perturbative regimes can be expected to apply whenever radiative interactions are much less than the intra atomic Coulomb forces or rather electric fields. The atomic unit (au) of Coulomb potential and electric field are derived from the hydrogen atom results

$$V_o = \frac{e^2}{a_o} = 43.6 \times 10^{-19}\,\text{joule} = 1\ \text{Hartree}\ (1\ \text{au}) = 27.25\ \text{ev} = 2\ \text{Rydbergs}$$

$$E_o = \frac{e}{a_o^2} = 5.15 \times 10^9\ \text{V/cm} \tag{1.237}$$

where $a_o = 1\,\text{Bohr} = 0.529 \times 10^{-8}$ cm. The corresponding electromagnetic intensity I_o for one au of electric field E_o is obtained from

$$I_o = \frac{cE_o^2}{8\pi} = 3.5 \times 10^{16}\ \text{W/cm}^2 \tag{1.238}$$

The electric dipole interaction for an electric field of magnitude E and a transition moment r can be expressed as the *Rabi frequency*, ω_R (i.e., the rate of radiative excitation [3, 6])

$$\hbar\omega_R = eEr \tag{1.239}$$

Using the value of one atomic unit of dipole interaction

$$eE_oa_o = \frac{e^2}{a_o} = 1\ \text{au} \tag{1.240}$$

one can reexpress Equation 1.239 as

$$eEr = \left(\frac{r}{a_o}\right)\left(\frac{I}{I_o}\right)^{1/2}\ (\text{au}) \tag{1.241}$$

This gives the following expressions for calculating the radiative interaction (or Rabi frequency ω_R) in terms of the incident laser intensity I (W/cm^2) and the transition moment r

$$eEr = r(au)\, I^{1/2}\, (W/cm^2) \times (5.4 \times 10^{-9})\, (au) \tag{1.242}$$

$$eEr = r(au)\, I^{1/2}\, (W/cm^2) \times (1.17 \times 10^{-3})\, cm^{-1} \tag{1.243}$$

At the field intensity $I_o = 3.5 \times 10^{16}$ W/cm^2, one sees that the radiative interaction is equal to the Coulomb potential energy of one atomic unit (27.2 ev) per atomic unit of transition moment.

In view of the applicability of the semiclassical approximation at high intensities (see Section 1.7), one can interpret these intensities in terms of photon energies (Equation 1.79). The photon density can thus be defined as

$$\frac{N}{V} = \frac{I\,(W/cm^2)}{\hbar\omega c} \tag{1.244}$$

For a CO_2 laser, $\omega = 1000\ cm^{-1}$, $\hbar\omega = 2 \times 10^{-20}$ joule, one obtains for the atomic unit intensity I_o a photon density $N_o/V = 6 \times 10^{25}$ photons/cm^3. Clearly at this intensity, the semiclassical approximation is therefore most appropriate.

One of the effects of such highly intense radiation on particles is to displace the particles from their quantum orbits giving rise to a ponderomotive energy (Equation 1.204). Thus for an arbitrary intensity I (W/cm^2) and frequency ω, one can evaluate this energy as

$$\frac{e^2E^2}{4m\omega^2} = \frac{(I/I_o)}{[2\hbar\omega(au)]^2}\,(au) \tag{1.245}$$

This gives for the atomic unit intensity I_o and frequency $\omega = 1000\ cm^{-1}$, an energy of 10^4 au (2.7×10^5 eV). Clearly, at such high intensity, an electron will see the Coulomb potential only as a perturbation as it follows the electromagnetic field. This observation can be confirmed by calculating the field-induced particle displacement $\alpha(t)$ described in the BN representation (Equation 1.180). Such a calculation can be done in the quantum BN representation [32] or classically. Thus integrating Equation 1.180 for a sinusoidal vector potential $A(t) = A \sin\omega t$, and using the relation $A = Ec/\omega$, one obtains the average displacement variance as

$$<\alpha(t)^2> = \frac{e^2E^2}{2m^2\omega^4} \tag{1.246}$$

Thus for an electron in the presence of a CO_2 laser ($\omega = 1000\ \text{cm}^{-1}$), of intensity $I_o = 3.5 \times 10^{16}\ \text{W/cm}^2$, one obtains $<\alpha(t)^2> = 10^{-8}\ \text{cm}^2$, or a displacement amplitude $\alpha \sim 10^{-4}$ cm. One concludes that at the atomic unit intensity I_o, an electron is becoming decoupled from the nuclear Coulomb potential as it vibrates essentially in the electromagnetic field, thus acquiring a "wiggling" or ponderomotive energy (Equation 1.245).

Hence at laser field intensities approaching the internal electric field strengths of atoms and molecules, considerable distortion of particle orbits will occur. In particular, since the dipole approximation gives rise to the simple radiative interaction (Equation 1.241), one must reexamine this approximation in light of the above calculation of laser-induced displacements $\alpha(t)$. The dipole approximation (Section 1.11) is based on the negligible electromagnetic plane wave phase factor, which for one atomic unit of photon energy $\hbar\omega$ and atomic size a_o can be evaluated to be

$$ka_o = \frac{\omega}{c} \cdot \frac{e^2}{\hbar\omega} = \frac{e^2}{\hbar c} = \alpha_F \tag{1.247}$$

where α_F is the fine structure constant $\simeq 10^{-2}$. This approximation is therefore valid for low excitations where electron orbits are restricted to atomic units in size. On the other hand, an electron driven by a circularly polarized field of frequency ω will acquire a velocity $v = \omega r$ with $r = eE/m\omega^2$. This gives the relation

$$kr = \frac{v}{c} = \frac{eE}{m\omega c} = \left(\frac{\lambda_c}{a_o}\right)\left(\frac{I}{I_o}\right)^{1/2} \frac{1}{\hbar\omega(\text{au})} \tag{1.248}$$

where $\lambda_c = \hbar/mc$ is the Compton wavelength (2.4×10^{-10} cm). For a CO_2 laser, $\hbar\omega_{CO_2} \sim 5 \times 10^{-3}$ (au), then at the atomic unit field intensity $I = I_o$

(Equation 1.248) is of order one and the electric dipole approximation is no longer valid. Concomitantly, the electron motion must be treated relativistically, thus necessitating a more complete treatment of the field-particle interactions at intensities $I \geq I_0 = 10^{16}$ W/cm^2 [49].

REFERENCES

1. A. D. Bandrauk (Ed.), *Atomic and Molecular Processes with Short Intense Laser Pulses*, NATO ASI, vol. 171B, Plenum Press, New York, (1988); A. D. Bandrauk, S. C. Wallace, (Eds.), *Coherence Phenomena in Atoms and Molecules in Laser Fields*, NATO ASI, vol. 287B, Plenum Press, New York, (1992).
2. E. A. Power, *Introductory Quantum Electrodynamics*, Elsevier, New York, (1965).
3. R. Loudon, *The Quantum Theory of Light*, Oxford University Press, London, (1973).
4. T. D. Lee, *Particle Physics and Introduction to Field Theory*, Harwood Academic, New York, (1981).
5. M. H. Mittleman, *Introduction to the Theory of Laser-Atom Interactions*, Plenum Press, New York, (1982).
6. F. H. M. Faisal, *Theory of Multiphoton Processes*, Plenum Press, New York, (1987).
7. C. Cohen-Tannoudji, J. Dupont-Roc, G. Grynberg, *Photons and Atoms: Introduction to Quantum Electrodynamics*, J. Wiley & Sons, New York, (1989).
8. R. Shen, *Principles of Nonlinear Optics*, J. Wiley & Sons, New York (1984).
9. G. L. Lamb, Jr., *Elements of Soliton Theory*, J. Wiley & Sons, New York, (1980).
10. K. Kulander, B. W. Shore, *J. Opt. Soc. Amer.* **B7**, 502 (1990).
11. S. Chelkowski, A. D. Bandrauk, *J. Chem. Phys.* **89**, 3618 (1988).
12. A.D. Bandrauk, S. Chelkowski, H.B. Wang, *Physica* **D50**, 31 (1991).
13. R. J. Glauber, *Phys. Rev.* **131**, 2766 (1963).
14. F. Bloch, A. Nordsieck, *Phys. Rev.* **52**, 54 (1937).
15. J. van de Ree, J. Z. Kaminski, M. Gavrila, *Phys. Rev.* **37A**, 4536 (1988).
16. T. T. Nguyen-Dang, A. D. Bandrauk, *J. Chem. Phys.* **79**, 3256 (1983); **80**, 4926 (1984); **83**, 2840 (1985).
17. E. Schrödinger, *Naturwiss.* **14**, 664 (1926).
18. E. Goldin, *Waves and Photons - An Introduction to Quantum Optics*, J. Wiley & Sons, New York, (1982).
19. A. Perelomov, *Generalized Coherent States and their Applications*, Springer Verlag, Berlin, (1986).
20. S. M. Barnett, D.T. Pegg, *J. Phys.* **A**19, 3849 (1986).
21. A. D. Bandrauk, O. Atabek, in *Lasers, Molecules and Methods*, (J. O. Hirschfelder, R. Wyatt, R. D. Coalson, Eds.), J. Wiley & Sons, New York, 1989, Adv. Chem. Phys., vol. 73, chap. 19.

22. A. D. Bandrauk, O. Atabek, *J. Phys.* Chem. **91**, 6469 (1987).
23. A. D. Bandrauk, J. F. McCann, *Comm. Atom. Molec. Phys.* **22**, 325 (1989).
24. M. Abramowitz, I. A. Stegun, *Handbook of Mathematical Functions*, Dover Publications. Inc., New York, (1972).
25. P. M. Morse, H. Feshbach, *Methods of Theoretical Physics*, Part I, McGraw-Hill, New York, (1953), page 557.
26. K. M. Watson, J. Nuttall, *Topics in Several Particle Dynamics*, Holden-Day Pub., San Francisco, (1964), chap. 1.
27. A. D. Bandrauk, G. Turcotte, *J. Chem. Phys.* **77**, 3867 (1982); A. D. Bandrauk, G. Turcotte, *J. Phys. Chem.* **89**, 3039 (1985).
28. R. D. Levine, *Quantum Mechanics of Molecular Rate Processes*, Oxford Univ. Press, Oxford, (1969), Appendix 2B.
29. A. Lami, N. K. Rahman, *Phys. Rev.* **A26**, 3360 (1982).
30. H. Yu, A. D. Bandrauk, in *Coherence Phenomena in Atoms and Molecules in Laser Fields*, (A. D. Bandrauk, S. C. Wallace, Eds.), Plenum Press, New York, 1992, NATO ASI, vol. 287B, pp. 31-43.
31. H. R. Reiss, *Phys. Rev.* **A19**, 1140 (1979).
32. A. D. Bandrauk, O. F. Kalman, T. T. Nguyen-Dang, *J. Chem. Phys.* **84**, 6761 (1986).
33. M. Göppert-Mayer, *Ann. Phys. (Leipzig)* **9**, 273 (1931).
34. E. A. Power, S. Zienau, *Phil. Trans. Roy. Soc.* **251**, 427 (1959).
35. E. A. Power, T. Thirunamachandran, *Phys. Rev.* **28A**, 2649 (1983).
36. W. E. Lamb, *Phys. Rev.* **85**, 259 (1952).
37. Z. Fried, *Phys. Rev.* **A8**, 2835 (1973).
38. T. Feuchtwang, E. Kazes, P. Cutter, H. Grotch, *J. Phys.* **A17**, 151 (1984).
39. W. Pauli, M. Fierz, *Nuovo Cimento* **15**, 167 (1938).
40. N. G. van Kampen, *Dan. Mat. Fys. Medd.* **26**, #15, 1, (1951).
41. C. Chanmugan, S. S. Schweber, *Phys. Rev.* **A1**, 1369 (1970).
42. T. A. Welton, *Phys. Rev.* **74**, 1157 (1948).
43. E. A. Power, T. Thirunamachandran, *Phys. Rev.* **A15**, 2366 (1977).
44. W. C. Henneberger, *Phys. Rev.* Lett. **21**, 838 (1968).
45. T. W. B. Kibble, *Phys. Rev.* **150**, 1060 (1966).
46. K. Husimi, *Prog. Theor. Phys.* **9**, 381 (1953).
47. J. R. Hiskes, *Phys. Rev.* **122**, 1207 (1961).
48. R. S. Mulliken, *J. Chem. Phys.* **7**, 20 (1939).
49. P. S. Krstic, M. H. Mittleman, *Phys. Rev.* **A42**, 4037 (1990).

2

H_2 in Intense Laser Fields

Anton Zavriyev* **and Philip H. Bucksbaum**

University of Michigan, Ann Arbor, Michigan

2.1 INTRODUCTION

High intensity laser-atom interactions can take place in molecules as well as in atoms. The laser field shifts and mixes molecular eigenstates just as for atoms, and multiphoton ionization (MPI) or above-threshold ionization (ATI; i.e., ionization by absorption of more photons than the minimum number needed to ionize) are readily observable in molecules [1–10]. In addition, molecules possess some degrees of freedom that are absent in atoms, due to the vibration and rotation of constituent atoms about the center of mass. This leads to a number of new and interesting high field phenomena. For example, nuclear motion can be affected by the field, causing molecules to change their rotational and vibrational quantum numbers and redistribute energy between different modes. Molecules can also dissociate in the field and

* *Current affiliation:* National Research Council of Canada, Ottawa, Ontario, Canada

above-threshold dissociation (ATD; i.e., dissociation by via absorption of more photons than the minimum number required for dissociation), analogous to ATI, has been observed [5, 6, 10, 11].

Due to its relative simplicity, H_2 is one of the most frequently studied molecules. In this chapter we use some of the experimental results involving molecular hydrogen as examples for different molecular processes which can take place in an intense laser field.

2.2 THEORETICAL BACKGROUND

A. Born-Oppenheimer Approximation

In order to explain molecular behavior in an intense laser field, we have to know how to describe the molecule when the field is off. We start with a Hamiltonian of the following form:

$$H_{tot} = \sum_i \frac{\mathbf{P}_i^2}{2M_i} + \sum_j \frac{\mathbf{p}_j^2}{2m} + \sum_{i,j} V(\mathbf{R}_i,\mathbf{r}_j) \tag{2.1}$$

Here

$$\sum_i \frac{\mathbf{P}_i^2}{2M_i}$$

is the total kinetic energy of nuclei,

$$\sum_j \frac{\mathbf{p}_j^2}{2m}$$

is the total kinetic energy of electrons, and $V(\mathbf{R}_i,\mathbf{r}_j)$ is the term containing all interactions. The behavior of the molecule is then given by the solution of the Schrödinger equation:

$$H_{tot}\Psi(\mathbf{R},\mathbf{r}) = E_T\Psi(\mathbf{R},\mathbf{r}) \tag{2.2}$$

where $\Psi(\mathbf{R},\mathbf{r})$ denotes the total wave function, a function of the coordinates for the nuclei ($\mathbf{R}_i$) as well as the electrons ($\mathbf{r}_j$).

Due to the number of particles involved, the problem is very complicated and a set of reasonable approximations must be made to approach it. The most important simplification usually made in the molecular description is the Born-Oppenheimer approximation (BOA), which decouples the nuclear and electronic wave functions. We seek approximate solutions of the form:

$$\Psi(\mathbf{R},\mathbf{r}) = \psi(\mathbf{R},\mathbf{r})\nu(\mathbf{R}) \tag{2.3}$$

Here $\psi(\mathbf{R},\mathbf{r})$ is a solution to:

$$\left[\sum_j \frac{p_j^2}{2m} + V(\mathbf{R}',\mathbf{r}_j)\right]\psi(\mathbf{R}',\mathbf{r}) = E'(\mathbf{R}')\psi(\mathbf{R}',\mathbf{r}) \tag{2.4}$$

In this equation the $\mathbf{R}'$ are not variables but parameters kept fixed, while the $\mathbf{r}_j$ change. $E'(\mathbf{R}')$ are the electronic eigenvalues, calculated for fixed nuclear positions. The Born-Oppenheimer approximation is based on the fact that electrons are much lighter and much more mobile than nuclei. As a consequence, we can freeze the nuclear motion while solving the Schrödinger equation for the electrons.

The calculation is repeated for a range of R values, so the electronic eigenvalues trace out curves E′(R). It is easy to show that the nuclear motion then satisfies the following equation:

$$\left[\sum_i \frac{\mathbf{P}_i^2}{2M_i} + E'(\mathbf{R}_i)\right]\nu(\mathbf{R}) \approx E\nu(\mathbf{R}) \tag{2.5}$$

where $\nu(\mathbf{R})$ is the nuclear wave function [12]. This result has a very important physical meaning: it implies that we can treat the nuclei as moving in an effective potential $E'(\mathbf{R})$.

The Born-Oppenheimer approximation provides us with a method for solving the Schrödinger equation for a molecular system. We begin by calculating the electronic eigenstates of the system as a function of the nuclear coordinates; then we force the nuclei to move along these calculated electronic potentials and solve the Schrödinger equation for the nuclear motion.

B. Vibrations and Rotations of Diatomic Molecules

We now apply the Born-Oppenheimer approximation to the case of a diatomic molecule. The molecule is in an electronic eigenstate, like the one shown in Figure 2.1. The task is to solve the Schrödinger equation for the nuclear motion. The kinetic energy of the nuclei can be written in the following form:

$$T_N = \frac{p_1^2}{2m_1} + \frac{p_2^2}{2m_2} \tag{2.6}$$

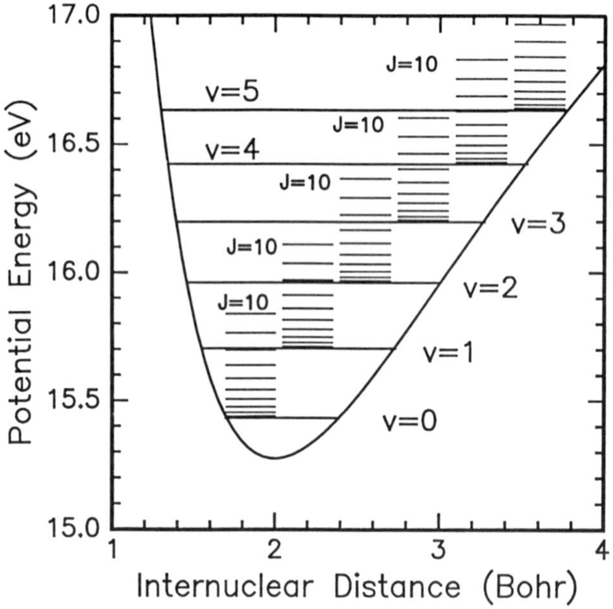

Figure 2.1 A bound electronic state supports a series of nuclear vibrational levels. A vibrational quantum number **v** = 0, 1, 2, 3, ... is assigned to each level. In addition, each nuclear vibrational level is divided into a series of rotational states. The separation between these states increases with the rotational quantum number **J**.

where the nuclear masses are $m_{1,2}$ and the momenta are $\mathbf{p}_{1,2}$. Part of this energy comes from the translational motion of the molecule as a whole and another part from the motion of nuclei with respect to each other. If we separate these two motions, we can rewrite the kinetic energy as follows:

$$T_N = \frac{\mathbf{P}^2}{2M} + \frac{\mathbf{p}^2}{2\mu} \tag{2.7}$$

where

$$\mu = \frac{m_1 m_2}{m_1 + m_2}$$

is the reduced mass of the system,

$$\mathbf{p} = \frac{m_1\mathbf{p}_2 - m_2\mathbf{p}_1}{m_1 + m_2},\ \mathbf{P} = \mathbf{p}_1 + \mathbf{p}_2, \text{ and } M = m_1 + m_2.$$

The intrinsic properties of the molecule depend on the relative position of the nuclei and not on the position of the center of mass. Therefore, we can discard the first term in Equation 2.7 as uninteresting. The potential energy term $E'(\mathbf{R})$ in Equation 2.5 depends on the coordinates of the nuclei $\mathbf{r}_{1,2}$. We introduce a new coordinate system: $\mathbf{r} = \mathbf{r}_1 - \mathbf{r}_2$ and $\mathbf{R} = \mathbf{r}_1 + \mathbf{r}_2$ and neglect the translation of the molecule. The Schrödinger equation for the motion of nuclei with respect to each other can then be written in the following form:

$$-\left(\frac{\hbar^2}{2\mu}\right)\nabla^2\psi(\mathbf{r}) + V(r)\psi(\mathbf{r}) = E\psi(\mathbf{r}) \tag{2.8}$$

where $V(r) = E'(r_1, r_2)$ and $\psi(\mathbf{r})$ is the nuclear wave function.

This equation describes the motion of a particle of mass μ in the spherically symmetric potential $V(r)$. In spherical coordinates we can rewrite it in the following form:

$$\left(\frac{h^2}{2\mu}\right)\left[\frac{-1}{r^2}\frac{\partial}{\partial r}\left(r^2\frac{\partial\psi(\mathbf{r})}{\partial r}\right)+\frac{\mathbf{L}^2}{r^2}\psi(\mathbf{r})\right]+V(r)\psi(\mathbf{r})=E\psi(\mathbf{r}) \tag{2.9}$$

where L is the angular momentum operator. The angular momentum is conserved in the spherically symmetric potential, therefore the eigenstates of the Hamiltonian are the simultaneous eigenstates of $\mathbf{L}^2$. We will look for solutions of the following form:

$$\psi(\mathbf{r})=\psi_{\text{vibrational}}\times\psi_{\text{rotational}}=\frac{1}{r}P(r)Y_{JM}(\theta,\varphi) \tag{2.10}$$

where $Y_{JM}(\theta,\varphi)$ are spherical harmonics with

$J = 0, 1, 2, ...;$ and $M = -J, -J + 1, ..., J - 1, J$

Since spherical harmonics are the eigenstates of angular momentum squared, with the eigenvalues $h^2J(J + 1)$, the Schrödinger equation reduces to an ordinary differential equation for the vibrational wave function:

$$\frac{-h^2}{2\mu}\frac{d^2P}{dr^2}+\left[V(r)+\frac{h^2}{2\mu}\frac{J(J+1)}{r^2}\right]P(r)=EP(r) \tag{2.11}$$

This equation is identical to the Schrödinger equation of a particle moving in a one-dimensional potential well

$$V(r)+\frac{\hbar^2}{2\mu}\frac{J(J+1)}{r^2}$$

The second term is the rotational *centrifugal* energy. Equation 2.11 may be solved for the vibrational eigenstates $[\psi_{\text{vibrational}}\equiv P(r)/r]$ and eigenvalues (E_v) associated with nuclear vibration in the central potential, modified by the centrifugal term.

Since diatomic molecules usually vibrate much faster then they rotate, it is possible to introduce the *mean internuclear distance* r_0, which is the average internuclear separation over one rotational period. The average rotational energy of the nuclei is then:

$$E_{rot} = \frac{\hbar^2}{2I} J(J+1) \tag{2.12}$$

where $I = \mu r_0^2$ is the *mean moment of inertia* of the nuclei around the axis of rotation. This average rotational energy does not depend on the internuclear separation and is a constant of motion. The energy of the nuclear rotation then may be written as:

$$E_{rot}(cm^{-1}) = F(J) = B_v J(J+1) \tag{2.13}$$

where B_v is the *rotational constant.* It is customary to express the molecular energy in units of cm^{-1}. Then B_v is given by

$$B_v = \frac{\hbar}{4\pi I c} \tag{2.14}$$

where c is the speed of light. Values of B_v are tabulated for different diatomic molecules in the literature [13].

The total kinetic energy of the nuclei can be written as the sum of the rotational energy and the vibrational energy, where the latter is calculated from Equation 2.11:

$$T_{N,tot} = E_{rot} + E_{vib} \tag{2.15}$$

Thus a molecule in a bound electronic state can occupy one of the allowed vibrational levels; each supporting a series of rotational sublevels whose separation increases with J (Figure 2.1). For an unbound electronic state we have to consider a continuous distribution of repulsive states instead of discrete vibrational levels. Still, similar to the case of bound vibrational levels, each of these repulsive states is divided into an infinite number of rotational substates.

It must be remembered that this hierarchy is a just an approximation, which ignores higher-order terms representing coupling between the vibrational and rotational modes. However, this simplification is good enough for many purposes and is widely used as such. For a more rigorous description of the vibrational-rotational structure in diatomic molecules, a reader can consult special literature on the subject [13].

C. Electronic States of Diatomic Molecules. Spectroscopic Notations

As established in the previous section, each molecular eigenstate has both vibrational and rotational quantum numbers. (Strictly speaking, there is no vibrational quantum number associated with unbound electronic states. However, one still can associate a certain energy with the radial motion of the nuclei.) To completely identify the molecular states, we must also classify the electronic eigenstates. This is done using standard spectroscopic notation.

Unlike atoms, molecules generally lack central symmetry. The electronic angular momentum is therefore not conserved. In diatomic molecules, however, the potential is symmetric around the internuclear axis, so the projection of the electronic angular momentum on this axis is conserved and can be used to classify the electronic states of a diatomic molecule. It is convenient to fix the z-axis parallel to the internuclear axis of the molecule. Then the projection of the electronic orbital angular momentum assumes discrete values $L_z\hbar$, where

$$L_z = \ldots -2, -1, 0, 1, 2, \ldots \tag{2.16}$$

States with $L_z = |L_z|$ and $L_z = -|L_z|$ have the same energy. (A simultaneous reversal of the direction of rotation of all electrons does not change the energy of the molecule.) Thus, it is appropriate to classify the states according to the absolute value of the projection of the electronic orbital angular momentum on the internuclear axis $\Lambda = |L_z|$. According to this classification, states with $\Lambda = 0, 1, 2, 3, \ldots$ are called $\Sigma, \Pi, \Delta, \Phi \ldots$ states (this is analogous to the atomic designation). Apparently, all the states with $\Lambda > 0$ are doubly degenerate, while Σ states are nondegenerate.

We can also characterize the molecular states by the total spin S of the electrons. Due to the different possible orientations of the spin vector, a state with the total spin S has a $2S+1$-fold degeneracy. The number $2S+1$ is called the *multiplicity* of the state. By convention, the multiplicity of a molecular state is denoted by a superscript placed before the letter showing the value Λ. For example, $^1\Pi$ denotes a state with $\Lambda = 1$ and total spin of 0.

Next, we should distinguish between two kinds of Σ states. If we reflect the molecule in any plane containing the internuclear axis, the energy of the Σ state does not change; hence the wave function can only be multiplied by a constant. However, a double reflection should bring the original state back. Thus, the multiplication constant for a single reflection must be either 1 or -1: the states either do not change or just change their sign. Those Σ states that do not change are called Σ^+ states, while the other kind are called Σ^- states.

One more special gradation is particular to homonuclear diatomic molecules: the energy of state should not change with simultaneous change of all electronic coordinates (this corresponds to the reflection in a plane bisecting the internuclear axis). Double reflection should again bring back the original function. Therefore, there must be two kinds of states in the diatomic homonuclear molecules: those that are invariant to the transformation and those that change their sign. The former are called *even* states (German: *gerade*) and the latter are called *odd* (*ungerade*). The even (odd) states are denoted by a subscript g (u) after the letter designating the value of Λ: for example, Π_g, Π_u.

Finally, we must mention one more spectroscopic notation that is used in this chapter. Sometimes in order to calculate the electronic eigenstates of a diatomic molecule, it is convenient to single out a particular electron, orbiting the molecular core consisting of the nuclei and the rest of the electrons. (Examples are molecular Rydberg states or the single electron in the H_2^+ molecular ion.) If the nuclei are close together compared to the electron orbital dimensions, the potential becomes more centro-symmetric. Then we can define atom-like principal and orbital quantum numbers n, l, and $\lambda = |l_z|$. Although the first two numbers are defined only for relatively small internuclear separations, λ is well-defined no matter how far apart the nuclei are, so long as $\hat{z}$ is taken along the internuclear axis. The single electronic states in diatomic molecules are therefore designated as follows: the quantum number n, followed by the Roman letter corresponding to the quantum number l, followed by the Greek letter describing the quantum number λ. The states with l = 0, 1, 2, 3, ... are designated by s, p, d, f, ..., and similarly, the states with λ = 0, 1, 2, 3, ... by $\sigma, \pi, \delta, \varphi$, ... In addition, for homonuclear diatomic molecules, we can define even (g) or odd (u) parity with respect to a reflection in a plane bisecting the internuclear axis. For example, the lowest lying two electronic states of the hydrogen molecular ion H_2^+ using this nomenclature are called $1s\sigma_g$ and $2p\sigma_u$.

D. Transitions in Diatomic Homonuclear Molecules

Thus far, we have not included a laser field in the discussion. Irradiation of a molecule by laser light can lead to molecular excitation, ionization, and dissociation. The latter process has no analog in laser-atom interactions.

Most lasers operate at visible or near-visible wavelengths. The separation between the rotational and vibrational sublevels of an electronic manifold in a molecule is usually much smaller than the photon's energy, whereas transitions between different electronic states typically require at least one photon. Selection rules for dipole transitions require that, for a system not possessing a permanent dipole moment, only transitions coupling the electronic states of the opposite symmetry are possible. Thus single photon transitions within a single electronic manifold are strongly forbidden. In a homonuclear diatomic molecule, one photon (or any odd number of photons) may only couple odd (u) states with even (g) states and vice versa. For example, a one-photon transition between the H_2 ground state (spectroscopical notation $X^1\Sigma_g^+$) and the excited state $I^1\Pi_g$ is forbidden. On the other hand, the H_2^+ ground state, $1s\sigma_g$, is strongly coupled to the first repulsive state, $2p\sigma_u$. As we will see later, this last transition strongly affects the behavior of hydrogen in intense laser fields.

Transitions between electronic states in a molecule usually involve changes of the vibrational and rotational quantum numbers as well. The *Franck-Condon* principle [14, 15] governs the change of the vibrational quantum number during an electronic transition in the molecule. This principle states that due to the difference in mass between nuclei and electrons, the time required for a radiative transition between two electronic levels is much shorter than a vibrational period of the molecule. Consequently, the nuclear positions and velocities do not change during an electronic transition. In a plot of the potential energy of the molecule versus internuclear separation, a transition satisfying the Franck-Condon principle is a vertical arrow, as shown in Figure 2.2. Such a transition is called a *vertical* transition.

A corollary of the Franck-Condon principle is that radiative transitions between bound electronic states in a diatomic molecule can take place only at certain internuclear separations. Consider, for example, the one-photon absorption shown in Figure 2.2. Since the positions and velocities of the nuclei remain constant during the transition, the electronic energy eigenvalues must be precisely one photon apart. This in turn determines allowed values

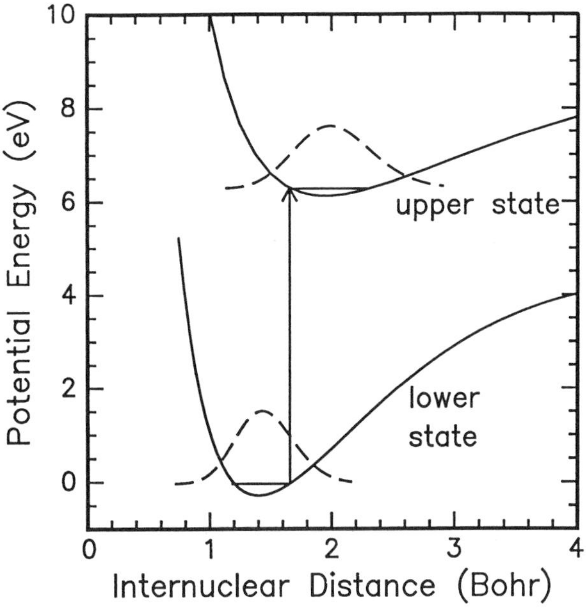

Figure 2.2 According to the Franck-Condon principle, the nuclear positions do not change during an electronic transition. Thus, such a transition is called a *vertical transition*. The excitation shown on the diagram does not change the nuclear vibrational quantum number (v = 0 for both states). The broken-line curves represent the vibrational wave functions of the initial (lower) and final (upper) states of the molecule.

for the internuclear separation. These points where the electronic states are exactly n photons apart are called *n-photon resonances*. Photoionization does not have the same restriction, since the departing electron may carry away some of the energy.

The quantum-mechanical statement of the Franck-Condon principle is that the transition rate between molecular states is proportional to the overlap of the initial and final wave functions. Consider a transition between two states: $\Psi_{initial}$ and Ψ_{final}. We will ignore rotation for the present. In the Born-Oppenheimer approximation these states can be written as

$$\Psi_{initial} = \psi_i \nu_i \qquad (2.17)$$

and

$$\Psi_{final} = \psi_f \nu_f \tag{2.18}$$

where $\psi_{i,f}$ are the final and initial electronic wave functions of the molecule calculated at each equilibrium internuclear separation, and $\nu_{i,f}$ are the initial and final nuclear vibrational wave functions. The probability for the transition is proportional to square of the matrix element coupling the states:

$$<\Psi_f|T|\Psi_i> = <\psi_f\nu_f|T_e + T_\nu|\psi_i\nu_i>$$

where T is the transition operator. This expression can be rewritten:

$$<\Psi_f|T|\Psi_i> = <\psi_f|\psi_i> <\nu_f|T_\nu|\nu_i> + <\nu_f|\nu_i> <\psi_f|T_e|\psi_i> \tag{2.19}$$

Since $\Psi_{initial}$ and Ψ_{final} represent two different electronic states, their electronic wave functions are orthogonal:

$$<\psi_f|\psi_i> = 0 \tag{2.20}$$

Hence, the first term in Equation 2.19 is zero and the matrix element coupling the states is

$$<\Psi_f|T|\Psi_i> = <\nu_f|\nu_i> <\psi_f|T_e|\psi_i> \tag{2.21}$$

The transition probability is proportional to the square of the overlap between the vibrational wave functions: $|<\nu_f|\nu_i>|^2$. This overlap is sometimes called the *Franck-Condon integral*; its square is the *Franck-Condon factor.*

2.3 H_2 IN INTENSE LASER FIELDS: POSSIBLE DECAY SCENARIOS

There are a number of ways a molecule may disintegrate in an intense laser field. Figure 2.3 shows a simplified potential energy diagram for molecular hydrogen (after reference 16). Consider the v = 0 level of the X $^1\Sigma_g^+$ ground state as the initial state of the molecule. Absorption of multiple photons to the dissociative state b $^3\Sigma_u^+$ is unlikely because of electric dipole

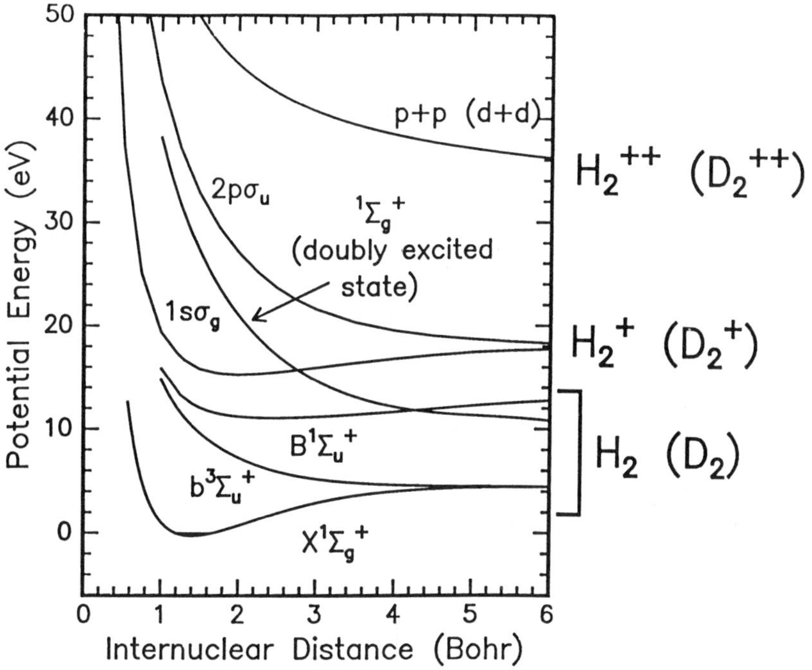

Figure 2.3 A simplified potential energy diagram for H_2 (D_2). Only a few relevant potentials are shown. They include four molecular states: the X $^1\Sigma_g^+$ ground state, the first repulsive state **b**, the long-range **B** state, and the lowest-lying doubly excited $^1\Sigma_g^+$ state. Also shown are three ion states: the $1s\sigma_g$ ground state and the $2p\sigma_u$ first repulsive state of molecular ion as well as the coulomb repulsion state of two bare nuclei (the uppermost curve in the diagram).

selection rules; however, multiphoton absorption to higher lying states is possible, leading to dissociation into a ground state neutral and an excited neutral. There are also states corresponding to the excitation of both electrons simultaneously, where the molecule could decay into two excited H atoms. The dissociated atoms can subsequently ionize and the photoelectron spectrum can be used to identify the excited states. The energy of a doubly excited state often exceeds that of the molecular ion so that autoionization is also possible. This process competes with ordinary multiphoton ionization.

If multiple ionization is very rapid, both electrons may leave in less time than it takes the molecule to dissociate. The *Coulomb explosion* afterwards produces very fast protons [2].

A final possibility is a single-electron multiphoton ionization of H_2 to form an H_2^+ molecular ion in the $1s\sigma_g$ ground state. In experiments carried out with visible and near-visible lasers with pulselengths from subpicoseconds to nanoseconds, this is the dominant mode of decay.

The H_2^+ molecular ion is extremely simple, containing only a single electron. It is also a very useful test laboratory for the study of molecules in intense laser fields. The ground state has a very large transition dipole moment to the $2p\sigma_u$ first excited state, which is identical to the ground state except for a parity change about the plane bisecting the protons. This state dissociates into a neutral hydrogen atom and a proton. The strong coupling between these two states leads to a number of interesting high field phenomena, such as bond-softening and the formation of light-induced bound states. Alternatively, the molecular ion can be ionized, resulting in production of two bare protons.

2.4 EXPERIMENTAL METHODS AND RESULTS

Several experimental techniques are available to explore the behavior of molecules in the laser focus. All of these methods involve collecting the products of molecular decay and analyzing their spectra. One of the techniques is based on detecting photons produced via emission from the molecular excited states. The others include the photoelectron and photoion spectroscopy. These last two methods are very frequently used in modern high-intensity molecular photophysics. In fact, most of the results described in this chapter were obtained using these techniques.

A. Electron Time-Of-Flight Spectroscopy

Suppose that we subject an atomic system (i.e., a molecule) to a burst of intense laser light (Figure 2.4). Furthermore, assume that the laser frequency ω is insufficient for single-photon ionization:

$$\hbar\omega << \mathrm{IP} \tag{2.22}$$

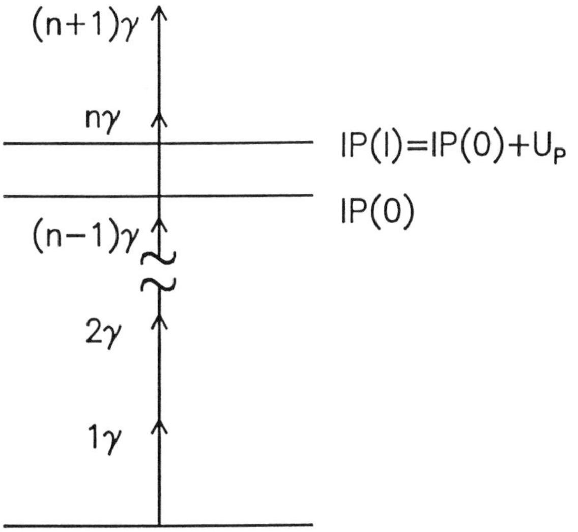

Figure 2.4 The ionization threshold shifts in an intense laser field. The magnitude of this shift is approximately equal to the ponderomotive potential at the instance of ionization, U_P.

where IP is the ionization potential of the molecule. These conditions rule out single-photon ionization. Hence, in order for photoionization to take place, the molecule must absorb more than one photon. A naive way to treat multiphoton ionization is as a sequential absorption of many photons, each increasing the molecule's energy by $\hbar\omega$. Thus, after absorption of m photons, the molecular energy is equal to $m\hbar\omega$. If there is no real eigenstate at this energy, the molecule will be in a virtual state where it only can spend a very limited time. According to the Heisenberg uncertainty principle, this time Δt is inversely proportional to the energy difference ΔE between the virtual state and the nearest eigenstate:

$$\Delta t_{virtual} = \frac{\hbar}{\Delta E} \tag{2.23}$$

Unless another photon is absorbed during time Δt, the molecule will decay back into a lower-energy state. Therefore, multiphoton ionization requires high photon flux which, in turn, implies high laser intensity. In fact, the required laser intensity is high enough to shift the ionization threshold of the molecule. If we take this shift into account, the electrons produced via n-photon ionization should have characteristic energies:

$$E_n(I) = n\hbar\omega - IP(I) \tag{2.24}$$

where I is the intensity at the moment of ionization and IP(I) is the intensity-dependent ionization potential, which may be expressed as the difference between the ground state energies of the atom and (ion + free electron).

We can calculate the shift of the ionization threshold with intensity using the *ponderomotive potential.* A free electron in a monochromatic light field oscillates with the field frequency ω. The time-averaged energy associated with this rapid oscillatory motion can be treated as a potential energy for the electron in the field and is called the ponderomotive potential U_P. (For a detailed review of ponderomotive potential and its effects on ionization dynamics see [17].) The total energy of an electron in the laser field is then:

$$E_{total} = T_e + E_{pot} + U_P \tag{2.25}$$

where T_e, E_{pot} are the kinetic and potential energies of the electron. The ponderomotive potential can be written as a function of the electric field:

$$U_P = \frac{e^2E^2}{4m_e\omega^2} \tag{2.26}$$

In practical units, this becomes:

$$U_P\,(\text{in eV}) = 9.3 \times 10^{-14}\, I\lambda^2 \tag{2.27}$$

where I is the laser intensity (in W/cm^2) and λ is the wavelength in μm. For the fundamental of an Nd:YAG laser (≈ 1.064 μm, a wavelength frequently used in experiments) this is approximately 1 eV of ponderomotive energy per 10^{13} W/cm^2 of laser intensity.

In order to ionize an atom (or a molecule) with an intense laser, the laser field must supply ponderomotive energy to the exiting electron in addition to overcoming the binding energy. Therefore, the ionization threshold of an atomic system in a light field is increased by the laser ponderomotive potential. Equation 2.24 can be rewritten as follows:

$$E_n(I) = n\hbar\omega - U_P - [E_{ion}(I) - E_{atom}(I)]$$
$$\approx n\hbar\omega - U_P \qquad (2.28)$$

If there is a gradient in the light intensity, such as that created by the inhomogeneity of a laser focus, photoelectrons will experience a force in the direction of lower intensity. The force is the negative gradient of U_P. For long pulses, the ponderomotive potential is conservative and the electron's ponderomotive potential energy is converted into the kinetic energy. In this case the final electron energy following n-photon ionization is given by:

$$E_n(0) = n\hbar\omega - IP(0) - U_P + U_P = n\hbar\omega - IP(0) \qquad (2.29)$$

Hence, experiments employing relatively long laser pulses should show no shift of the ionization threshold.

A typical *long-pulse* photoelectron spectrum is shown in Figure 2.5. This is the kinetic energy distribution of electrons detected during irradiation of molecular hydrogen by 100 ps pulses of intense 532 nm light. The light was focused to a small spot (10 μm) in order to obtain a peak intensity of 2×10^{13} W/cm^2. For this focal geometry the electrons move out of the focus before the laser pulse is over. Consequently, the molecular ionization potential appears unshifted. Most of the detected electrons were produced via multiphoton ionization of the neutral H_2 to various vibrational levels of the $1s\sigma_g$ ground state of the H_2^+ molecular ion. Some electrons appear to have absorbed many more photons than the minimum number required for ionization. This phenomenon is called *above-threshold ionization* (ATI).

Intermediate resonances in molecules can play a very important role during multiphoton ionization. If there is an m-photon resonance (i.e., an eigenstate with energy $m\hbar\omega$), then the multiphoton absorption cross-section is significantly enhanced. Since the ponderomotive shift of excited levels in the molecule can be comparable to the photon energy, such resonances are practically inevitable, and greatly affect the ionization rate.

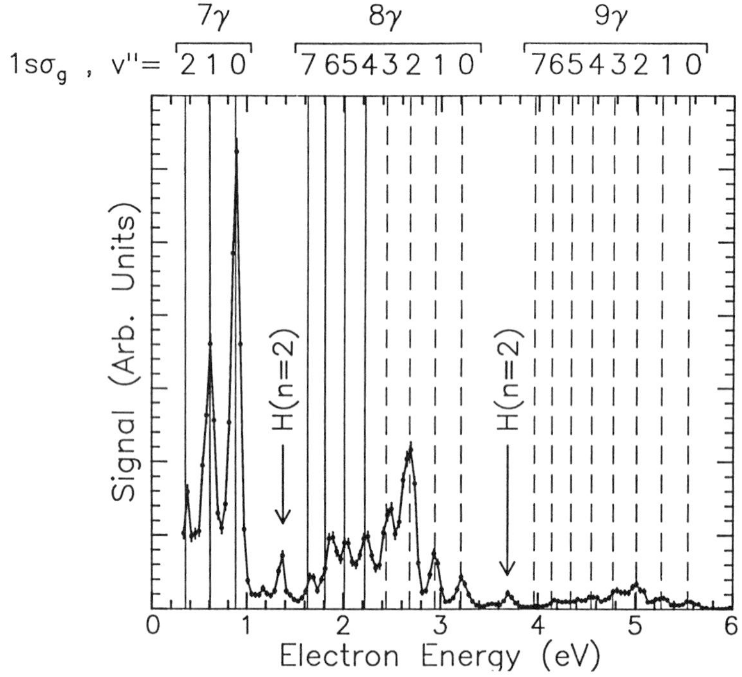

Figure 2.5 The diagram shows the kinetic energy distribution of the electrons detected during irradiation of molecular hydrogen by long pulses (100 ps) of intense (2×10^{13} W/cm^2 at the peak) 532 nm light. Vertical lines indicate expected electron energies for ionization into different vibrational levels of the H_2^+ ground state (vibrational quantum number v″ values are marked on top of the figure). Solid lines indicate threshold ionization peaks (i.e. those with the minimum possible number of photons). Dashed lines show the expected positions of ATI peaks. These peaks are clustered by the total number of photons absorbed (marked on top of the figure). Also seen are the electrons produced via ionization of exited (n = 2) hydrogen atoms. Peaks due to the ionization of H atoms in n = 1 ground state coincide with those from molecular ionization.

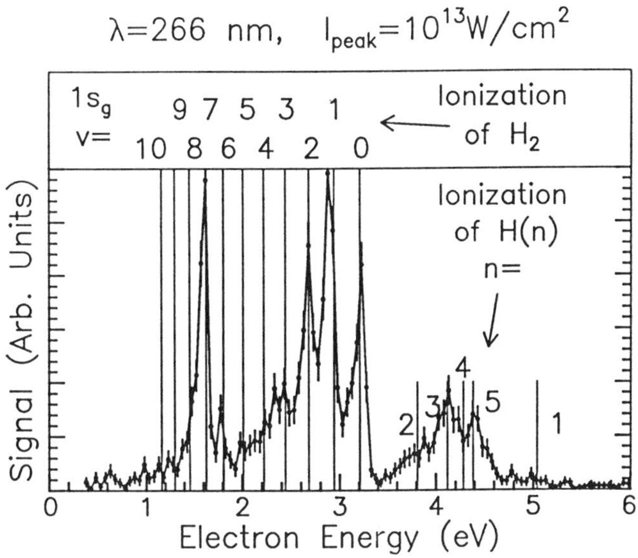

Figure 2.6 Intermediate resonances can play a very important role in multiphoton ionization. The spectrum shown in this figure was collected with 100 ps pulses of intense 266 nm light. Ionization of H_2 into different vibrational levels of the H_2^+ ground state, as well as ionization of hydrogen atoms in different n states is observed. Ionization into v = 1 and v = 7 vibrational states is enhanced via intermediate resonances at three-photon level with B′ and C molecular states. *See also* [4].

The photoelectron energy spectrum in Figure 2.6 illustrates the effects of resonance on ionization. These electrons were produced during irradiation of H_2 by 100 ps pulses of intense (10^{13} W/cm^2 peak intensity at the focus) 266 nm light. The pulse duration is long enough to be in the *long-pulse regime*. The spectrum contains a sharp peak at 1.65 eV and a series of peaks centered around 3 eV. The former can be assigned to ionization of molecular hydrogen into the seventh vibrational level of the $1s\sigma_g$ ground state of H_2^+; however, the prominence of this peak is due to a three-photon resonance in H_2 at this wavelength: the energy of the seventh vibrational level of the molecular hydrogen C state is nearly $3\hbar\omega$. The electron configuration of C state ($^1\Pi_u$) allows three-photon absorption from the ground state of molecular hydrogen. In addition, the C state has a potential very similar to the molecular ion ground state. This is a characteristic of Rydberg states, whose wavefunctions lie well beyond the molecular core. Consequently, the Franck-Condon principle favors $\Delta v = 0$ transitions between these two states, resulting in the production of copious amounts of hydrogen molecular ions in the $v = 7$ vibrational state.

The most prominent peak around 3 eV is due to ionization into the first vibrational state of the molecular ion ground state. The enhancement of this peak is due to the three-photon resonance with the $v = 1$ vibrational level of the $B'\,{}^1\Sigma_u^+$ state of molecular hydrogen. This is also one of the Rydberg states converging to the $1s\sigma_g$ state of the molecular ion. Hence, ionization of H_2 via this intermediate resonance will favor the production of H_2^+ ions in the $1s\sigma_g$ ($v = 1$) state.

It is not always easy to identify the dominant ionization path contributing to a particular spectral feature; an example is the spectrum shown in Figure 2.5. The peak ponderomotive potential for this data was approximately 500 meV. Hence, one can expect the intermediate state shifts to be comparable to the separation between the vibrational levels in H_2 molecule. This significantly complicates the analysis of the problem. In addition, the ionization rate is a steep function of the laser intensity. Thus, the states shifted to resonance at higher laser intensities will produce more signal than those that are at resonance when no light is present. Autoionization of doubly excited molecular states also might affect the distribution of ionic population among the different vibrational levels.

Finally, the electron energy spectrum is complicated by the photoelectrons produced via ionization of atomic hydrogen. Both spectra (Figures 2.5 and 2.6) have peaks due to ionization of H atoms in $n \geq 1$ states. In fact, for ionization at 532 nm (as in Figure 2.5), the photoelectrons from atomic ground state hydrogen have energies around 0.4 eV, 2.73 eV, . . ., so that they are hardly distinguishable from the photoelectrons from molecular ionization.

The situation is complicated further when the light pulse is short enough to turn off before electrons leave the laser focus. The initial energy of an electron upon ionization will still be given by Equations 2.24 and 2.28. However, unlike for a long pulse, the electron will not be able to convert the ponderomotive energy into kinetic energy. Hence, the electron will leave the focus with less energy. In the extreme limit of very short pulses, the ponderomotive component of the electron's energy simply disappears. This regime is called a *short-pulse regime*. The kinetic energy distribution of the electrons produced via ionization by a short pulse laser generally depends on the laser wavelength and the pulse duration.

Several types of lasers produce pulses in the 0.5–5 psec range. This is short enough to neglect ponderomotive acceleration after ionization but still much longer than a vibrational period in simple molecules. Experiments in this regime often show that ionization takes place via multiphoton resonances [7, 8]. The electron kinetic energy spectrum then often contains sharp peaks attributed to the presence of intermediate resonances.

B. Ion Time-Of-Flight Spectroscopy; Bond Softening

Intense light can not only ionize molecules but can dissociate them as well. Figures 2.5 and 2.6 show peaks from ionization of both atomic and molecular hydrogen. It is clear that dissociation may occur prior to ionization. Additional information about dissociation can be obtained from direct measurements of the fragment kinetic energy spectrum. Combined with photoelectron spectroscopy, this ion spectroscopy is a very powerful tool for investigating molecular structure.

Figure 2.7 illustrates how the ion fragment kinetic energy spectrum can yield information about the molecular processes inside a laser focus. Consider H_2^+ molecular ions subjected to intense 532 nm laser light. There is a very large optical coupling between the ground $1s\sigma_g$ state and the first excited $2p\sigma_u$ state. The molecular ion's initial state (v^{th} vibrational level of the $1s\sigma_g$ ground state) extends beyond the point where $1s\sigma_g$ and $2p\sigma_u$ states are exactly one photon apart. Every time the vibrating nuclei pass the one-photon resonance point, the ion can absorb a photon, transfer to the upper repulsive state and dissociate into a hydrogen atom and a proton. The kinetic energy shared between the dissociating fragments is

$$KE_{tot} = E_v + \hbar\omega - E_D \tag{2.30}$$

where E_v is the nuclear energy in the lower state and E_D is the dissociation limit of the upper state. Since the masses of a proton and a hydrogen atom are almost the same, both fragments share the available kinetic energy equally and each emerges with $KE = KE_{tot}/2$.

A convenient way to look at this process is to consider the eigenstates of a *dressed* system of the molecule combined with a perfectly monochro-

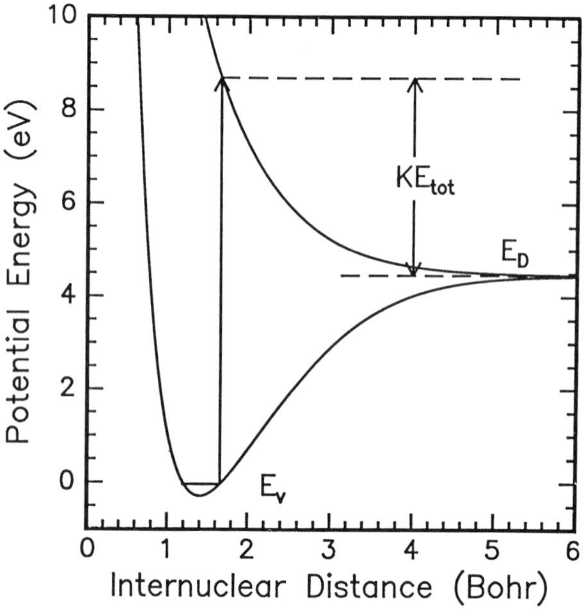

Figure 2.7 A molecule initially confined in the vth vibrational level of a bound electronic state can be photoexcited into a repulsive state. The total energy available for dissociation is then equal to $E_v + \hbar\omega - E_D$. In the case of a homonuclear diatomic molecule (like H_2, H_2^+, D_2, or D_2^+) this energy is equally shared by the dissociation fragments.

matic laser field. In this picture the energy of the system is equal to the sum of molecular energy, interaction energy, and $n\hbar\omega$, where n is the total number of photons absorbed by the molecule. There exists several ways to solve the Schrödinger equation for this system. One of them, based on the Floquet methods [18] to find electronic eigenvalues as a function of the field strength is outlined below.

The Hamiltonian of the system may be divided into three parts: molecule, field, and interaction:

$$H = H_{mol} + H_{field} + e\mathbf{r}\cdot\mathbf{E}(t) \tag{2.31}$$

An eigenstate can be expressed as a product of a molecular wave function and a *photonic* state |n> (a state with n photons). The actual value of n in an

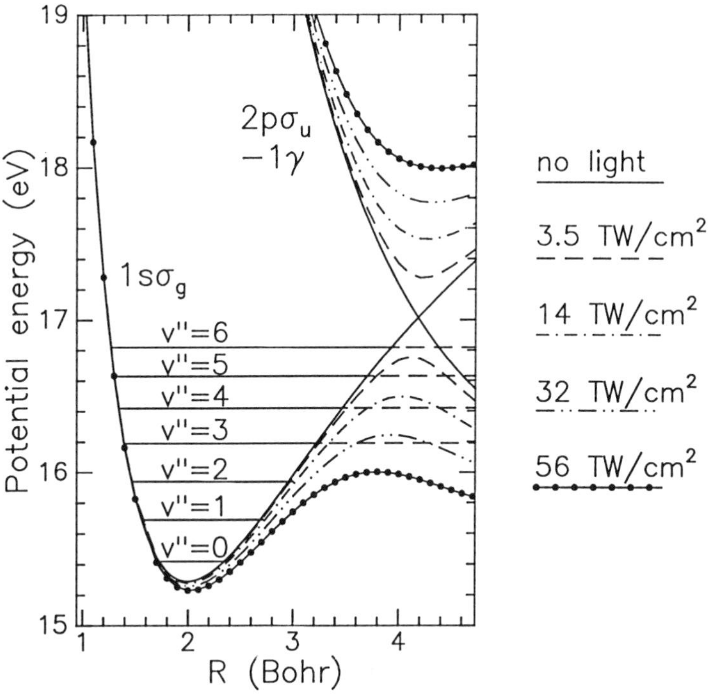

Figure 2.8 The one-photon resonance for 532 nm light. Solid lines show the potential curve of the unperturbed $1s\sigma_g$ state, and the $2p\sigma_u$ state lowered by $\hbar\omega$. The component of the laser field parallel to the internuclear axis perturbs the states and shifts them onto new curves. As the parallel polarization grows, the gap between the states opens, the binding energy of the H_2^+ ion decreases, and the vibrational states become unbound. *See also* [6].

experiment is huge; however, it does not appear directly in the calculations. The Hamiltonian matrix $H_{ij} = <\psi_i|H|\psi_j>$ contains two different kinds of nonzero elements: 1) elements in the diagonal blocks where $n_i = n_j$, and 2) off-diagonal blocks where the interaction Hamiltonian couples states with $\Delta n = \pm 1$ and different parity. The matrix diagonalization results in a series of *quasi-eigenvalues* for molecular states in the field as is shown in Figure 2.8. (See also Chapter 3.)

Although the number of the resultant eigenstates depends on the number of initial basis states, only the eigenvalues representing $1s\sigma_g$ and $2p\sigma_u$ states are shown in the figure. The other states are removed in energy by many photons, so populating them requires absorption of many more photons and only becomes important at still higher laser intensities. The solid lines in the figure represent the so-called *diabatic* states, which are the electronic eigenstates in the limit of very small laser field. (The diabatic eigenvalues still include the photon energies $n\hbar\omega$). When the ion in the ground $1s\sigma_g$ state passes the one-photon resonance point it can either absorb a photon and switch onto the $2p\sigma_u$ curve, lowered by a photon, or stay on the $1s\sigma_g$ curve (no photon exchange takes place in this case).

The strength of the coupling between the $1s\sigma_g$ and $2p\sigma_u$ states depends on the magnitude of the laser field component along the internuclear axis of the ion. It is therefore convenient to introduce the notion of *parallel intensity*, I_{pl}:

$$I_{pl} = I_{tot} \times (\cos\theta)^2 \tag{2.32}$$

where θ is the angle between the laser polarization and the internuclear axis.

The curves in Figure 2.8 are plotted for several different values of I_{pl}. As I_{pl} increases, the $1s\sigma_g - 2p\sigma_u$ coupling deforms the potential at points where two curves are separated exactly by an integer number of photons of energy. This deformation leads to the opening of the gap (or avoided crossing) between the states at these point. In the weak field limit, vibrational levels whose energy lies above the one-photon crossing point are able to switch onto the $2p\sigma_u$ curve (by absorption of a photon) and dissociate. On the other hand, the states whose energy lies below the crossing point are bound and stable against dissociation. (This includes all the vibrational levels shown in Figure 2.8.) However, as the parallel light intensity increases, the gap is opening wider and ultimately drops below these vibrational levels, thereby allowing them to dissociate. This effect had been named *bond softening*. (In fact, a small amount of dissociation may occur, even if the vibrational state is below the bottom of the one-photon crossing gap, due to quantum mechanical tunneling [11].)

Figure 2.9 shows the kinetic energy spectra of protons produced via dissociation of H_2^+ via bond softening. The characteristic signature of bond

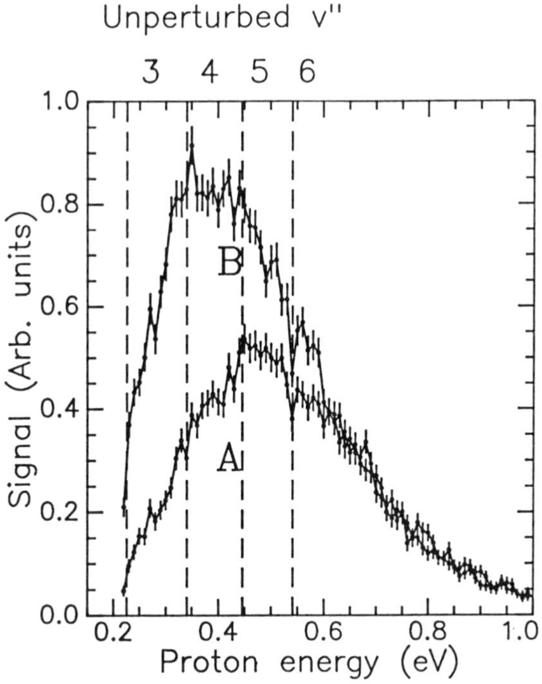

Figure 2.9 Kinetic energy distribution of protons produced during irradiation of H_2 by long pulses (100 ps) of intense 532 nm light. Only the protons emerging along the direction of the laser polarization were detected. Vertical lines indicate expected proton energies for dissociation of H_2^+ molecular ions in different vibrational levels of the $1s\sigma_g$ ground state (the vibrational quantum numbers are marked on top). When the laser intensity is increased, the gap at the one-photon crossing gets wider, allowing lower vibrational states to dissociate. This results in broadening of the energy distribution to the left as is seen in the figure. *See also* [6].

softening is the low energy of the ions: The distribution is centered around 0.4 eV, which approximately corresponds to dissociation of the fourth or fifth vibrational levels. Figure 2.8 shows that were it not for the distortion of the potential (i.e., the bond softening effect), these levels would never come into one-photon resonance with the repulsive state, and would therefore never dissociate via this path. Additional evidence comes from the comparison of fragment energy spectra at different intensities, as shown in the figure. As the laser intensity increases, the distribution broadens to the left,

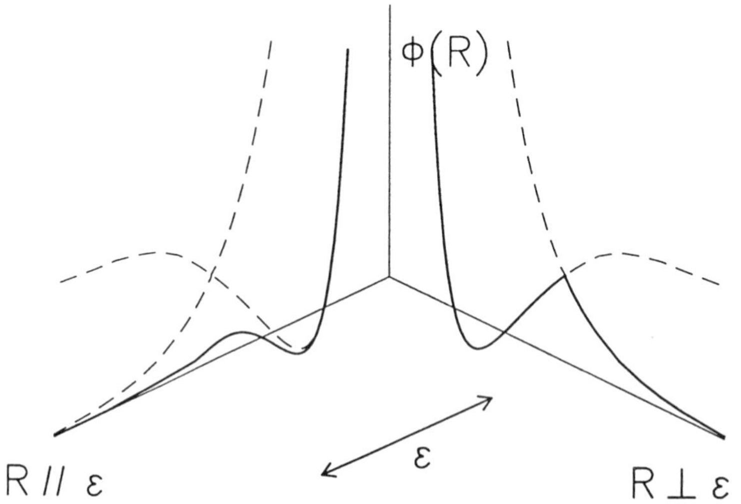

Figure 2.10 The effective potential well formed by the adiabatic crossing of the $1s\sigma_g$ state and the $2p\sigma_u$ state, is a function of the relative orientation of the internuclear axis and the laser polarization. As is seen from the figure, most of dissociation is expected along the direction of the laser polarization.

indicating dissociation of lower vibrational states (as is implied by the bond softening mechanism).

In addition to affecting the proton spectra, deformation of potential curves with intensity affects electron spectra as well. As the intensity grows, the bottom of the potential well becomes more shallow, so that vibrational eigenvalues change. As a consequence, the electrons produced via photoionization of neutral molecules increase (see Equation 2.29). This is one of the reasons for the relative width of the vibrational peaks in the photoelectron spectra seen in Figures 2.5 and 2.6.

If the molecule is viewed as a classical vibrating rotator, this lowering of the vibrational eigenvalues as the parallel intensity increases is equivalent to an angle-dependent potential in the dressed Hamiltonian. This potential produces a torque on the molecule in the direction of alignment with the laser polarization. We can imagine the molecular ion moving in a three-dimensional potential well, with sides that become quite low along the laser polarization axis, as pictured in Figure 2.10. As the molecular ion rotates

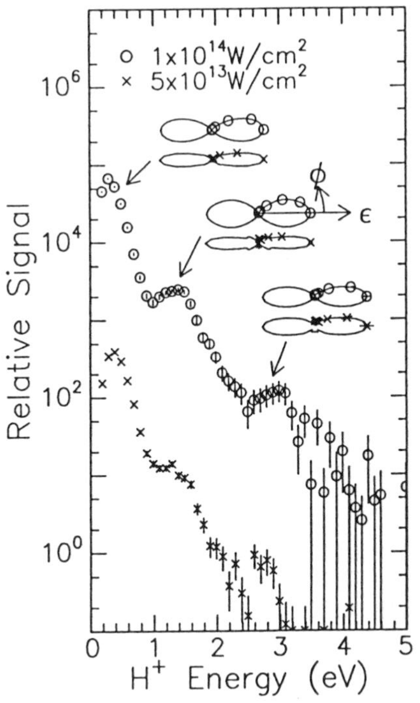

Figure 2.11 The energy spectrum and angular distribution of ions detected during irradiation of H_2 by 532 nm linearly polarized light. Above 5×10^{13} W/cm^2 the proton spectra develop a series of peaks separated by $\hbar\omega/2 = 1.165$ eV. These peaks are due to the dissociation via different multiphoton crossings. Most of the detected protons are emitted along the laser polarization $\bar{\varepsilon}$ in narrow distributions that become broader as the intensity increases. This is expected from the general bond softening picture: Since dissociation probability is a function of I_{pl} only, increase of the total laser intensity forces the ions misaligned with the field to dissociate. *See also* [6].

into the alignment with the field, its binding energy decreases and it eventually dissociates. Therefore, the molecular ions are likely to dissociate along the direction of the laser polarization, creating a narrow angular distribution of as seen in Figure 2.11. When the laser intensity is increased, I_{pl} grows proportionally, thereby lowering the binding energy of the ions and causing a broader angular distribution of the dissociating fragments (Figure 2.11). Any initial rotation of the molecule will have a very strong influence on its

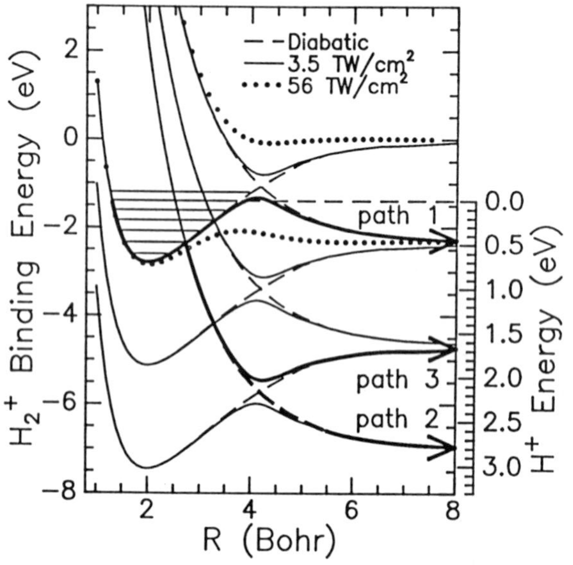

Figure 2.12 The results of Floquet calculations for 532 nm light. Solid curves: adiabatic potential for $I_{pl} = 3.5 \times 10^{13}$ W/cm^2. Dashed curves: diabatic (unperturbed) $1s\sigma_g$ and $2p\sigma_u$ states shifted by $n\hbar\omega$. Different paths are available for dissociation, resulting in production of different energy protons. The probability of the different pathways can be estimated by Landau-Zener theory. *Source*: from [6].

motion in the potential, since the component of angular momentum along the laser polarization should be conserved.

Figure 2.11 also shows multiple peaks in the kinetic energy spectra of dissociating ions. The separation between the peaks is $\hbar\omega/2$. These peaks result from the dissociation via the higher-order crossings, representing multiphoton transitions between $1s\sigma_g$ and $2p\sigma_u$. These states can be coupled by any odd number of photons. An H_2^+ molecular ion passing through an odd-photon crossing can adiabatically switch between these two states, corresponding to emission or absorption of extra photons. In Floquet dressed state formalism this leads to the gaps appearing in place of odd-number-of-photon resonances (Figure 2.12). These gaps become wider with increasing laser field, forcing the ion to take the adiabatic path. The molecular fragments from these events emerge faster since more photons are absorbed.

Figure 2.12 shows three different dissociation paths; each path results in a single peak in the observed proton energy distribution. The slowest peak in Figure 2.11 (centered around 0.4 eV) contains the protons produced via one-photon excitation into $2p\sigma_u$ state, followed by dissociation. In the Floquet picture, this corresponds to adiabatic passage through the one-photon gap (Path 1 in Figure 2.12). An alternative path, three-photon absorption followed by one-photon emission, produces the peak $\hbar\omega/2$ higher in energy. This is Path 3 in Figure 2.12 (adiabatic transition at both crossings). The highest energy peak in the spectra in Figure 2.11 corresponds to dissociation via three-photon absorption. In the Floquet picture, this corresponds to an adiabatic passage through the three-photon gap, followed by a diabatic transition through the one-photon gap (Path 2 in Figure 2.12).

As is seen from Figure 2.12, the dissociation via higher-order multiphoton crossings is also possible. However, it requires much higher laser intensities and has not been observed.

A careful look at Figure 2.11 reveals that the peaks are not separated by exactly $\hbar\omega/2$. This is caused by the fact that probability of different dissociation paths depends on the initial vibrational state of the ion. For example, as is seen from Figure 2.12, only the classically unbound levels can dissociate through the one-photon gap (neglecting the tunneling), but all vibrational states except $v' = 0$ can dissociate through the three-photon gap. Thus, the one-photon peak contains higher vibrational states than the other two peaks. Consequently, the separation between first two peaks is less than $\hbar\omega/2$.

The appearance of multiple peaks in the fragment dissociation spectrum is analogous to the additional peaks in the electron spectra in ATI. Thus, this phenomenon has been called *above-threshold dissociation* (ATD) [11].

The distribution of the signal over the different peaks in the spectrum depends on the adiabatic/diabatic branching ratios at the crossings; these ratios can be calculated by applying Landau-Zener theory [19]. The basic idea may be stated in semiclassical terms, where we imagine that the ion is a particle moving along the internuclear coordinate along the potential curves shown in the diagram. As the molecule moves through a crossing point, it will follow the adiabatic curve so long as its rate of passage through the gap is slow compared to the Rabi rate $V_{12}/\hbar$ for transitions between the interacting states [19]. The Rabi rate is related to the size of the gap Δ:

$$\Delta = 2|V_{12}| \tag{2.33}$$

Landau-Zener theory (see Chapter 3) predicts the probability W for the diabatic passage of a particle through a curve-crossing, in terms of the relative slopes of the diabatic (unperturbed) potentials and the Rabi frequency:

$$W = e^{(-2\pi V^2)/(\hbar v|F_1 - F_2|)} \tag{2.34}$$

where $F_{1,2}$ are the slopes $d\varphi/dR$ of the diabatic curves (φ is the potential), and v is the velocity of internuclear motion at the crossing. It follows from this equation that the slower ions are more likely to move adiabatically through the crossing. This dependence can be understood if we recall that an adiabatic transition involves either absorption or emission of photons. The photoexchange process can only take place when the ion passes the corresponding multiphoton resonance. The slower ions spend more time in the vicinity of the resonance, therefore increasing the probability of photon absorption or emission.

C. Above-Threshold Dissociation With Ultrashort Laser Pulses

The distribution of the above-threshold dissociation signal among the different peaks depends on the laser wavelength and intensity, as well as the time history of the laser pulse. Figure 2.13 shows kinetic energy spectra of ions detected during irradiation of H_2 and D_2 by intense 769 nm light. The general shape of the deuterium spectra is similar to that of hydrogen. This is not surprising at all, since both isotopes share the same potential curves; however, they have different nuclear masses which results in slightly different spectra.

Figure 2.13 (opposite) The kinetic energy spectra of the dissociation fragments produced during irradiation of H_2 and D_2 by 100 fs pulses of intense 769 nm light. In addition to the peaks due to above-threshold dissociation via bond softening, the spectra show the evidence for the vibrational population trapping in the light-induced bound states: The Franck-Condon overlaps between these states and the repulsive two proton (or two deuteron) Coulomb state form a series of maxima separated by the same intervals as the observed modulations in the spectra. The overlap integrals were calculated for the lowest eight (ten) vibrational states of the hydrogen (deuterium) molecular ion in the adiabatic potential well above the three-photon crossing at $I_{pl} = 9 \times 10^{13}$ W/cm^2. *See also* [20].

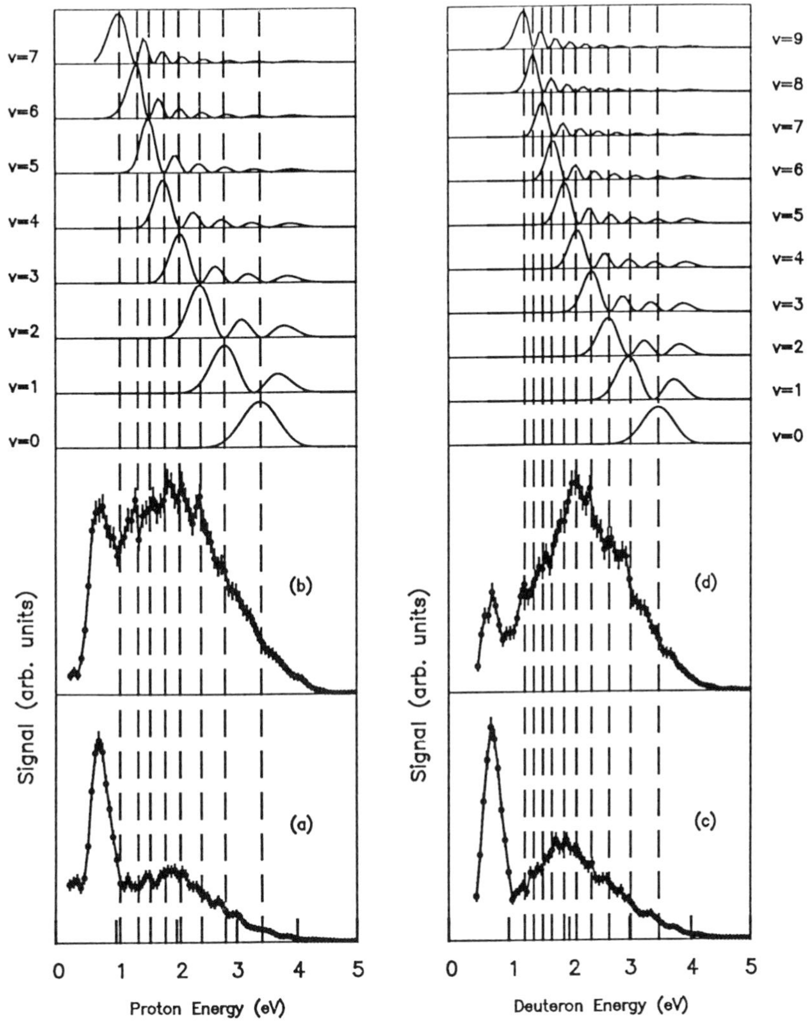
v=7
v=6
v=5
v=4
v=3
v=2
v=1
v=0
(b)
(a)
Signal (arb. units)
0 1 2 3 4 5
Proton Energy (eV)
v=9
v=8
v=7
v=6
v=5
v=4
v=3
v=2
v=1
v=0
(d)
(c)
Signal (arb. units)
0 1 2 3 4 5
Deuteron Energy (eV)

As is seen in Figure 2.13, the general shape of the spectra is quite different from that seen in Figures 2.9 and 2.11. What causes this difference? At least three laser parameters have been changed. First, the laser peak intensity was increased to 10^{15} W/cm^2 (for the traces a and c) and to 2×10^{15} W/cm^2 (for the traces b and d). This is an order of magnitude higher than the intensity used to produce the energy distributions of Figures 2.9 and 2.11. Second, the wavelength was increased from 532 nm to 769 nm. Finally, the most important change was laser pulse duration; it was decreased by three orders of magnitude: from ~100 ps to ~100–150 fs!

To understand the importance of this change we have to consider the characteristic time scales for internuclear motion in the hydrogen and deuterium molecular ions. The rotational period of the H_2^+ (D_2^+) molecular ion is ~500 fs (~1 ps) [13]. Therefore, in the experiments with the 100 fs laser the ions were essentially rotationally frozen. The vibrational period of the ions (~20 fs for hydrogen and ~30 fs for deuterium) is shorter than the laser pulse. However, unlike in the experiments with 100 ps laser pulses, the duration of the nuclear vibration is comparable to the time scale on which the laser intensity changes. This last difference has the most profound effect on the dissociation fragment energy spectra.

We begin by identifying the features of the spectra in Figure 2.13. The spectra consist of two parts: the slow ions, grouped into the peaks separated by approximately $\hbar\omega/2$, and fast ions, those with energies above 2 eV. We will concentrate on the low energy part first, and discuss the fast ions later.

The slow ions can be attributed to above-threshold dissociation via bond softening. However, unlike experiments with longer laser pulses, most of the ions emerge with higher kinetic energy, which is characteristic of dissociation at multiphoton crossings (such as Paths 2 and 3 in Figure 2.12).

This pulse-width dependence of ATD may have several causes. Most simply, the shorter laser pulse requires higher intensity in order to saturate the dissociation. Higher intensity tends to favor the multiphoton channels over the single photon channel.

The preponderance of multiphoton dissociation, when bond softening would permit single photon dissociation, may also be influenced by transient behavior during ionization. The dressed state picture assumes a monochromatic and continuous laser; however, especially for ultrashort pulses, the finite turn-on time of the laser may be significant. If the excitation rate increases more rapidly than the vibrational period of the initial state, the molecule may form a vibrational wave packet. The ionization potential of neutral H_2 or D_2 molecule is ~15.4 eV. At 769 nm, $\hbar\omega$ is 1.612 eV. Thus, the initial ionization is a high-order multiphoton process whose rate increases rapidly with intensity. For the ultrashort laser pulses used in the experiments, the ionization rate may approach or exceed the ion's vibrational period, resulting in the creation of a vibrational wave packet in the $1s\sigma_g$ ionic state.

To analyze how the formation of the wave packet affects the vibrational motion, we can model the process as follows: we assume 10^{15} W/cm^2 for the laser peak intensity and 1.4×10^{13} W/cm^2 for the intensity at which ionization of a neutral deuterium molecule takes place. Ionization then occurs on the rising edge of the laser pulse, producing a molecular ion near the inner turning point of the $1s\sigma_g$ ground state. The Franck-Condon principle favors wave-packet states centered around the v = 5 vibrational level.

Most of the essential features of the ensuing time evolution of the state can be seen in a model where the ion is a classical particle vibrating in a slowly changing potential. About 15 fs after the ion is born, it reaches its outer turning point. The momentary laser peak intensity at this time is $\approx 2.9 \times 10^{13}$ W/cm^2 and the potential curves are moderately perturbed, as can be seen in Figure 2.14. However, the one-photon gap is still not wide enough for the ion to dissociate. Consequently, the ion just turns around, goes back to the inner turning point and starts a new vibrational cycle. However, by the time the ion reaches the three-photon gap (about 40 fs into the ion's life), the laser intensity is 9×10^{13} W/cm^2 and the gap is about 200 meV wide. The Landau-Zener formula predicts a significant probability for adiabatic passage through this gap. As a result, the ion might absorb three photons and shift onto the dissociating potential curve. As the ion continues to dissociate, it reaches the one-photon gap which is now wide open, forcing the ion to emit a photon in the adiabatic trajectory. The resultant deuterons emerge with the kinetic energy characteristic of a *net two-photon absorption* (three-photon absorption followed by one-photon emission). This effect is responsible for reshaping of the ion spectra observed in the experiments with ultrashort laser pulses.

D. Light-Induced Bound Vibrational States Of Molecular Ions

Figure 2.8 shows how the deformation of a molecular bond in the presence of an intense laser field leads not only to bond softening but also to the

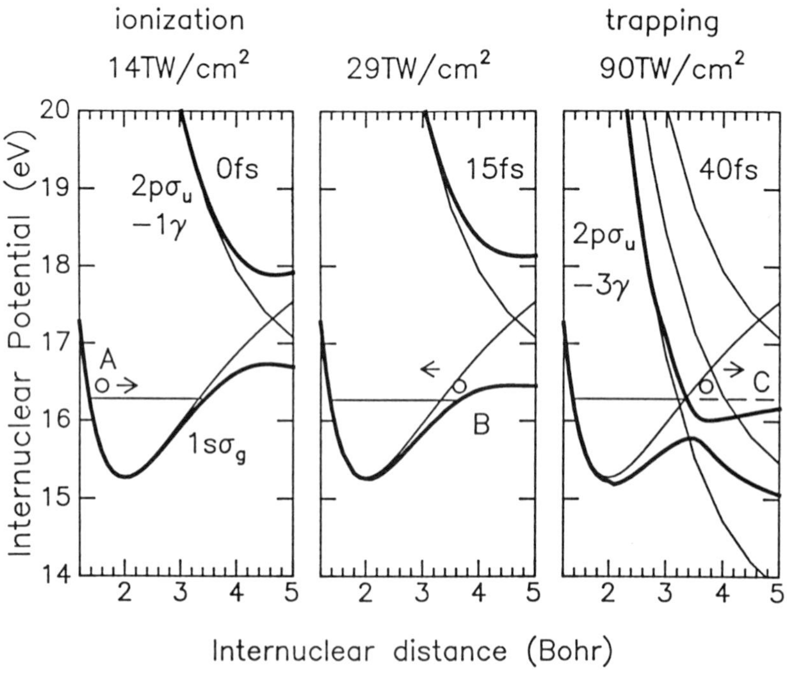

Figure 2.14 *A possible dissociation scenario:* during the second vibrational cycle the ion dissociates via three-photon avoided crossing between $1s\sigma_g$ and $2p\sigma_u$. *An alternative possibility:* A diabatic passage through the same gap leads to the trapping of the ion in the new light-induced vibrational bound state. This state is created in the adiabatic well above the three-photon avoided crossing between $1s\sigma_g$ and $2p\sigma_u$ states of molecular ion. *See also*: [20].

formation of a new potential well above the avoided crossing. The vibrational states supported in this well depend on the presence of intense laser light for their stability. Each vibrational cycle in these states involves absorption and emission of photons. Such a state has a $2p\sigma_u$ character on the left side of the adiabatic well ($R < R_{crossing}$) and $1s\sigma_g$ character on the right ($R > R_{crossing}$). These states are stable against dissociation as long as the value of I_{pl} is in the range where the potential well exists and is deep enough.

These light-induced bound states have been predicted by Fedorov [21] and independently by Bandrauk [22] (see also Chapter 3). Evidence for them has been seen in experimental work by Allendorf and Szoke [7] and Decker

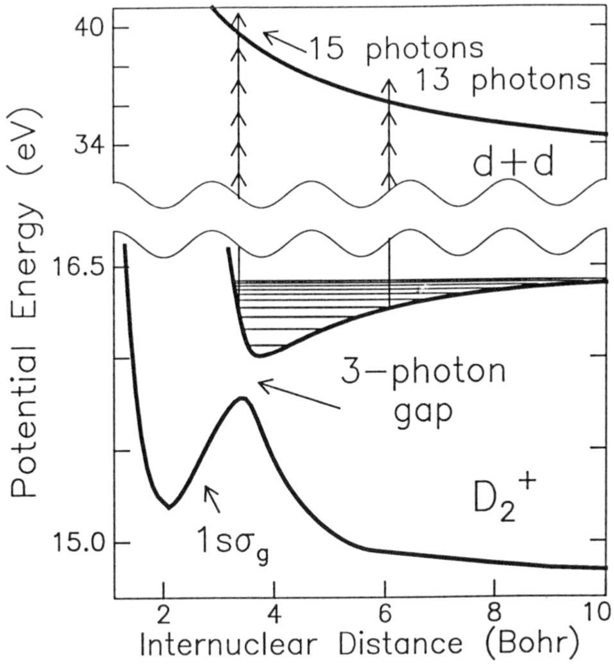

Figure 2.15 At laser intensities sufficient to open the three-photon gap, vibrational population can be trapped in the new adiabatic potential well. The vibrational levels are calculated for a D_2^+ molecular ion in intense laser field polarized along the internuclear axis (λ = 769 nm). Unable to dissociate, these states can be exposed to a very intense laser field and multiphoton ionize, resulting in production of two bare nuclei repelling each other. The light-induced bound vibrational states in the figure are calculated at $I_{pl} = 9 \times 10^{13}$ W/cm^2.

et al. [23]. The theoretical calculations on the dissociation of H_2^+ by intense laser fields found that a significant portion of population can be trapped in these light-induced vibrational bound states, depending on the initial distribution of vibrational population in $1s\sigma_g$ ground state and the laser pulse parameters [24–28].

Experimental evidence for similar light-induced states has been observed in experiments by Zavriyev et al. [20]. The ion kinetic energy spectra that they observed features a broad distribution of faster ions with kinetic energies up to 5 eV (Figure 2.13). These faster ions appear to be the result of photoionization out of the light-induced bound initial states, which are formed in the adiabatic well above the three-photon resonant coupling of the $1s\sigma_g$ and $2p\sigma_u$ states. This ionization produces two bare protons or deuterons, which accelerate due to their Coulomb repulsion as shown schematically in Figure 2.15.

The trapping of vibrational population in light-induced states is quite natural within the context of the wave-packet model described earlier. Landau-Zener theory predicts a significant probability for adiabatic crossing at the three-photon gap. As a result, instead of dissociation through this gap, the ion can stay on the $1s\sigma_g$ curve and continue traveling to the right until it reaches the new outer turning point (point C in Figure 2.14). Consequently, it turns around and starts moving back to the three-photon gap. However, by the time it reaches the gap, the gap is wide open. Thus, the ion takes the adiabatic pass and becomes trapped in the new adiabatic potential well. Since the ion in this state cannot dissociate, it becomes exposed to the extremely high intensity at the peak of the laser pulse. This light intensity can become high enough for second ionization: $D_2^+ + 1\gamma = d + d + e$. In this case two deuterons forcibly repel each other and produce the fast ions seen in the data.

The spectra observed by Zavriyev et al. [20] actually show the evidence for bound state vibrational structure: each spectrum has slight periodic modulations superimposed on the broad, generally featureless distribution. The period of the modulations is different for hydrogen and deuterium. These modulations are due to the ionization of several vibrational states formed in the adiabatic light-induced potential well. The positions of these states depend on the shape of the potential well, which in turn depends on the parallel intensity at the moment of ionization. Figure 2.13 features the Franck-Condon overlaps between these states and the repulsive two proton (or two deuteron) Coulomb state. The light-induced vibrational eigenvalues were calculated at $I_{pl} = 9 \times 10^{13}$ W/cm^2.

The Franck-Condon overlaps form a series of maxima separated by the same intervals as the observed modulations in the spectra. This suggests that the ionization of molecular ions saturates at approximately 9×10^{13} W/cm^2. The data also provide qualitative information about the change in the ionization probability as a function of the number of photons needed to ionize. Figure 2.13 shows a relative abundance of slower fragments. The ionization probability should be relatively higher for lower energy fragments than for higher energy ones, since faster ions come from the inner turning point of the light-induced bound state, where several more photons are required to ionize.

This Franck-Condon overlap model provides only a qualitative explanation of the data. For example, it does not include any homogeneous broadening of the peaks expected due to the finite lifetime of the light-induced vibrational levels in the intense laser field. (Since the laser pulse is so short, the expected lifetime of these states is only a few vibrational periods at most.) A complete quantum-mechanical time-dependent calculation is really needed for a quantitative interpretation of the results. However, as is seen from Figure 2.13, even the simple qualitative model described in this chapter is useful in obtaining an insight on the molecular behavior in an intense laser field.

REFERENCES

1. C. Cornaggia, D. Normand, J. Morellec, G. Mainfray, and C. Manus, *Phys. Rev. A* 34, 207 (1986); S. T. Pratt, P. M. Dehmer, and J. L. Dehmer, *Chem. Phys. Lett.* **105**, 28 (1984).
2. M. H. Nayfeh, J. Mazumder, D. Humm, T. Sherlock, and K. Ng in *Atomic and Molecular Processes with Short Intense Laser Pulses,* (A.D. Bandrauk, Ed.), Plenum Press, New York, 1988, p.177.
3. T. S. Luk and C. K. Rhodes, *Phys. Rev. A*, **38**, 6180 (1988).
4. J. W. J. Verschuur, L. D. Noordam, and H. B. van Linden van den Heuvell, *Phys. Rev. A*, **40**, 4383 (1989).
5. P. H. Bucksbaum, A. Zavriyev, H. G. Muller, and D. W. Schumacher, *Phys. Rev. Lett.* **64**, 1883 (1990).
6. A. Zavriyev, P. H. Bucksbaum, H. G. Muller, and D. W. Schumacher, *Phys. Rev. A* **42**, 5500 (1990).
7. S.W. Allendorf and A. Szoke, *Phys. Rev. A* **44**, 518 (1991).
8. H. Helm, M. J. Dyer, and H. Bissantz, *Phys. Rev. Lett.* **67**, 1234 (1991).
9. G. N. Gibson, R. R. Freeman, and T. J. McIlrath, *Phys. Rev. Lett.* **67**:1230 (1991).
10. B. Yang, M. Saeed, L. F. DiMauro, A. Zavriyev, and P. H. Bucksbaum, *Phys. Rev. A* **44**, R1458 (1991).
11. A. Guisti-Suzor, X. He, O. Atabek, and F. H. Mies, *Phys. Rev. Lett.* **64**, 515 (1990).
12. See, for example, M. Weissbluth, *Atoms and Molecules*. Academic Press, Inc., New York, 1978.
13. See, for example, G. Herzberg, *Molecular Spectra and Molecular Structure. I. Spectra of Diatomic Molecules,* Van Nostrand Reinhold Ltd, New York, 1950.

14. J. Franck, *Trans. Faraday Soc.* **21**, 536 (1925).
15. E. U. Condon, *Phys. Rev.* **32**, 858 (1928).
16. T. E. Sharp, *Atomic Data,* **2**, 119 (1971).
17. R. R. Freeman and P.H. Bucksbaum, *Jour. Phys. B* **24**, 325 (1991).
18. See, for example, F. H. M. Faisal, *Theory of Multiphoton Processes*, Plenum Press, New York, 1987.
19. See, for example, L. D. Landau and E. M. Lifshitz, *Quantum Mechanics, Course of Theoretical Physics 3,* Pergamon Press, New York, 1965.
20. A. Zavriyev, P. H. Bucksbaum, J. Squier, and F. Saline, *Phys. Rev. Lett.*, **70**, 1077 (1993).
21. M. V. Federov, O. V. Kudrevatova, V. P. Makarov, and A. A. Samokhin, *Opt. Comm.* **13**, 299 (1975).
22. A. D. Bandrauk and M. L. Sink, *J. Chem. Phys.* **74**, 1110 (1981); *Chem. Phys. Lett.*, **57**, 569 (1978).
23. J. E. Decker, G. Xu, and S. L. Chin, *J. Phys. B* **24**, L281 (1991).
24. A. Guisti-Suzor and F. H. Mies, *Phys. Rev. Lett.* **68**, 3869 (1992).
25. G. Yao and S.-I. Chu, *Chem. Phys. Lett.*, **197**, 413 (1992).
26. E. E. Aubanel and A. D. Bandrauk, *Chem. Phys. Lett.*, **197**, 419 (1992).
27. A. D. Bandrauk, E. E. Aubanel, and J. M. Gauthier, *Laser Phys.*, **3**, 381 (1993).
28. See Chapter 3 for more details.

3

Theory Of Molecules in Intense Laser Fields

André D. Bandrauk, Eric E. Aubanel, and Jean-Marc Gauthier

Université de Sherbrooke, Sherbrooke, Quebec, Canada

3.1 INTRODUCTION

Great progress in our understanding of the electronic structure of molecules has come from the introduction of the molecular orbital concept by Mulliken in the 1940s to 1960s. Thus, as in atoms, electrons in molecules occupy orbitals which envelope the whole nuclear space, creating stable molecular species if the molecular orbitals are bonding and unstable species if these are antibonding [1–3]. The bonding characteristics of molecular orbitals can be inferred from photoelectron spectroscopy [4]. Improvements in this method has even led to the determination of the electron momentum distribution in these orbitals [5]. A concomitant structure which appears often in the photoelectron spectrum is the vibronic structure of the remaining molecular ion after photoionization. This structure which is created by the coupling of the ionized electron to the core of the ion reveals the vibrational structure of the molecular ion and the degree of coupling between both electron and ion [6].

The advent of intense lasers has spurred interest in examining the nonlinear behavior of atomic electrons in intense laser fields [7–10]. Similar nonlinear phenomena, such as *above-threshold ionization* (ATI) has been observed in molecules [9–16]. In particular, experiments on the nonlinear (multiphoton) photoionization of molecules has revealed that the vibronic structure of the molecular ion is considerably altered with respect to the free ion [10–18]. In the present chapter, we will endeavor to show that these results can be explained in terms of a theoretical model, the *dressed molecule*, which enables one to span the perturbative (Franck-Condon, Fermi-Golden rule) to the nonperturbative, highly nonlinear regime of molecule-photon interactions. The concept of dressed states, states modified by coupling to an external field, goes back to classical physics, solid state physics, and renormalization theory in quantum electrodynamics (see Chapter 1). It has been exploited in the atomic context by Cohen-Tannoudji et al. [19]. It will be shown in the present chapter that such states are also very useful in allowing one to perform numerical calculations of multiphoton transition amplitudes in molecules, from the weak field, perturbative to intense field, nonperturbative regime [20–25]. Such a dressed state approach is time-independent, describing the field-molecule interactions in terms of time-independent quantum stationary states of the field-molecule system. Current laser pulses are necessarily time-dependent in order to achieve high intensities [9–10]. Thus an extension of the above theory to time-dependent field-molecule interactions will also be described. For long pulses, longer than the relaxation times of the system (e.g., ionization or dissociation), the dressed state representation will still serve as a useful description to help understand the highly nonlinear, multiphoton processes induced by intense lasers.

The additional nuclear degrees of freedom in molecules (as compared to atoms) leads to new laser-induced phenomena. An example is *above-threshold dissociation* (ATD) in which a molecule dissociates within the same electronic state by multiphoton absorption [23–25]. Another new phenomenon is the process of *laser-induced avoided crossing* between resonant field-molecular potentials [7, 21, 26–31]. These molecular laser-induced avoided crossings offer the possibility of creating new laser-induced molecular quasi-bound states or resonances for which a semiclassical stability theory was first developed by Bandrauk et al. [28, 30, 31] in analogy to the semiclassical theory of molecular predissociation [3, 32, 33].

Since electronic transition moments in molecules are multiplied by nuclear Franck-Condon factors [3], strong fields are required in order to produce noticeable field-induced nonlinear effects in molecules. In both the atomic and molecular case, large transition moments are therefore required. Such large moments are of course encountered in Rydberg transitions. Nevertheless molecules can exhibit intense transitions which as pointed out first by Mulliken [34], have no counterpart in the spectra of atoms. These are the *charge-resonance* and *charge-transfer* spectra, where the transition moment is proportional to the internuclear distance (i.e., the distance across which charge resonates or is transferred) [2, 34]. In molecules, the simultaneous transfer of more than one electron can occur with considerable intensity, as emphasized by Mulliken. This phenomenon is analogous to the *plasmon* effects encountered in macroscopic systems [35, 36], with the main difference being that dissociation in the molecular case leads to large nuclear separations and therefore diverging transition moments. Hence large radiative couplings will be encountered in multiphoton dissociation processes mediated by these electronic resonance or transfer transitions. Clearly, a nonperturbative treatment of such transitions in the presence of laser fields is essential and this is the main subject of this chapter.

One can classify the coupling regimes between the laser and the molecular system according to the nature of the process induced. The first regime which can be treated perturbatively corresponds to low intensity lasers which couple weakly with the system. The radiative coupling can be expressed as the Rabi frequency or frequency of radiative transitions [7, 8],

$$\omega_R(\mathrm{cm}^{-1}) = \frac{\mathbf{d} \cdot \mathbf{E}}{\hbar} = 1.17 \times 10^{-3}\, \mathrm{d(au)}\, \mathrm{I}\, (\mathrm{W/cm}^2)^{1/2} \tag{3.1}$$

$$\mathrm{I} = \frac{\mathrm{c}E^2}{8\pi} \tag{3.2}$$

where $\mathbf{d}$ is the transition moment, I the intensity for a field $\mathbf{E}(t) = \mathbf{E}\cos\omega t$. In the perturbative regime this must be less than the natural frequencies (rotational $\nu_r \sim 10^{+9}\ \mathrm{s}^{-1}$, vibrational $\nu_v \sim 10^{+12}\ \mathrm{s}^{-1}$, electronic $\sim \nu_e \sim 10^{+15}\ \mathrm{s}^{-1}$) of a molecule. Consequently the excitation processes are well described by leading order perturbation theory, such as Fermi's Golden rule. For molecules this leads to a Franck-Condon picture of electronic radiative transitions [1–3]. With increasing intensity, one encounters a domain in which multiphoton processes begin to take effect. This is signaled by a nonlinear behavior of transition probabilities as a function of intensity. In particular two or more states may be strongly coupled together as a result of being near resonant. An example of this is the Rabi oscillations of a two level system with the frequency defined in Equation 3.1 [7, 8] or an n-level molecule [37]. These Rabi oscillations introduce energy shifts of same magnitude as ω_R. These are referred to usually as dynamic Stark shifts since they are frequency dependent. In this chapter we will discuss in detail the nonlinear interaction between vibrational manifolds of different electronic molecular states induced by sufficiently intense laser fields so that the Rabi frequency ω_R approaches the vibrational frequency of the molecular system. One can establish the upper limit of this regime as the intensity $I \leq 10^{13}$ W/cm^2. Thus for a 1 a.u. transition moment d, one has $\omega_R \simeq 3000\,\text{cm}^{-1}$ at this intensity (Equation 3.1). Furthermore at intensities $I > 10^{13}$ W/cm^2, ionization rates of atoms [9, 10] and molecules [17, 18] exceed 10^{12} s^{-1} (i.e., lifetimes become less than a picosecond, 10^{12} s). At even higher intensities, available with current superintense lasers, Rabi frequencies become comparable to electronic and photon frequencies. In such cases ionization (see Chapter 2) and Coulomb explosions (see Chapter 4) occur.

We will focus in this chapter on the intermediate to high intensities $I \leq 10^{13}$ W/cm^2, just below the ionization threshold of molecular ions (this usually involves ionization of the neutral molecule first [18]). We will show that in this particular radiative interaction regime where Rabi frequencies ω_R approach vibrational frequencies ω_v, the incident laser can create *dressed adiabatic* states as a result of laser-induced avoided crossings. From a semiclassical analysis of the problem [28, 30, 31], one can predict a *stabilization* of the new dressed molecular states for particular intensities. This stabilization stems from the molecule resonating between the two bound states, diabatic (unperturbed, zero-field) and the adiabatic (field-induced), dressed molecular states. Such stabilization of electronic states at high intensities is currently being discussed extensively in the atomic case [38, 39]. Finally we also point out that at intensities where Rabi energies (Equation 3.1) exceed rotational energies, laser-induced orientational effects or alignment are expected to predominate in the angular distribution of photodissociation fragments [40, 41].

3.2 MOLECULAR HAMILTONIANS IN LASER FIELDS

Multiphoton infrared molecular photodissociation was observed in the 1970s with long pulses (~ 1–100 nanosecond) and peak powers of ~ 1 GW/cm^2 (10^9). Already the effects of these fields were considered to be nonperturbative, and most importantly, it was recognized that a proper theoretical treatment of multilevel systems, as encountered in molecules, was necessary [42]. Short subpicosecond pulses, ($<10^{-12}$ sec) and much higher intensities TW/cm^2 (10^{12} W/cm^2) are now available [9, 10, 43], thus making a multilevel nonperturbative approach imperative [44]. In principle one should treat the field-molecule system as a complete system, and investigate all the possible adequate approximations spanning the low-field, perturbative regime to the intense nonperturbative limit. In fact, we have previously emphasized that the stationary states of the field-molecule system, called *dressed* states, can provide a useful representation to describe multiphoton processes in molecules [28, 30, 31, 40, 41].

One of the basic approximations in molecular dynamics is the adiabatic or Born-Oppenheimer separability of electronic and nuclear motion, due to the various time scales, τ_{el} ~ femtosecond (10^{-15} s) versus τ_v ~ picosecond (~10^{-12} s). In intense laser fields, dynamic Stark shifts accompanied by intrastate and interstate couplings are expected to influence the photodissociation dynamics [42]. In the field-free case, the Equations for the general Born-Oppenheimer separation into electronic and nuclear coordinates have been examined rigorously and coupled Equations have been obtained for collision processes (continuum to continuum transitions) involving heavy-particles [45], with emphasis being made on two representations, a *diabatic* and an *adiabatic* representation [3]. In this chapter we will show that it is also possible to treat electromagnetic-field-molecule transitions of the bound-bound, bound-continuum, and continuum-continuum type in a unified way, both in a time-independent (dressed state) approach and for time- dependent problems, the latter being more appropriate for short laser pulses. Both approaches lead to coupled equations in the two representations. We therefore first examine the various forms of the molecular Hamiltonian in the presence of electromagnetic fields in the various gauges introduced in Chapter 1 [i.e., the Coulomb, electric-field (EF), and Block-Nordsieck (BN) gauges] [46–48]. Coupled equations for calculating multiphoton transitions in the different gauges will depend on the adiabatic approximations appropriate to each gauge.

We consider a molecular system in an electromagnetic field described by a Coulomb-gauge vector potential $\mathbf{A}(\mathbf{r},t)$. Denoting the nuclear coordinates and momenta by $\mathbf{R}_\alpha$, $\mathbf{P}_\alpha$, $\alpha = 1, 2, \ldots N$ and the electronic coordinates and momenta by $\mathbf{r}_i$, $\mathbf{p}_i$, $i = 1, 2, \ldots n$, one can define the Hamiltonian for the complete system, field + molecule as:

$$\hat{\mathbf{H}} = \hat{\mathbf{H}}_R + \hat{\mathbf{H}}_M + \mathbf{V}_I \tag{3.3}$$

where $\hat{\mathbf{H}}_M$ is the isolated (zero-field) molecular Hamiltonian,

$$\begin{aligned} \hat{\mathbf{H}}_M &= \hat{\mathbf{T}}_N + \hat{\mathbf{H}}_{el} \\ &= \sum_\alpha \frac{\hat{\mathbf{P}}_\alpha^2}{2M_\alpha} + \sum_i \frac{\mathbf{p}_i^2}{2m} - \sum_\alpha \frac{Z_\alpha e^2}{\mathbf{r}_{i\alpha}} + \frac{1}{2}\sum_{i\neq j} \frac{e^2}{\mathbf{r}_{ij}} + \frac{1}{2}\sum_{\alpha\neq\beta} \frac{Z_\alpha Z_\beta e^2}{\mathbf{R}_{\alpha\beta}} \end{aligned} \tag{3.4}$$

The operator $\hat{\mathbf{T}}_N$ represents the kinetic energy of the nuclei, while $\hat{\mathbf{H}}_{el}$ denotes the usual electronic Hamiltonian, which depends parametrically on the nuclear configuration $\{\mathbf{R}_\alpha\}$. The operator $\hat{\mathbf{H}}_R$ is the Hamiltonian for the free field. It is well known that $\hat{\mathbf{H}}_R$ can be written as the sum of harmonic oscillator Hamiltonians (Chapter 1),

$$\begin{aligned} \hat{\mathbf{H}}_R &= \frac{1}{2}\sum_{k,\lambda} (\hat{\mathbf{P}}_{k,\lambda}^2 + \omega_k^2 \hat{\mathbf{Q}}_{k,\lambda}^2) = \sum_{k,\lambda} \hbar\, \omega_k \hat{\mathbf{a}}_{k,\lambda}^+ \hat{\mathbf{a}}_{k,\lambda} \\ \hat{\mathbf{a}}_{k,\lambda} &= \frac{\omega_k \hat{\mathbf{Q}}_{k,\lambda} + i\hat{\mathbf{P}}_{k,\lambda}}{(2\hbar\omega_k)^{1/2}} \end{aligned} \tag{3.5}$$

The corresponding classical field dynamical variables $(\mathbf{P}_{k,\lambda}, \mathbf{Q}_{k,\lambda})$ are also related to the Fourier transform of the vector potential $\mathbf{A}(\mathbf{r},t)$ through the definitions

$$\mathbf{A}(\mathbf{r},t) = V^{-1/2} \sum_{k,\lambda} \exp(i\mathbf{k}\cdot\mathbf{r})\, \hat{\varepsilon}_{k,\lambda}[\mathbf{A}_{k,\lambda}(t) + \mathbf{A}_{-k,\lambda}^*(t)] \tag{3.6}$$

$$\mathbf{P}_{k,\lambda} = |k|\, [\mathbf{A}_{k,\lambda}(t) + \mathbf{A}_{k,\lambda}^*(t)] \tag{3.7}$$

$$\mathbf{Q}_{k,\lambda} = \frac{i}{c}\,[\mathbf{A}_{k,\lambda}(t) - \mathbf{A}^*_{k,\lambda}(t)] \tag{3.8}$$

Then $\dot{\mathbf{Q}}_{k,\lambda} = \mathbf{P}_{k,\lambda}$ showing the canonical relation between both variables. c is the speed of light, V is the quantization volume of the free field, and $\hat{\varepsilon}_{k,\lambda}$ is a unit polarization vector associated with one of the two independent polarizations ($\lambda = 1, 2$) for transverse radiation with wave vector $\mathbf{k}$ ($\omega_k = c|k|$). In terms of the canonical field dynamical variables $\mathbf{P}$ and $\mathbf{Q}$, we can also write

$$\mathbf{A}(\mathbf{r},t) = cV^{-1/2} \sum_{k,\lambda} \hat{\varepsilon}_{k,\lambda}\,[\sin(\mathbf{k}\cdot\mathbf{r})\,\mathbf{Q}_{k,\lambda}(t) + \omega_{,}^{-1}\cos(\mathbf{k}\cdot\mathbf{r})\,\mathbf{P}_{k,\lambda}(t)] \tag{3.9}$$

We adhere to the canonical representation (Equation 3.9) in what follows as it gives compact expressions. In quantizing the field, the dynamical variables ($\mathbf{P},\mathbf{Q}$) become operators which are time-independent in the Schroedinger representation, and which satisfy the usual commutation relations for canonical variables. Hence, $\mathbf{A}(\mathbf{r},t)$ becomes a time-independent operator, in terms of which the interaction term (Equation 3.3) is given by

$$\hat{\mathbf{V}}_I = \frac{e}{mc}\sum_i \hat{\mathbf{A}}(\mathbf{r}_i)\cdot\hat{\mathbf{p}}_i - \frac{e}{c}\sum_\alpha \frac{Z_\alpha}{M_\alpha}\hat{\mathbf{A}}(\mathbf{R}_\alpha)\cdot\mathbf{P}_\alpha +$$

$$\frac{e^2}{2c^2}\left\{\frac{1}{m}\sum_i |\hat{\mathbf{A}}(\mathbf{r}_i)|^2 + \sum_\alpha \frac{Z_\alpha^2}{M_\alpha}|\hat{\mathbf{A}}(\mathbf{R}_\alpha)|^2\right\} \tag{3.10}$$

In the present approach, we shall neglect the A^2 terms, on the basis that they only correspond to couplings between the radiation field's degrees of freedom, when the dipole approximation is made [48]. Indeed, in this approximation, one assumes that the molecule's constituents are localized on the scale of the wavelengths of the field (i.e., $\mathbf{k}\cdot\mathbf{r}_i \sim 0, \mathbf{k}\cdot\mathbf{R}_\alpha \sim 0$ for all i, α, and $\mathbf{k}$ admitted in the above equations). Hence in this approximation we have

$$\hat{\mathbf{A}}_{k,\lambda} = cV^{-1/2}\,\omega_k^{-1}\,\hat{\mathbf{P}}_{k,\lambda}\,\hat{\varepsilon}_{k,\lambda} \tag{3.11}$$

Using the classical canonical relation obtained from Maxwell's equations (Chapter 1), one obtains the quantum canonical relation,

$$\hat{\mathbf{E}}_{k,\lambda} = -\frac{1}{c}\dot{\hat{\mathbf{A}}} = -\frac{i}{c\hbar}[\hat{\mathbf{H}}_R,\hat{\mathbf{A}}] = V^{-1/2}\sum_{k,\lambda} \omega_k \hat{Q}_{k,\lambda}\,\hat{\varepsilon}_{k,\lambda} \tag{3.12}$$

Thus in the dipole approximation the interaction reduces to

$$\hat{\mathbf{V}}_1 = \left[\left(\frac{\theta}{m}\sum_i \hat{\mathbf{p}}_i - \sum_\alpha \frac{Z_\alpha}{M_\alpha}\hat{\mathbf{P}}_\alpha\right)\cdot \sum_{k,\lambda}\omega_k^{-1}\hat{\mathbf{P}}_{k,\lambda}\,\hat{\varepsilon}_{k,\lambda}\right] \tag{3.13}$$

where θ is a new radiative coupling parameter, $\theta = eV^{-1/2}$. Therefore

$$\hat{\mathbf{H}}_R + \hat{\mathbf{V}}_1 = \frac{1}{2}\sum_{k,\lambda}(\hat{\mathbf{P}}'^2_{k,\lambda} + \omega_k^2\hat{Q}^2_{k,\lambda}) - \frac{2\pi e^2}{V}\omega_k^{-2}\left[\frac{1}{m}\left(\sum_i \hat{\mathbf{p}}_i - \sum_\alpha \frac{Z_\alpha}{M_\alpha}\hat{\mathbf{P}}_\alpha\right)\cdot\hat{\varepsilon}_{k,\lambda}\right]^2 \tag{3.14}$$

where

$$\mathbf{P}_{k,\lambda}' = \hat{\mathbf{P}}_{k,\lambda} + \frac{\theta}{\omega_k}\left(\frac{1}{m}\sum_i \hat{\mathbf{p}}_i - \sum_\alpha \frac{Z_\alpha}{M_\alpha}\hat{\mathbf{P}}_\alpha\right)\cdot\hat{\varepsilon}_{k,\lambda} \tag{3.15}$$

Clearly Equation 3.15 corresponds to a unitary transformation leading to a displacement of the field canonical variables (**P**,**Q**). This transformation is described by the unitary translation operator

$$\hat{\mathbf{T}}_1(\{\hat{\mathbf{Q}}_{k,\lambda},\hat{\mathbf{p}}_i,\hat{\mathbf{P}}_\alpha\}) = \exp\left\{-\frac{i\theta}{\hbar}\cdot\sum_{k,\lambda}\omega_k^{-1}\left(\frac{1}{m}\sum_i \hat{\mathbf{p}}_i - \sum_\alpha Z_\alpha\frac{\hat{\mathbf{P}}_\alpha}{M_\alpha}\right)\cdot\hat{\varepsilon}_{k,\lambda}\,\hat{\mathbf{Q}}_{k,\lambda}\right\} \tag{3.16}$$

In the dipole approximation, the field momentum $\mathbf{P}_{k,\lambda}$ is displaced by quantities linear in particle momenta $\hat{\mathbf{p}}_i$ and $\hat{\mathbf{P}}_a$ whereas the field coordinate $\hat{\mathbf{Q}}_{k,\lambda}$ remains invariant. Going beyond the dipole limit introduces photon recoil corrections to $\hat{Q}_{k,\lambda}$ proportional to 1/c [48]. Similarly for particle variables we obtain

$$\hat{p}_i' = \hat{T}_1 \hat{p}_i \hat{T}_1^{+} = \hat{p}_i \tag{3.17}$$

$$\hat{p}_\alpha' = \hat{T}_1 \hat{P}_\alpha \hat{T}_1^{+} = \hat{P}_\alpha \tag{3.18}$$

$$\hat{r}_i' = \hat{T}_1 \hat{r}_i \hat{T}_1^{+} = \hat{r}_i - \frac{\theta}{m} \sum_{k,\lambda} \omega_k^{-1} \hat{Q}_{k,\lambda} \hat{\varepsilon}_{k,\lambda} \tag{3.19}$$

$$\hat{R}_\alpha' = \hat{T}_1 \hat{R}_\alpha \hat{T}_1^{+} = \hat{R}_\alpha + \sum_{k,\lambda} \omega_k^{-1} \hat{Q}_{k,\lambda} \hat{\varepsilon}_{k,\lambda} \tag{3.20}$$

for all i = 1, ... n, α = 1, ... N. We see therefore that the particle momenta are invariant but not the coordinates nor the velocities.

Taking these last equations into account, we rewrite the total Hamiltonian in terms of the transformed (primed) dynamical variables of the system

$$\hat{H}_1 = \hat{H}_R(\{P_{k,\lambda}', Q_{k,\lambda}\}) + \hat{H}_M(\{\hat{p}_i, \hat{P}_\alpha, \hat{r}_i', \hat{R}_\alpha' - \theta\sigma_\alpha \sum_{k,\lambda} \omega_k^{-1} \hat{Q}_{k,\lambda} \hat{\varepsilon}_{k,\lambda}\}) - \theta^2 \sum_{k,\lambda} \omega_k^{-1} \left[\left(\frac{1}{m}\sum_i p_i - \sum_\alpha \frac{Z_\alpha}{M_\alpha} P_\alpha\right) \cdot \hat{\varepsilon}_{k,\lambda}\right]^2 \tag{3.21}$$

where

$$\theta = eV^{-1/2}; \quad \sigma_\alpha = \frac{Z_\alpha}{M_\alpha} + \frac{1}{m} \tag{3.22}$$

Clearly, in Equation 3.21, the $\mathbf{A} \cdot \mathbf{p}$ couplings have now been eliminated (but not $\mathbf{A}^2$ terms).

The last term in Equation 3.21 (see also Equation 3.13) denotes interactions between the particles in the radiation vacuum state, interactions which give rise to and are of the same magnitude as the mass renormalizations of the particles. This last term can be incorporated into $\hat{\mathbf{H}}_M$ as a mass corrections, as in the case of free electrons [49-51]. In the following, we neglect these corrections in order to focus on the parametric dependence of the molecular term $\hat{\mathbf{H}}_M$ on the field coordinates $\mathbf{Q}_{k,\lambda}$. We emphasize that in the dipole approximation, the displacement of the particle coordinates $\mathbf{r}_i$ and $\mathbf{R}_\alpha$ depend on their charge/mass ratio (Equations 3.19–3.20), in such a manner that the interaction between like particles are invariant. This is a generalization to the molecular case of the result of Bergou et al. [52], that in the dipole approximation the radiation field has no effect on a pair of electrons or identical nuclei. As discussed in Chapter 1, the unitary transformation (Equation 3.16) was first introduced by Bloch and Nordsieck (BN) and the other early workers on quantum electrodynamics in connection with discussions of mass renormalization and of the Lamb shift [49–53]. We note finally that in view of the definition of the quantum electric field $\hat{\mathbf{E}}$ (Equation 3.12), we can reexpress the field perturbed particle coordinates (Equations 3.19–3.20) in the BN gauge as (see also Chapter 1),

$$\hat{\mathbf{r}}_i' = \hat{\mathbf{r}}_i - \frac{e}{m} \sum_{k,\lambda} \frac{\hat{\mathbf{E}}_{k,\lambda}}{\omega_k^2} \tag{3.23}$$

$$\hat{\mathbf{R}}_\alpha' = \hat{\mathbf{R}}_\alpha + \frac{eZ_\alpha}{M_\alpha} \sum_{k,\lambda} \frac{\hat{\mathbf{E}}_{k,\lambda}}{\omega_k^2} \tag{3.24}$$

This is consistent with the classical particle displacements expected in an electromagnetic field thus giving rise to ponderomotive forces (see Chapter 1).

Passage to the electric field or EF gauge is obtained by using the unitary operator

$$\hat{\mathbf{T}}_2 = \exp\left(-\frac{i}{\hbar} \theta\, \mathbf{d} \cdot \sum_{k,\lambda} \frac{\hat{\varepsilon}_{k,\lambda}}{\omega_k^2} \hat{\mathbf{P}}_{k,\lambda} \right) \tag{3.25}$$

where $\mathbf{d}$ is the total dipole moment of the molecular system,

$$\mathbf{d} = -\sum \mathbf{r}_i + \sum Z_\alpha \mathbf{R}_\alpha \tag{3.26}$$

The resulting coordinate translations are obtained,

$$\hat{\mathbf{r}}_i' = \hat{\mathbf{r}}_i; \quad \hat{\mathbf{R}}_\alpha' = \hat{\mathbf{R}}_\alpha \tag{3.27}$$

$$\mathbf{p}_i' = \hat{\mathbf{p}}_i + \frac{e}{c}\hat{\mathbf{A}}; \quad \hat{\mathbf{P}}_\alpha' = \hat{\mathbf{P}}_\alpha - \frac{eZ_\alpha}{c}\hat{\mathbf{A}} \tag{3.28}$$

$$\hat{\mathbf{P}}_{k,\lambda}' = \hat{\mathbf{P}}_{k,\lambda}; \quad \hat{\mathbf{Q}}_{k,\lambda}' = \hat{\mathbf{Q}}_{k,\lambda} + \frac{\theta}{\omega_k}\hat{\varepsilon}_{k,\lambda} \cdot \mathbf{d} \tag{3.29}$$

As compared to the BN transformation the particle coordinates and hence velocities are now invariant but not the momenta. The contrary applies to the photon coordinates and momenta (Equation 3.29). Recoil corrections of order v/c remain beyond the dipole approximation invoked here. In this approximation, particle coordinates are invariant, particle momenta are displaced by quantities linear in the vector potential thus canceling it in the Coulomb gauge Hamiltonian (3.4). The displacement of the field (photon) coordinate $\mathbf{Q}_{k,\lambda}$ which by Equation (3.2.10) is the transversal electric field component $\mathbf{E}_{k,\lambda}$, gives rise upon squaring this coordinate in $\mathbf{H}_R$, the dipolar coupling $-e\mathbf{d} \cdot \hat{\mathbf{E}}$ and a vacuum correction proportional to $|\mathbf{d}|^2$. Neglecting this vacuum correction for large fields (or large photon numbers), we obtain the EF molecular Hamiltonian,

$$\hat{\mathbf{H}}_2 = \hat{\mathbf{H}}_R\,(\hat{\mathbf{P}}_{k,\lambda}, \hat{\mathbf{Q}}_{k,\lambda}' - \frac{\theta}{\omega_k}\hat{\varepsilon}_{k,\lambda} \cdot \mathbf{d}) + \hat{\mathbf{H}}_M \tag{3.30}$$

$$= \hat{\mathbf{H}}_R\,(\hat{\mathbf{P}}_{k,\lambda}, \hat{\mathbf{Q}}_{k,\lambda}) + \mathbf{H}_M\,(\hat{\mathbf{p}}_i, \hat{\mathbf{P}}_\alpha, \mathbf{r}_i, \mathbf{R}_\alpha) - e\,\mathbf{d} \cdot \hat{\mathbf{E}} \tag{3.31}$$

Recapitulating, the unitary operator $\hat{\mathbf{T}}_1$ transforms the Coulomb gauge (minimal coupling) Hamiltonian $\hat{\mathbf{H}}$ (Equation 3.3) in which photons and particles are coupled by $\mathbf{A} \cdot \mathbf{p}$ terms, to $\hat{\mathbf{H}}_1$ (Equation 3.21) where nuclear particle coordinates are displaced by a term proportional to the field coordinate $\hat{\mathbf{Q}}_{k,\lambda}$. In this representation, the electronic Hamiltonian $\hat{\mathbf{H}}_{el}$ parametrically depends on the nuclear coordinates $\hat{\mathbf{R}}_\alpha$ and the field coordinates $\hat{\mathbf{Q}}_{k,\lambda}$. Mutual electronic and nuclear repulsion are left invariant. This is a specific nature of the dipole approximation since new electron-electron field-dependent couplings will appear as one goes beyond the dipole approximation [48]. In the EF gauge, only the field coordinates $\hat{\mathbf{Q}}_{k,\lambda}$ are displaced by a term proportional to the total molecular dipole moment $\mathbf{d}$. This gives rise to a simple total field-molecule Hamiltonian $\hat{\mathbf{H}}_2$ (Equation 3.31) where the radiative interaction separates as the term $-e\mathbf{d} \cdot \hat{\mathbf{E}}$ in the dipole approximation.

3.3 COUPLED EQUATIONS AND ADIABATIC EXPANSIONS

Quantum coupled equations are useful to study numerically simple models of collisions and spectroscopy of molecules in electromagnetic fields. The advantage of the coupled Equations arises from the fact that one can extend calculations in a unified manner from weak fields (perturbative multiphoton spectroscopy) to intense fields where perturbative methods are no longer applicable [20–25, 54–57]. In these calculations the appropriate electron-photon bases were taken to be the products of zeroth-order eigenstates of $\hat{\mathbf{H}}_R$ and $\hat{\mathbf{H}}_{el}$ $(\{\mathbf{R}_\alpha\})$ (Equations 3.4 and 3.5), coupled via the EF semiclassical radiative interaction $\mathbf{d} \cdot \mathbf{E}$, thus successively generating one-photon transitions between different electronic states. In the following we shall derive the coupled Equations appropriate to the various gauge representations derived in the previous section.

Let $|\Psi\rangle$ be an eigenstate of the total field-molecule Coulomb gauge Hamiltonian $\hat{\mathbf{H}}$ (Equation 3.3). By inspection of Equations 3.3 and 3.31, the appropriate basis functions for both Coulomb and EF gauge is the basis of unperturbed states (i.e., the direct products of eigenstates of $\hat{\mathbf{H}}_R$ and $\hat{\mathbf{H}}_M$) defined in Equations 3.4–3.5. In contrast, in the BN representation, $\mathbf{H}_M$ is obviously parametrically dependent on the field coordinate $\mathbf{Q}_{k,\lambda}$ (Equation 3.21), so that the adiabatic states and potentials will also depend parametrically on $\mathbf{Q}_{k,\lambda}$ [46–48, 58]. Thus the basis appropriate to the minimal coupling (Coulomb gauge) form of the Hamiltonian (Equation 3.3) is the set of all product states $|n\rangle\, |\Psi_i; \mathbf{R}_\alpha\}$ where $|n\rangle$ is a free photon eigenstate of $\hat{\mathbf{H}}_R$ and $|\Psi_i; \mathbf{R}_\alpha\}$ is an eigenstate of $\hat{\mathbf{H}}_{el}$ $\{(\mathbf{R}_\alpha)\}$ evaluated at a given nuclear configuration $\{\mathbf{R}_\alpha\}$. The expansion of the total wavefunction $|\Psi\rangle$ in terms of this basis set reads:

$$|\Psi\rangle = \sum_{n,I} |\chi_{n,I}\rangle\, |n\rangle\, |\Psi_I; \{\mathbf{R}_\alpha\}\rangle \tag{3.32}$$

The form of this expansion indicates that electrons are *adiabatic* (fast variables) with respect to the nuclei since the electronic wave function depends parametrically on the nuclear positions. The photon states $|n\rangle$ being independent of electronic and nuclear coordinates would correspond to a *diabatic* basis [3, 45].

The following coupled equations for the nuclear states $|\chi_{n,I}\rangle$ are obtained after projection of the time-independent Schroedinger Equation associated with $\hat{\mathbf{H}}$ over the basis states defined in Equation 3.32, with the $\hat{\mathbf{A}}^2$ term neglected in $\hat{\mathbf{V}}_I$ (Equation 3.10):

$$E|\chi_{n,I}\rangle = \left\{\hat{\mathbf{T}}_N + \hat{\mathbf{E}}_I^{el}\,\{(\mathbf{R}_\alpha)\} + \left(n + \frac{1}{2}\right)\hbar\omega_k\right\} |\chi_{n,I}\rangle + \frac{\theta}{\omega_k}\,\hat{\varepsilon}_k \cdot$$

$$\left\{\sum_{n',I'} \langle\Psi_I; \{\mathbf{R}_\alpha\}|\, \frac{1}{m}\sum_i \hat{\mathbf{p}}_i - \sum \frac{Z_\alpha}{M_\alpha}\,\hat{\mathbf{P}}_\alpha\, |\Psi_{I'}; \{\mathbf{R}_\alpha\}\rangle\, \langle n|\mathbf{P}_k|n'\rangle |\chi_{n',I'}\rangle - \right.$$

$$\sum_{n'} \langle n|\hat{\mathbf{P}}_k|n'\rangle \sum_\alpha \frac{Z_\alpha}{M_\alpha}\,\hat{\mathbf{P}}_\alpha\, |\chi_{n',I'}\rangle + \sum_{I',\alpha} \left\{\langle\Psi_I; \{\mathbf{R}_\alpha\}| \cdot \frac{\hat{\mathbf{P}}_\alpha}{M_\alpha}\, |\chi_{n',I'}\rangle\right\} +$$

$$\left. \langle\Psi_I; \{\mathbf{R}_\alpha\}|\, \frac{\hat{\mathbf{P}}_\alpha}{2M_\alpha}\, |\Psi_{I'}; \{\mathbf{R}_\alpha\}\rangle |\chi_{n,I}\rangle \right\} \tag{3.33}$$

and θ, the radiative coupling parameter is defined in Equation 3.22.

There are two distinct groups of coupling terms on the left-hand side of Equation 3.33. The terms involving the radiative matrix elements $\langle n|\mathbf{P}_k|n'\rangle$ induce single photon transitions, $\Delta n = n' - n = \pm 1$, while the remaining terms are associated with *nonradiative* nonadiabatic interactions. One further distinguishes radiative couplings between different electronic manifolds, I and I′ (i.e., nondiagonal and diagonal ones, $I = I'$). The nondiagonal radiative matrix elements correspond to electronic transitions. The diagonal electronic terms vanish exactly by the symmetry of $\mathbf{p}_i$ or by virtue of the commutation relation,

$$\frac{1}{m}<\Psi_I;\{\mathbf{R}_\alpha\}|\hat{\mathbf{p}}_i|\Psi_I;\{\mathbf{R}_\alpha\}> = <\Psi_I;\{\mathbf{R}_\alpha\}|[\hat{\mathbf{H}}_{el},\mathbf{r}_i]|\Psi_I;\{\mathbf{R}_\alpha\}> = 0 \tag{3.34}$$

Furthermore for nondegenerate electronic bound states, we have $<\Psi_I|\hat{\mathbf{P}}_\alpha|\Psi_I> = 0$, so that in the Coulomb gauge, the only direct contributions to intramanifold ($I = I'$), nuclear transitions are represented by the term $<n|\hat{\mathbf{P}}_k|n'>\ \hat{\mathbf{P}}_\alpha|\chi_{n',I}>$. Electrons can contribute to these intramanifold (within one electronic state only) nuclear (normally infrared IR), transition via nonadiabatic corrections (i.e., the last two terms in Equation 3.34), as shown by explicit calculation of nonadiabatic corrections to IR nuclear (vibrational transitions) [47,59,60]. In the case of intermanifold (i.e., electronic transitions $I \neq I'$), simultaneous electron and nuclear contributions will arise in the Coulomb gauge [47].

The basis for the EF gauge is the set of functions

$$\hat{\mathbf{T}}_2|n>|\Psi_I;\{\mathbf{R}_\alpha\}> = |n;\mathbf{d}>|\Psi_I;\{\mathbf{R}_\alpha\}> \tag{3.35}$$

where $\hat{\mathbf{T}}_2$ is the EF gauge translation unitary operator defined in Equation 3.25. This operator displaces the field (photon) coordinates $\hat{\mathbf{Q}}_{k,\lambda}$, the displacement being proportional to the projection of the total dipole vector $\mathbf{d}$ (Equation 3.26) on the field polarization direction $\hat{\varepsilon}_{k,\lambda}$. The corresponding photon wavefunction (see Chapter 1) which results from the action of $\hat{\mathbf{T}}_2$ on $|n>$ is denoted by $|n;\mathbf{d}>$ and is a displaced harmonic oscillator wave function [47–49]. Clearly, on this basis, the photon and electronic variables can be considered as fast variables, due to the parametric dependence of the respective Hamiltonians $\hat{\mathbf{H}}_R$ and $\hat{\mathbf{H}}_{el}$ on the nuclear coordinates $\mathbf{R}_\alpha$, so that the photons and electrons are the proper adiabatic subsystems [61]. We note also that the photons are adiabatic simultaneously with respect to electrons and nuclei, as indicated by the parametric $\mathbf{d}(\{\mathbf{r}_i,\mathbf{R}_\alpha\})$ dependence of the photon states.

Expanding the total wave function in terms of this double adiabatic basis and writing

$$|\Psi> = \sum_{n,I} |\chi_{n,I}>\ |n;\mathbf{d}>\ |\Psi_I;\{\mathbf{R}_\alpha\}> \tag{3.36}$$

and using the EF Hamiltonian (Equation 3.30), we find the following coupled equation for the nuclear states $|\chi_{n,I}>$,

$$
\begin{aligned}
|E\chi_{n,I}\rangle = & \left[\hat{\mathbf{T}}_N + \mathbf{E}_I^{el}(\{\mathbf{R}_\alpha\}) + (n+\tfrac{1}{2})\,\hbar\omega_k\right]|\chi_{n,I}\rangle + \\
& \sum_{n'} \langle n;\mathbf{d}|\hat{\mathbf{T}}_e + \hat{\mathbf{T}}_N|n';\mathbf{d}\rangle\chi_{n',I}\rangle + \\
& \sum_{I',n'}\left(\sum_i \langle n,\mathbf{d}|\hat{\mathbf{p}}_i|n';\mathbf{d}\rangle \left\langle \Psi_I;\{\mathbf{R}_\alpha\}\left|\frac{\hat{\mathbf{p}}_i}{m}\right|\Psi_{I'};\{\mathbf{R}_\alpha\}\right\rangle|\chi_{n',I'}\rangle + \right. \\
& \left. \sum_\alpha \langle n;\mathbf{d}|\hat{\mathbf{P}}_\alpha|n';\mathbf{d}\rangle \left\langle \Psi_I;\{\mathbf{R}_\alpha\}\left|\frac{\hat{\mathbf{P}}_\alpha}{M_\alpha}\right|\Psi_{I'};\{\mathbf{R}_\alpha\}\right\rangle|\chi_{n',I'}\rangle\right) + \\
& \sum_{n'\alpha} \langle n;\mathbf{d}|\hat{\mathbf{P}}_\alpha|n';\mathbf{d}\rangle \cdot \frac{\hat{\mathbf{P}}_\alpha}{M_\alpha}|\chi_{n',I'}\rangle + \\
& \sum_{I'}\Big[\langle\Psi_I;\{\mathbf{R}_\alpha\}|\hat{\mathbf{T}}_N|\Psi_I;\{\mathbf{R}_\alpha\}\rangle|\chi_{n,I'}\rangle + \\
& \sum_\alpha \langle\Psi_I;\{\mathbf{R}_\alpha\}|\hat{\mathbf{P}}_\alpha|\Psi_{I'};\{\mathbf{R}_\alpha\}\rangle \cdot \frac{\hat{\mathbf{P}}_\alpha}{M_\alpha}|\chi_{n,I'}\rangle\Big]
\end{aligned}
\tag{3.37}
$$

The parametric dependence of the photon states $|n;\mathbf{d}\rangle$ on the electronic and nuclear coordinates $\{\mathbf{r}_i;\mathbf{R}_\alpha\}$ generate nonadiabatic coupling terms, namely nonadiabatic photon-nuclear and photon-electron couplings. On the other hand, the field energy $(n+½)\hbar\omega_k$ remains invariant as the state $|n;\mathbf{d}\rangle$ is related to the free photon state $|n\rangle$ by the unitary transformation $\hat{\mathbf{T}}_2$.

Let us now evaluate some of these field-particle nonadiabatic couplings. As an example, for matrix elements,

$$
\langle n;\mathbf{d}|\hat{\mathbf{p}}_i|n';\mathbf{d}\rangle = \langle n|\hat{\mathbf{T}}_2^{+}\,\hat{\mathbf{p}}_i\,\hat{\mathbf{T}}_2|n'\rangle \tag{3.38}
$$

and using the relation in Equation 3.28,

$$
\hat{\mathbf{T}}_2\hat{\mathbf{p}}_i\hat{\mathbf{T}}_2 = \hat{\mathbf{p}}_i + \frac{\theta}{\omega_k}\,\hat{\varepsilon}_k\hat{\mathbf{P}}_k
$$

we obtain

$$<n;\mathbf{d}|\hat{\mathbf{p}}_i|n';\mathbf{d}> = \frac{\theta}{\omega_k}\hat{\varepsilon}_k <n|\hat{\mathbf{P}}_k|n'> \tag{3.39}$$

Similarly one can show that:

$$<n;\mathbf{d}|\hat{\mathbf{P}}_\alpha|n';\mathbf{d}> = \frac{Z_\alpha\theta}{\omega_k}\hat{\varepsilon}_k <n|\hat{\mathbf{P}}_k|n'> \tag{3.40}$$

$$<n;\mathbf{d}|\hat{\mathbf{T}}_e + \mathbf{T}_N|n';\mathbf{d}> = \frac{e^2}{2c^2} < \left| \frac{1}{m}\sum_i |\hat{\mathbf{A}}(\mathbf{r}_i)|^2 + \sum_\alpha \frac{Z_\alpha}{M_\alpha} |\hat{\mathbf{A}}(\mathbf{R}_\alpha)|^2 \right| n'> \tag{3.41}$$

Comparing term by term the Coulomb gauge coupled Equation 3.33 and the EF gauge Equation 3.39 and taking into account the results (Equations 3.39–3.41), we arrive at the conclusion that the two sets of field-molecule coupled equations are equivalent. Thus in the former gauge, transitions occur between zeroth-order (unperturbed) electronic-field states whereas in the latter, the same transitions occur between unperturbed electronic states but displaced field states.

Finally, using the EF gauge Hamiltonian in the form of Equation 3.31 where now the field-particle interaction $\mathbf{d} \cdot \hat{\mathbf{E}}$ is made explicit (and neglecting the vacuum correction term $(\hat{\varepsilon}_k \cdot \mathbf{d})^2$ in the case of strong fields), we see that the appropriate basis expansion is the zeroth-order unperturbed field-electronic states (Equation 3.32). All radiative couplings are dipolar and can be diagonal, $I = I'$, and nondiagonal, $I \neq I'$, in the adiabatic electron space. We reemphasize that in the Coulomb gauge (Equation 3.33), radiative couplings are always nondiagonal in the adiabatic electronic space. The equivalence between diagonal electronic dipole transition moments and nondiagonal momentum transition moments is obtained only after using the exact, nonadiabatic electronic states (i.e., only after total momentum conservation has been restored) [47].

The Block-Nordsieck (BN) transformation (Equations 3.16–3.20) and the resulting Hamiltonian (Equation 3.21) suggest another adiabatic basis, $|n>|\Psi_I;\{\mathbf{R}_\alpha\},\hat{\mathbf{Q}}_k>$, where $|\Psi_I;\{\mathbf{R}_\alpha\},\hat{\mathbf{Q}}_k>$ is an eigenstate of

$$\hat{\mathbf{H}}_{el}\left[\{\mathbf{R}_\alpha - \frac{\theta\sigma_\alpha}{\omega_k}\hat{\mathbf{Q}}_k\,\hat{\varepsilon}_k\}\right]$$

for a given nuclear configuration $\{\mathbf{R}_\alpha\}$ and field oscillator coordinates $\hat{\mathbf{Q}}_k$. The BN electronic wavefunctions are related to the unperturbed adiabatic electronic functions $|\Psi_I;\{\mathbf{R}_\alpha\}$ introduced in Equation 3.32, by

$$|\Psi_I;\{\mathbf{R}_\alpha\},\hat{\mathbf{Q}}_k> = \exp\left\{-\frac{i}{\hbar}\frac{\theta}{\omega_k}\hat{\mathbf{Q}}_k\cdot\hat{\varepsilon}_k\sum_\alpha \sigma_\alpha \hat{\mathbf{P}}_\alpha\right\}|\Psi_I;\{\mathbf{R}_\alpha\}> \tag{3.42}$$

The field-molecule equations associated with this adiabatic expansion have been derived previously in [46], giving coupled Equations for the nuclear states $|\chi_{n,I}>$,

$$\begin{aligned}
E|\chi_{n,I}> = {} & \left[\hat{\mathbf{T}}_N + \left(n+\frac{1}{2}\right)\hbar\omega_k + <n|\mathbf{E}_I^{el}\{\mathbf{R}_\alpha - \frac{\theta\sigma_\alpha}{\omega_k}\hat{\mathbf{Q}}_k\,\hat{\varepsilon}_k\}|n>\right]|\chi_{n,I}> + \\
& \sum_{n'\neq n} <n|\mathbf{E}_I^{el}|n'>\chi_{n',I}> + \sum_{I',n'} <n|<\Psi_I|\left\{\frac{\theta^2}{2\omega_k^2}\left(\sum_\alpha \sigma_\alpha \hat{\mathbf{P}}_\alpha\cdot\hat{\varepsilon}_k\right)^2 + \right. \\
& \left.\sum_\alpha \frac{\hat{\mathbf{P}}_\alpha^2}{2M_\alpha}\right\}|\Psi_{I'}>|n'>|\chi_{n',I'}> + \sum_{I',n',\alpha} <n|<\Psi_I\left|\frac{\hat{\mathbf{P}}_\alpha}{M_\alpha}\right|\Psi_{I'}>|n'>\cdot\hat{\mathbf{P}}_\alpha|\chi_{n',I'}> - \\
& \frac{\theta}{\omega k}\sum_{I',n'} <n|<\Psi_I|\sum_\alpha \sigma_\alpha \hat{\mathbf{P}}_\alpha\cdot\hat{\varepsilon}_k|\Psi_{I'}>\hat{\mathbf{P}}_k|n'>|\chi_{n',I'}>
\end{aligned} \tag{3.43}$$

In the coupled Equation 3.43 one now encounters averages over photon states of electronic matrix elements which are parametrically dependent on the photon coordinate (field) $\hat{\mathbf{Q}}_k$ (i.e., these matrix elements denote double integrals: an integral first over electronic coordinates followed by an integration over the field coordinate $\hat{\mathbf{Q}}_k$).

The main difference between the BN representation and the Coulomb and EF gauge representations, as seen by comparing Equation 3.43 with 3.33 and 3.37 is that in the BN representation, the adiabatic electronic-field potential energy surfaces are now field-intensity dependent, being represented by the quantum field average $<n|\mathbf{E}_I^{el}|n>$, whereas in the last two (traditional) gauges, these surfaces are merely the field-free surfaces shifted uniformly by the unperturbed field energy $(n+½)\hbar\omega_k$. The dynamics of nuclear motion in the BN representation is governed by the field-modified surfaces as well as by the field-modified nonadiabatic couplings appearing in Equation 3.43. These field-dependence electronic surfaces become quite distorted at high intensities ($I > 10^{13}$ W/cm^2) [46, 58]. Furthermore, in contrast to the traditional descriptions which allow for no field influence on the nonadiabatic couplings [3], the BN representation predicts that all couplings, including the nonadiabatic couplings (last three sums in Equation 3.43) are strongly affected by the field [47]. We also note the appearance of new coupling matrix elements $\langle n|\mathbf{E}_I^{el}|n'\rangle$ which involve direct multiphoton transitions since $\mathbf{E}_I^{el}$ is in principle a nonlinear function of the field coordinate $\hat{Q}_k$. These multiphoton transitions have been investigated in the hydrogen atom case by Gavrila et al. [62, 63]. No detailed calculation exists yet for these non-diagonal multiphoton radiative and nonradiative transitions in molecules. We finally emphasize the multiphoton nature of these matrix elements by pointing out that for zero-field, $\langle 0|\mathbf{E}_I^{el}|0\rangle$ gives the *Lamb shift* for a molecule [46] (i.e., the vacuum radiative correction due to the zero-point energy of the quantum field).

3.4 DRESSED MOLECULE MODEL FOR THE PHOTODISSOCIATION OF A_2^+ SYSTEMS

In the previous section we have examined the field-molecule coupled equations for the nuclear motion in various gauges. These fully quantum equations are completely time-independent and the corresponding eigenstates are stationary states of the system, called *dressed* states, which correspond to nuclear (vibrational, rotational) states evolving on coupled zeroth-order electronic-field states appropriate to each gauge or representation. As emphasized in Chapter 1, the EF gauge is the most physical of the gauges. Thus, as seen already in Equation 3.31, the radiative interaction takes on the classical form $-\mathbf{d}\cdot\mathbf{E}$, where $\mathbf{d}$ is the total dipole moment of the system (Equation 3.26). Another advantage of the EF gauge is that the particle momenta in the EF molecular Hamiltonian, $\{\hat{\mathbf{p}}_i, \hat{\mathbf{P}}_a\}$, are equal to the mechanical velocities $\{m\dot{\hat{\mathbf{r}}}_i, M_\alpha\dot{\mathbf{R}}_\alpha\}$ (i.e., these quantities are gauge invariant), whereas in the Coulomb and BN gauges, this equivalence is no longer true.

Therefore we shall examine the field-molecule equations in detail in the EF gauge as it will be the basis for our discussion of numerical results and their relation to the dressed molecule model. Furthermore we shall specialize to diatomic systems A_2^+. We have originally treated the Ar_2^+

molecule [30, 56, 64] and the H_2^+ system [22–25] by coupled equations; other coupled-equation calculations on the last system have been reported also [24, 65, 66].

We digress first on the choice of Ar_2^+ and H_2^+ as simple systems in which one might expect nonlinear electromagnetic effects to occur in bound-continuum transitions as a result of photodissociation induced by a strong laser field. In both cases one is dealing with a photodissociation process which is the first electronic transition with a large electronic transition moment, and higher electronic states are well separated from this strong transition. Both bound and continuum electronic states are of $\Sigma_u^+(\Sigma_g^+)$ and $\Sigma_g^+(\Sigma_u^+)$ for $Ar(H)_2^+$. The electronic transitions correspond in both cases to electronic excitation from a σ_g to a σ_u molecular orbital, such that both orbitals, bonding (σ_g) and antibonding (σ_u), are degenerate at infinite nuclear separation, leading therefore to the same atomic fragments. Such an electronic transition as emphasized first by Mulliken will lead to an intense electronic absorption and has been named by him a *charge-resonance* spectrum [34]. Another intense absorption to be expected is that due to *charge-transfer* where, as a consequence of electron transfer between the separating atoms, a redistribution of charge occurs in the products. Thus in the charge-resonance case discussed for Ar_2^+ and H_2^+, the transition can be viewed as an $A + A^+ \rightarrow A^+ + A$ symmetric electronic rearrangement, whereas in a charge-transfer case, one could view the process as $A + A^+ \rightarrow A^+ + A^{++}$ (i.e., a nonsymmetric one electron transfer). For the first type of transfer, taking the molecular orbitals to have the form

$$\sigma_{g(u)} \simeq \frac{\varphi_a \pm \varphi_b}{[2(1 \pm S)]^{1/2}}$$

where $\varphi_{a(b)}$ are the same atomic orbitals on either atom ($1s$ for H_2^+ and $3p\sigma$ for Ar_2^+), and +(−) signs correspond to g(u) symmetries, one can show that the one electron transition moment is for large distances,

$$\mathbf{d}_{gu} = \langle\sigma_g|\mathbf{r}|\sigma_u\rangle = \frac{R}{2}(1 - S)^{-1/2} \simeq \frac{R}{2} \tag{3.44}$$

where

$$S = \int \varphi_a \varphi_b dv$$

is the atomic overlap integral. The corresponding one-electron transfer transition moment [34] is

$$<AA^+|r|A^-A^{++}> \simeq SR \tag{3.45}$$

Of note is that in the case of electronic-resonance transfer, transition moments diverge since the two possible dissociation products are identical. In the charge-transfer case, the moment now vanishes at infinity but nevertheless remains large near the equilibrium distance ($S \to 1$ as $R \to 0$, $S \to 0$ as $R \to \infty$). The latter mechanism has been shown to be operative in the intense field dissociation of HCl^+ [18].

We now derive the EF coupled Equations for the transition from a bound σ_g electronic state to a dissociative (continuum) electronic σ_u state induced by a single-mode, monochromatic, linearly polarized radiation field. The electronic Hamiltonian in this two-state model can be therefore written as [30, 66]

$$\hat{\mathbf{H}}_{el}(R) = \mathbf{E}_g(R)\,|\sigma_g><\sigma_g| + \mathbf{E}_u(R)\,|\sigma_u><\sigma_u| \tag{3.46}$$

where $\mathbf{E}_g(R)$ is a bound state potential and $\mathbf{E}_u(R)$ is a repulsive potential (Figure 3.1). The solutions for the time-independent Schroedinger Equation $\hat{\mathbf{H}}|\Psi> = E|\Psi>$ determine the stationary or *dressed* molecular states which are written for the present model as:

$$|\Psi> = \sum_n [|\chi_{g,n}(R)>\,|\sigma_g> + |\chi_{u,n}(R)>\,|\sigma_u>]\,|n> \tag{3.47}$$

$|n>$ are the photon eigenstates of the free radiation field $\hat{\mathbf{H}}_R$, with corresponding eigenvalues $(n + ½)\hbar\omega$. Substituting Equation 3.47 in the total Hamiltonian (Equation 3.31) (neglecting the vacuum correction, last term) gives the *infinite* set of coupled equations for the nuclear amplitudes $|\chi>$

$$[\hat{\mathbf{T}}_N+\mathbf{E}_g(R)+n\hbar\omega-E]\,|\chi_{gn}(R)> = V_{gu}(R)\,[(n+1^{½}\,|\chi_{u,n+1}(R)>+n^{½}\,|\chi_{u,n-1}(R)>]$$

$$V_{gu}(R) = \frac{d_{gu}(R)}{2}\left(\frac{\hbar\omega}{2\mathbf{V}}\right)^{½} \tag{3.48}$$

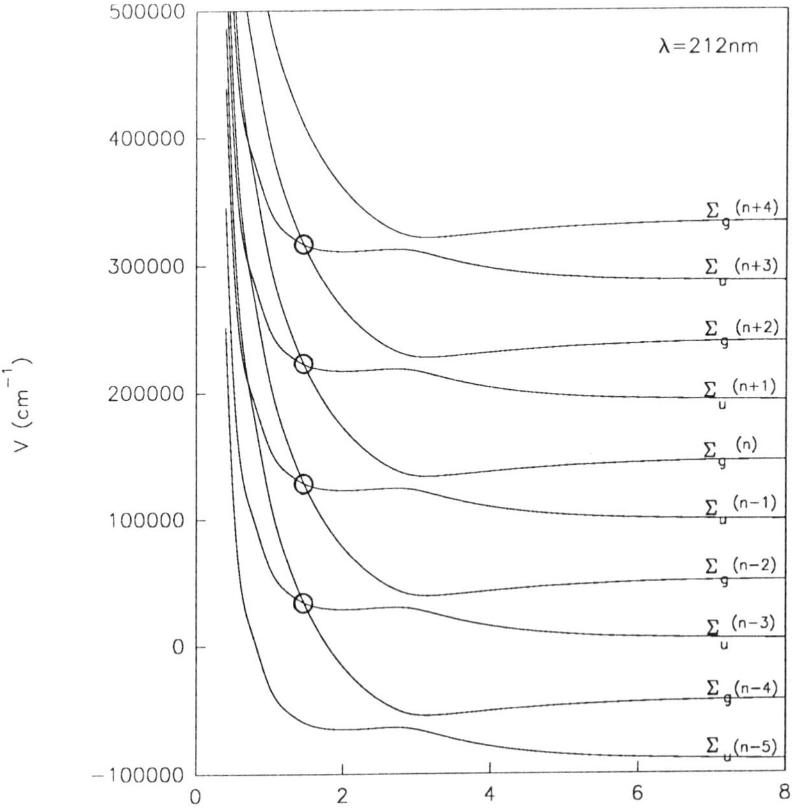

Figure 3.1 Adiabatic potentials (in cm^{-1}) versus R (in a.u.) obtained from the diagonalization of a ten-channel (five Floquet blocks) diabatic field-molecule potential matrix for the dressing of $^2\Sigma_g^+$ and $^2\Sigma_u^+$ electronic states of H_2^+ by a field of wavelength $\lambda = 532$ nm and intensity $I = 10^{14}$ W/cm^2. Avoided crossings also occur on a smaller scale in the encicled (O) regions.

$$[\hat{\mathbf{T}}_N + \mathbf{E}_u(R) + (n-1)\hbar\omega - E]\,|\chi_{u,n-1}(R)\rangle =$$
$$V_{gu}(R)\,[(n^{1/2}\,|\chi_{g,n}(R)\rangle + (n-1)^{1/2}\,|\chi_{u,n-2}(R)\rangle] \tag{3.49}$$

For large intensities, one can replace $(n-1)^{1/2}$ by $n^{1/2}$ so that the effective radiative coupling becomes, using the relation $I = cn\hbar\omega/V$

$$\overline{V}_{gu}(R) = \frac{1}{2}\left(\frac{n\hbar\omega}{2V}\right)^{1/2} \hat{\varepsilon}\cdot\mathbf{d}_{gu} = 5.85\times10^{-4}\,[I(W/cm)^2]^{1/2}\,\mathbf{d}_{g,u}\ (\text{a.u.}) \tag{3.50}$$

We see therefore that $\overline{V}_{gu}$ is half the electronic Rabi frequency introduced in Equation 3.1 as a measure of the frequency of laser-induced electronic transitions at laser intensity I.

The structure of the coupled Equations 3.48–3.49 deserves some comment as it is the consequence of the quantum nature of the electric field $\hat{\mathbf{E}}$, which could have been written as

$$\hat{\mathbf{E}} = i\left(\frac{\hbar\omega}{2V}\right)^{1/2}(\hat{\mathbf{a}} - \hat{\mathbf{a}}^{+}) \tag{3.51}$$

where $\hat{\mathbf{a}}$ and $\hat{\mathbf{a}}^{+}$ are the annihilation and creation harmonic oscillator operators (Equation 3.5, also introduced in Chapter 1). (Alternatively one can use the equivalent Equation 3.12 in terms of the field coordinate $\hat{\mathbf{Q}}$ which has matrix elements $\langle n|\hat{\mathbf{Q}}|n'\rangle$ with $n' = n \pm 1$). Of note is that each nuclear state, $|\chi_{g(u),n}\rangle$ is coupled to two other states [i.e., with photon quantum numbers (n + 1) and (n − 1)]. The first coupling corresponds to an emission of a photon whereas the second coupling represents absorption of a photon. Thus two different photon transitions occur simultaneously from any given state as a result of the composite form of the electric field operator $\hat{\mathbf{E}}$ (Equation 3.57). Assuming that we are dealing with H_2^+, then the ground bound state is a σ_g electronic state from which photodissociation occurs resonantly by transition to the σ_u excited state. The field-molecule state coupling described by Equations 3.54–3.55 is illustrated in Figure 3.2. The horizontal solid arrows correspond to *resonant* transitions [absorption ($\hat{\mathbf{a}}$) or emission ($\hat{\mathbf{a}}^{+}$)]. The dotted arrows correspond to *nonresonant* (virtual) transitions which always accompany a resonant transition and thus gives rise to the double state couplings in Equations 3.54–3.55. All these transitions differ in photon

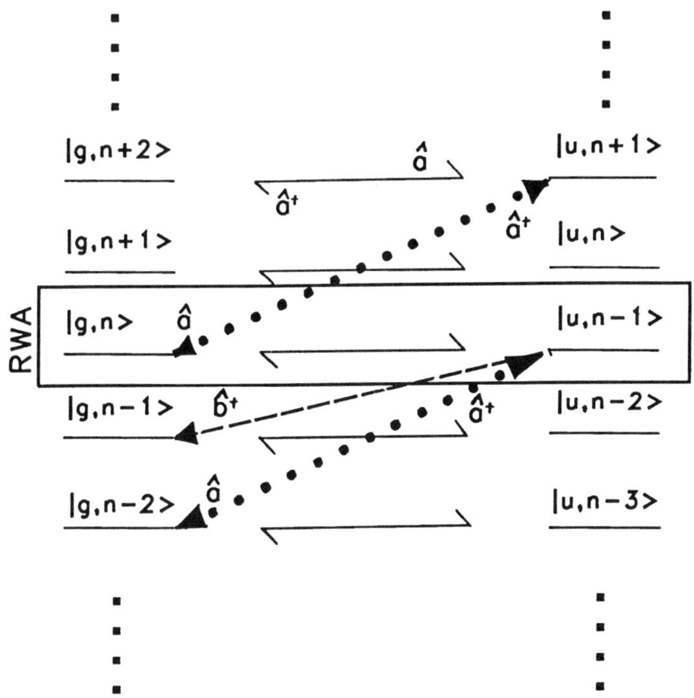

Figure 3.2 Field-molecule states for a g to u electronic transition. n is the photon quantum number. a is an annihilation operator for absorption; a^+ is the creation operator for emission, for an incident field E (Equation 3.51). b^+ is the creation operator for free spontaneous emission. —> resonant transition; · · · > virtual transition; – – –>- spontaneous transition. RWA: Rotating wave approximation neglects virtual transitions.

quantum number by ± 1 as they are laser-induced (stimulated). We have also added to Figure 3.2 transitions between states of same quantum number (e.g., |u,n − 1> → |g,n − 1>; hyphenated line arrow). These are spontaneous emissions creating new photons (with creation operator $\hat{\mathbf{b}}^+$). If their frequency $\omega_s = \omega$, then one has resonance fluorescence. These spontaneous emissions can be used to measure the dressed spectrum of the system illustrated in Figure 3.2 [56]. Since the spontaneous emissions are independent of the incident laser intensity, they are therefore neglected in the present model and calculations. (These become important in the X-ray region since the spontaneous emission (scattering) probability varies as ω_s^4). Finally, we note that retaining only the resonant transitions, $|g\rangle|n\rangle \leftrightarrow |u\rangle|n-1\rangle$, corresponds to the rotating wave approximation, RWA [7, 8]. Since corrections to this approximation involve the virtual transitions at photon energies $\pm 2\hbar\omega$ from the resonant process, we clearly see the limit of applicability of RWA, as being the condition that radiative interaction (Rabi frequency, Equation 3.1), is much less than the photon energies,

$$h\omega_R < 2\,\hbar\omega \tag{3.52}$$

The coupled Equations 3.48–3.49 can be expressed succinctly in the matrix form

$$[\hat{\mathbf{T}}_N + E + \mathbf{W}^d(R)]\,|\chi^d(R)\rangle = 0 \tag{3.53}$$

where **I** is the identity matrix, $\mathbf{W}^d(R)$ is the *diabatic* potential matrix

$$\mathbf{W}^d(R) = \begin{bmatrix} & \cdot & \cdot & & & & \\ & \cdot & \cdot & 0 & 0 & 0 & \cdots \\ & \cdot & \cdot & & & & \\ \cdots & W_{gu}^T(R) & W^{(1)}(R) & W_{gu}(R) & 0 & 0 & \cdots \\ \cdots & 0 & W_{gu}^T(R) & W^{(0)}(R) & W_{gu}(R) & 0 & \cdots \\ \cdots & 0 & 0 & W_{gu}^T(R) & W^{(-1)}(R) & W_{gu}(R) & \cdots \\ \cdots & 0 & 0 & 0 & & & \\ & \cdot & \cdot & \cdot & \cdot & \cdot & \cdots \\ & \cdot & \cdot & \cdot & \cdot & \cdot & \\ & \cdot & \cdot & \cdot & \cdot & \cdot & \end{bmatrix} \tag{3.54}$$

made up of 2×2 submatrices

$$\mathbf{W}_{gu}(R) = \begin{bmatrix} 0 & 0 \\ \bar{V}_{gu}(R) & 0 \end{bmatrix} \tag{3.55}$$

$$\mathbf{W}^{n}(R) = \begin{bmatrix} E_g(r) + (2n+1)n\omega & \overline{V}_{gu}(R) \\ \overline{V}_{gu}(R) & E_u(R) + (2n)\hbar\omega \end{bmatrix} \tag{3.56}$$

and $|\chi^d(R)>$ is a vector column consisting of the diabatic nuclear states

$$|\chi^d(R)> = \begin{bmatrix} \cdot \\ \cdot \\ \cdot \\ \chi_{u,2n+2}(R) \\ \chi_{g,2n+1}(R) \\ \chi_{u,2n}(R) \\ \chi_{g,2n-1}(R) \\ \cdot \\ \cdot \\ \cdot \end{bmatrix} \tag{3.57}$$

The matrix in Equation 3.53 is written in a *diabatic* representation since the zeroth order states $|\sigma_g> |n>$ and $|\sigma_u> |n \pm 1>$ in Equations 3.48–3.49 are coupled by potential radiative couplings (i.e., there are no nuclear momentum couplings typical of nonadiabatic corrections to adiabatic representations) [3, 32, 45]. The above basis set constitutes a proper diabatic basis set since the photon states are **R** independent. Furthermore the matrix Equations 3.53–3.56 are equivalent to the dressed state description illustrated in Figure 3.1. The 2 × 2 matrix $\mathbf{W}^{(n)}(R)$ (Equation 3.55) is the resonant matrix corresponding to the rotating wave approximation, RWA, which approximation is useful whenever Rabi frequencies are much less than the photon frequency (Equation 3.53). The radiative matrix $\mathbf{W}_{gu}(R)$ corresponds therefore to the virtual (nonresonant) transitions coupling, as an example the states $|g>|2n> \leftrightarrow |u>|2n+1>$ or $|u>|2n-1> \leftrightarrow |g>|2n>$. This distinction between resonant and nonresonant transition applies when the transition under study is truly resonant. It is possible to have of course nonresonant transitions (i.e., $\hbar\omega < E_u - E_g$). In that case one must retain both types of transitions, absorption and emission, simultaneously. An example of this is the nonresonant Raman effect [67]. The same Equations are obtained in the semiclassical time-dependent approach as shown in Chapter 1. Thus the two-dimensional resonant matrix $\mathbf{W}^{(n)}(R)$ (Equation 3.56) becomes a single Floquet block in that formalism. Thus for weak fields such that

Rabi frequencies are less than photon frequencies (Equation 3.52), a single Floquet block will suffice. This can be augmented for nonresonant transitions by the matrix in Equation 3.55 in order to preserve the equal importance of absorption and emission in the latter case.

Alternatively, an electronic-field basis can be defined in which the diabatic potential matrix $\mathbf{W}^d(R)$ is diagonal. This can be achieved by subjecting the diabatic states to an orthogonal transformation

$$|\chi^{ad}(R)\rangle = \mathbf{C}(R)|\chi^d(R)\rangle \tag{3.58}$$

such that an adiabatic potential is obtained

$$\mathbf{W}^{ad}(R) = \mathbf{C}(R)^+ \mathbf{W}^d\mathbf{C}(R) \tag{3.59}$$

This new representation is called an *adiabatic* representation [3, 32, 45], so that the new field-molecule states are explicitly parametrically dependent on the nuclear coordinate R. Figure 3.1 illustrates the adiabatic channel potentials obtained from the diagonalization of a five Floquet-block (ten dressed potentials) diabatic potential matrix $\mathbf{W}^d$ for H_2^+, neglecting rotational energies. We note first of all *laser-induced avoided crossings* at the one-photon resonant (crossing) points

$$\mathbf{E}_g(R) + n\hbar\omega = \mathbf{E}_u(R) + (n+1)\,\hbar\omega \tag{3.60}$$

Further crossing points occur at the three-photon resonant points

$$\mathbf{E}_g(R) + n\hbar\omega = \mathbf{E}_u(R) + (n-3)\,\hbar\omega \tag{3.61}$$

and so on. This last crossing point also becomes a laser-induced avoided crossing at high intensities due to three-photon processes. It has been detected experimentally by Bucksbaum et al. [68] and is discussed further in Chapter 2.

In this new adiabatic representation, obtained by diagonalizing the potential matrix for fixed nuclear positions R, laser-induced avoided crossings as seen in Figure 3.2 are a new feature. The new adiabatic potentials $\mathbf{W}_\pm(R)$ in the single photon resonance or RWA approximation are obtained by diagonalizing Equation 3.56,

$$W_{\pm}(R) = \frac{E_g(R) + \hbar\omega + E_u(R)}{2} + 2n\hbar\omega \pm$$

$$\frac{1}{2}\{[E_g(R) + \hbar\omega - E_u(R)]^2 + 4\bar{V}_{gu}^2(R)\}^{1/2} \tag{3.62}$$

Clearly, the upper adiabatic well $W_+(R)$ can support new quasi-bound states called *laser-induced resonances*. The properties of these new laser-induced states were *first* predicted and described in detail by Bandrauk et al. [28, 30, 31, 33, 40, 41], and these will be reported below.

One important aspect of these representations must not be forgotten however. Both the diabatic states, the original bound vibrational states of the ground state potential $E_g(R)$ of H_2^+, and the new adiabatic states created by the new adiabatic potential $W_+(R)$ are quasi-bound states in the presence of the laser field. In the diabatic representation, the initial states photodissociate via the diabatic radiative interaction $\bar{V}_{gu}(R)$ coupling the bound nuclear functions $\chi_{g,n}^d(R)$ to the unbound dissociative nuclear functions $\chi_{u,n'}^d(R)$ where $n' = n \pm 1, n \pm 3$, etc. Alternatively, one can also consider the new adiabatic bound nuclear states χ_+^{ad} of $W_+(R)$ to dissociate into the adiabatic continuum states $\chi_-^{ad}(R)$ of $W_-(R)$. In fact, the latter lower adiabatic potential can also support bound states in the shallow well induced by the laser. These were first detected by Bucksbaum et al. [13, 14] and were interpreted as tunneling states due to "bond softening" by the laser field. Thus for the lower adiabatic states χ_-^{ad}, tunneling through the barrier created by the laser-induced avoided crossing is a dominant dissociation channel. For the upper adiabatic states, χ_+^{ad}, nonadiabatic interactions mediated by the nuclear momenta (velocities), is the dominant photodissociation mechanism as we show next. Thus in both cases one is dealing with quasi-bound states (i.e., states which acquire a finite lifetime due to photodissociation).

Neglecting rotational couplings and introducing new radial nuclear functions, $F(R) = \chi(R)/R$, the coupled equations become in the adiabatic representation, coupled adiabatic radial matrix equations [3, 32, 45, 66, 69, 70]

$$\left[\frac{d^2}{dR^2} + Q(R)\frac{d}{dR} + W^{ad}(R)\right]F^{ad}(R) = 0 \tag{3.63}$$

where

$$Q(R) = 2\, C(R)^{+} \frac{d}{dR} C(R) \tag{3.64}$$

$$W^{ad}(R) = \frac{2M}{\hbar^2} [EI - W_{\pm}(R)] + C(R)^{+} \frac{d^2}{dR^2} C(R) \tag{3.65}$$

M is the reduced mass of the molecular system. In general, the nonadiabatic coupling matrix **Q** is a Lorentzian shaped function centered about the crossing point R_c and becomes more localized (i.e., narrower) as the diabatic interaction increases, or equivalently, the system becomes more adiabatic [3, 32].

We note that because of the different symmetries of the ground ($^2\Sigma_g^+$) and first excited ($^2\Sigma_u^+$) state of H_2^+, and since the nuclear coordinate R is symmetric, no nonadiabatic couplings exist between these two zeroth-order electronic states. However in the presence of a laser field, nonadiabatic couplings now appear between the new dressed adiabatic electronic states obtained from the transformation (Equation 3.58)

$$\begin{bmatrix} |\Psi_+(R)\rangle \\ |\Psi_-(R)\rangle \end{bmatrix} = \begin{bmatrix} |\sigma_g(R)\rangle|n\rangle \\ |\sigma_u(R)\rangle|n-1\rangle \end{bmatrix} C(R) \tag{3.66}$$

As with gauge transformations, the diabatic and adiabatic-coupled equations derived above are equivalent provided complete expansions are used. In particular, this involves infinite dimension matrices, involving complete sets of photon and electronic states. A limited expansion over electronic states, such as the two-state electronic model, $|\sigma_g\rangle$ and $|\sigma_u\rangle$, will suffice provided the electronic Rabi field frequency ω_R (Equation 3.1) is much less than higher electronic excitation frequencies. Similarly, provided the RWA condition (Equation 3.52) is satisfied, the two-state electron-field representation (Equation 3.66) will suffice.

The adiabatic electronic-field representation offers useful new concepts (i.e., laser-induced avoided crossings and laser-induced quasibound states or resonances). For numerical simulations, the diabatic-coupled equation approach is easier to carry out since one does not have to calculate derivatives of nuclear wavefunctions [20–25]. Progress in the adiabatic-coupled equation integration methodology [66, 70] has enabled one to identify the adiabatic laser-induced resonances more rigorously. Thus in the language of Feshbach resonance theory [69, 71], diabatic resonances involve coupling between open and closed-channel wavefunctions, and are called Feshbach resonances. In the adiabatic basis, on the other hand, the adiabatic uncoupled open-channel potential, $\mathbf{W}_-(R)$, may support resonances of the shape type, such as the tunneling states mentioned above, whereas the closed channel potential, $\mathbf{W}_+(R)$ will support *Feshbach* resonances.

3.5 PHOTODISSOCIATION AND PHOTOIONIZATION OF H_2

In the previous section we have introduced two representations, diabatic and adiabatic, to describe multiphoton molecular transitions in the field-molecule or dressed state representation. In the present section we use the dressed representation to discuss qualitatively multiphoton transition amplitudes for the ten-photon ionization of H_2 measured by van Linden van den Heuvell [12] and Bucksbaum [13, 14]. This will allow us to evaluate the influence of laser-induced avoided crossings on the proton yield; these findings gave the first indication that the laser-induced avoided crossing mechanism is operative at high laser intensities [14]. In the following section we will present numerical results obtained by diabatic-coupled equations.

The ten-photon absorption of H_2 is illustrated in Figure 3.3 at the wavelength $\lambda = 532$ nm leading to photodissociation of H_2^+. As pointed out by van Linden van den Heuvell [12], this wavelength allows one to reach the $B^1\Sigma_u^+$ state of H_2 via a five-photon nonresonant transition. A sixth photon radiatively and resonantly couples the B state to the doubly excited $2p\sigma_u^2$ electronic state, the so-called F state which crosses in a diabatic representation the Rydberg type E electronic state [12]. In an adiabatic picture, the EF potential system actually forms a double well potential as does the GK system due to avoided crossings between the two systems which remain coupled by nondiabatic non-Born-Oppenheimer couplings of the type discussed in the previous section. One can adopt the equivalent diabatic representation discussed above where nondiabatic GF and EK curves cross as shown in Figure 3.3. In this diabatic picture, the crossing states are coupled non-radiatively by *nondiabatic* potential couplings due to the fact that in this representation the molecular electronic Hamiltonian, $\hat{\mathbf{H}}_{el}$, is not diagonal [3, 32, 45]. Thus a nondiabatic coupling $\langle EK|\hat{\mathbf{H}}_{el}|GF\rangle$ will transfer the $B^2\Sigma_u^+$

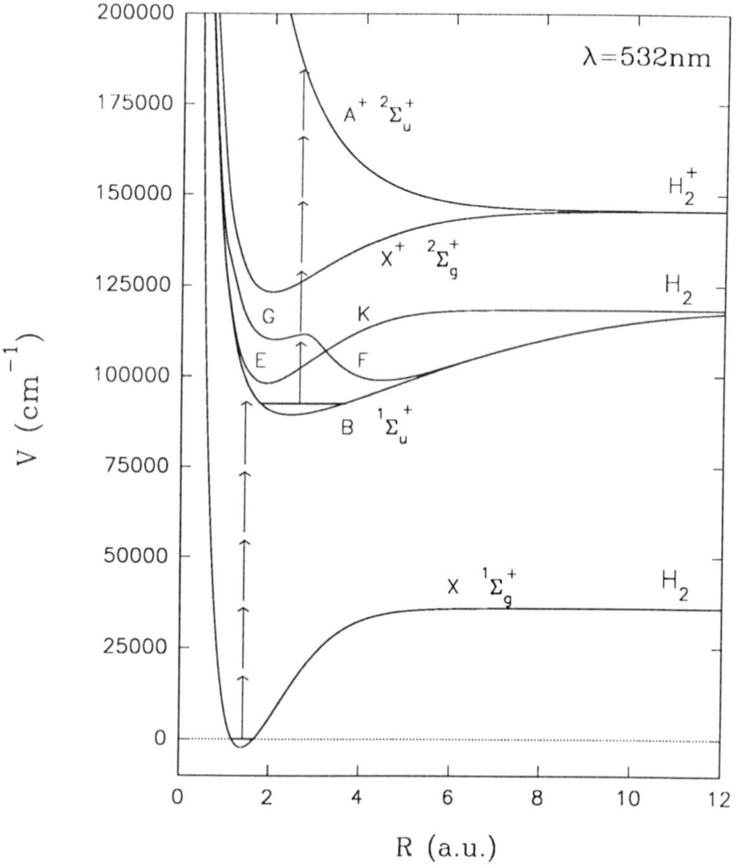

Figure 3.3 Ten-photon perturbative absorption scheme for $H_2 \rightarrow H_2^+$ photoionization at λ = 532 m through essential states. GF and EK are *diabatic* Rydberg-valence electronic potentials.

excitation to the Rydberg E and G state via the nondiabatic coupling with the F state. The sixth photon is resonant with the vibrational states of the GF and EK diabatic electronic potentials which further interact nondiabatically. A seventh photon induces a radiative transition between these last states to the $X^+(^2\Sigma_g^+)$ ground electronic state of H_2^+. In this process, a free electron is now created so that the electronic transition moment involves the Rydberg electrons of the E and G states and the ionizing electron in H_2^+ (we are assuming that the core is nearly the same for the E, G, and X^+ states; see Chapter 5 for further discussion).

We emphasize that the F state, which is doubly excited cannot interact radiatively directly with the X^+ ionized state since the electronic transition moment $<2p\sigma_u^2|r|1s\sigma_g,f_c>$, where f_c is the ionized electron wavefunction, is rigorously zero as radiative transitions involve one electron excitations in the present case [3]. The B → F transition moment being due to a $1\sigma_g \rightarrow 1\sigma_u$ molecular orbital transition has the value of the electronic resonance transfer reported in Equation 3.44 (i.e., R/2). We thus have the interesting case that the B state couples radiatively strongly to the F state, which then couples nonradiatively to the Rydberg E and G states. It is from these two Rydberg states that the seventh photon can now access resonantly the H_2^+ molecule, leading to ATI (i.e., above threshold ionization), a process whereby the ionized electron keeps absorbing further photons. This last process leads to dressing of the electron and various theoretical methods have been developed over the years to treat this problem for atoms [7, 8]; however, there is no rigorous treatment of this process for molecules at this time. The doubly excited F state of H_2 has been emphasized by van Linden van den Heuvell to be the principle *doorway* state for ATI in H_2^+ [12] due to the large transition moment between it and the B state of H_2. There is in fact the possibility of another excitation mechanism via the other highly doubly excited states of H_2, contributing to ionization via autoionization to H_2^+ [72]. This is examined further in Chapter 5.

Following photoionization of H_2, photons will interact further with the H_2^+ molecular ion, leading to an anomalous vibronic spectrum of H_2^+ in the ATI peaks [12] and an anomalous proton yield upon photodissociation of H_2^+ [13, 14]. A nonresonant three-photon transition induces direct photodissociation from the bound $X^+(^2\Sigma_g^+)$ state to the repulsive, dissociative $A^+(^2\Sigma_u^+)$ state of H_2^+. This is illustrated in Figure 3.3, the standard *perturbative* vertical image of multiphoton transitions. The more general, *nonperturbative*

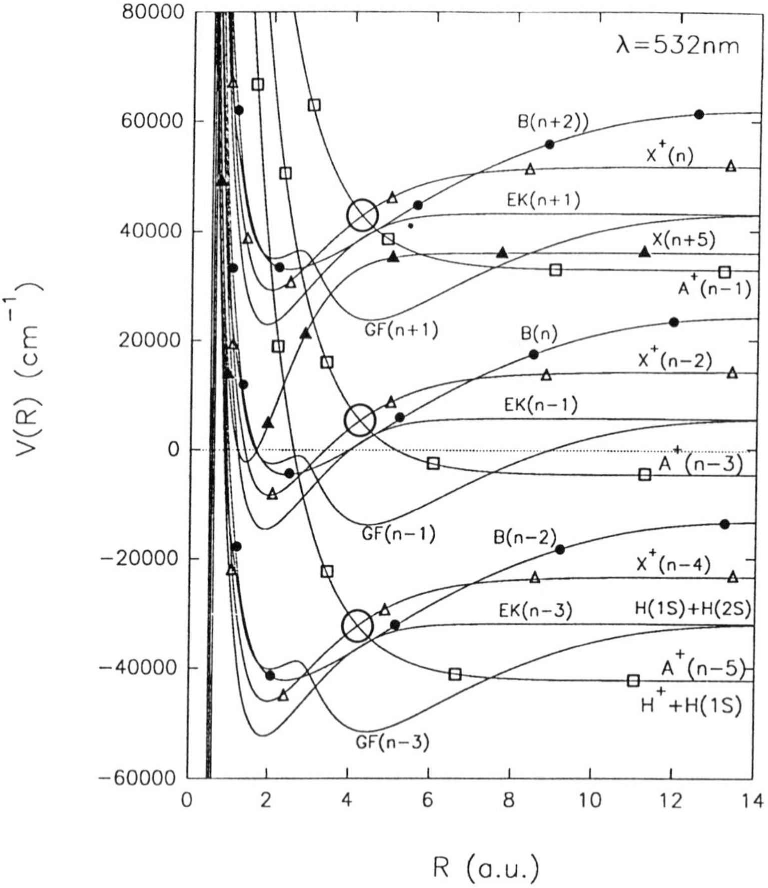

Figure 3.4 Field-molecule (dressed) representation of the ten-photon absorption of Figure 3.3, for zero electron kinetic energy. Circles (o) indicate resonant one-photon avoided crossings for the H_2^+ potentials.

representation is that of Figure 3.4, where we now use the field-molecular or dressed state description defined in the previous section. (The total wavefunctions are linear superpositions of products of photon and molecular states).

In the dressed state representation, we begin with the H_2 ground state, $X(^1\Sigma_g^+)$ with (n + 5) photons coupled nonresonantly radiatively to the $B(^1\Sigma_u^+)$ leaving only n photons after a five-photon transition. Since this transition is nonresonant, it will be weak and will have no effect on the ground nuclear (vibrational) states. The remaining radiative transitions being resonant couple strongly amongst themselves. Thus the B(n) state is coupled radiatively to two sets of states in keeping with the full quantum treatment (see Figures 3.1 and 3.2): the GF(n – 1) and GF(n + 1) field-molecule states. The first (n – 1) state corresponds to removal of one photon from the field and this transition is assigned as an absorption. The second (n + 1) state is reached by a *virtual* photon emission. Thus the first transition is due to the annihilation operator, $\hat{\mathbf{a}}$, whereas the second is due to the creation operator $\hat{\mathbf{a}}^+$, defining the quantum field operator (Equation 3.51). Since the B → F transition is resonant for λ = 532 nm excitation, then the GF(n – 1) state is seen to cross (resonantly) the B(n) state in the Franck-Condon region for that transition. The B(n) → GF(n + 1) virtual transition being nonresonant, the latter state is at energy $2\hbar\omega$ (λ = 532 nm) above the crossing of GF and B states. Neglect of the virtual transition corresponds to RWA. These virtual transitions are neglected only for the weak, nonresonant five-photon X → B transition but is retained for all remaining higher transitions (Figure 3.4).

We now continue to follow the photon paths. The GF diabatic states are coupled nondiabatically, as explained above, to the diabatic EK states with the same photon number. Since this is a nonradiative transition, the photon number cannot change. Next the Rydberg E(n – 1) and G(n – 1) give up a photon by resonant absorption to the $X^+(n-2)$ state and interact virtually with the $X^+(n)$ state of H_2^+. Subsequently $X^+(n-2)$ transits nonresonantly to $A^+(n-3)$ and $A^+(n-1)$. $A^+(n-3)$ couples radiatively to $X^+(n-4)$ and $X^+(n-2)$. The first transition corresponds to the nonresonant absorption of the ninth photon shown in Figure 3.4. The $A^+(n-3) \rightarrow X^+(n-2)$ serves to dress the H_2^+ $X^+(^2\Sigma_g^+)$ electronic state. As seen in Figure 3.4, these two field-molecule states cross at an energy above the v = 4 vibrational level of the ground state of H_2^+ [12]. The symmetric radiative coupling $\langle X^+|\mathbf{d}|A^+\rangle \cdot \hat{\mathbf{E}}$ gives rise to both absorption and emission between these two states. Similar crossings occur in the other field-molecule states corresponding to resonant absorption-emission processes. Finally, at the bottom of Figure 3.4, we have a transition from $X^+(n-4)$ to $A^+(n-5)$. This last state corresponds to the photodissociation-ionization of $H_2(X^1\Sigma_g^+)$ to $H_2^+(A^2\Sigma_u^+)$ after absorption of ten photons. As indicated in the previous section, in principle an infinite number of photon states should be included, since the classical electric field **E**(t) if coherent is a linear superposition of an infinite number of photon states with Poisson distribution (Chapter 1). In practice a finite number of states is used as determined by the numerical convergence of the numerical procedure to be described further on.

Figure 3.4 represents the minimal number of field-molecule states required for a proper treatment of the nonlinear photoelectron spectrum of H_2. Further it is a zero-kinetic energy representation of the ionized electron which in practice will carry with it some kinetic energy. Thus for nonzero kinetic energy electrons, all the X^+, A^+ electronic states of H_2^+ must be shifted up with respect to the H_2 states by an amount equal to the free electron energy. This allows one to map out the vibronic spectrum of H_2^+ observed in the photoelectron spectrum [12, 25]. As the intensity increases, more and more field-molecule states must be included until numerical convergence is achieved. In the weak field limit, the essential states are contained in the perturbative picture of Figure 3.3. In the strong field limit, such that Rabi frequencies ω_R (Equation 3.1), approach vibrational frequencies ω_v, many more photon pathways are allowed due to the virtual photon creation processes which are normally neglected in RWA. Thus the minimal state count illustrated in Figure 3.4 allows one to bridge the weak (perturbative) and strong (nonperturbative field) limits.

The field-molecule or dressed state representation depicted in Figure 3.4 allows us to make the following useful predictions. Crossings of field molecular states involving a one-photon resonant processes become *laser-induced avoided crossings* as one increases the field intensity I (Figure 3.1). Thus the crossing of the states $X^+(n)$, $A^+(n-1)$; $X^+(n-2)$, $A^+(n-3)$; $X^+(n-4)$, $A^+(n-5)$ (illustrated by circles in Figure 3.4) all undergo an avoided crossing with increasing intensity, the laser-induced gap at each crossing being equal to the Rabi frequency, ω_R (Equation 3.1). Similar laser-induced avoided crossings occur at the interactions of B(n), GF(n − 1); B(n + 2), GF(n + 1); B(n − 2), GF(n − 3). These last radiative avoided crossings are further perturbed by the nondiabatic interactions with the EK states. Figure 3.4 further demonstrates that at the wavelength 532 nm five open channels appear (i.e., channels which are below the initial zero energy line illustrated in the figure). All states above this line are physically unobservable (i.e., they are virtual), leading to renormalization of the spectrum of the system. All states below this line lead to real physically observable effects. Thus the open channels correspond to dissociation of H_2 and H_2^+ into neutral atoms and protons with kinetic energies equal to the distance from the zero-energy line of the product potential asymptotes. In particular, H_2^+ photodissociation will produce protons at three different kinetic energies. The lowest energy proton will emanate from the $A^+(n-3)$ channel as a result of tunnelling of the vibrational states of $X^+(n-2)$ by bond softening, under the barrier created by the field [13, 14]. A second proton with higher kinetic energy will emerge from the $X^+(n-4)$ channel. This is produced by absorption of the ninth photon in Figure 3.3, which is equivalent to the two-photon dissociation of H_2^+ via the nonresonant process $X^+(n-2) \rightarrow A^+(n-3) \rightarrow X^+(n-4)$. Thus the $A^+(n-3)$ channel allows this process via a virtual transition. The dressed state picture, Figure 3.4, as opposed to the perturbative representation, Figure 3.3, has the immediate advantage in predicting that the $X^+(n-4)$ channel (two-photon dissociation of H_2^+) remains radiatively coupled to $A^+(n-5)$, such that a laser-induced avoided crossing will occur between these two channels at the energy 32000 cm^{-1} (~ 4 eV) below the initial zero energy. Thus the yields of the lowest energy proton from $A^+(n-3)$ and the second and higher energy protons emanating from the $X^+(n-4)$ and $A^+(n-5)$ channels are all expected to be nonperturbatively influenced at high intensities by laser-induced avoided crossings. This was confirmed by the experimental results [13, 14] (see also Chapter 2). The high kinetic energy protons from the $A^+(n-5)$ channels are in fact the result of the three-photon nonresonant dissociation of H_2^+.

The two-photon photodissociation of H_2^+ through the virtual (nonresonant) process $X^+(n-2) \rightarrow A^+(n-3) \rightarrow X^+(n-4)$ is the molecular dissociation analog of ATI (i.e., the dissociating molecule keeps absorbing photons in its ground state). We have previously shown this process to important for molecules with large transition moments [23]. It has been called *above threshold dissociation*, ATD [14, 23–25]. The analogy with ATI stems from the fact that the electronic dipole moment, er, of the ionizing atom diverges as the electron recedes away, whereas in the H_2^+ photodissociation, the electronic transition moment diverges as R/2 due to the resonant exchange of the bonding electron between two receding nuclei [34]. Hence large transition moments in both cases facilitate multiphoton transitions. With increasing intensity, stimulated emission from the $A^+(n-5)$ state will also contribute, leading to the avoided crossing between $X^+(n-4)$ and $A^+(n-5)$.

We emphasize here that ATD should also occur in the B state of H_2. thus as illustrated in Figure 3.4, an open channel in H_2, B(n − 2) occurs between the $A^+(n-3)$ and $X^+(n-4)$ H_2^+ proton channels. The B(n − 2) is crossed by the GF(n − 3) channel to which it is also strongly radiatively coupled as a result of a similar electron-resonance transition moment, R/2, in H_2^+. Thus neutral H(1*s*) + H(n = 2) atoms are also expected to appear due to ATD in the B state. At high intensities, a laser-induced avoided crossing between B(n − 2) and GF(n − 3) should also result in abundant neutrals appearing at kinetic energies situated between the two first (lowest) proton energies discussed above. No evidence of these neutrals (or their ionized products) have been discerned to date [73]. Two possible reasons can be suggested for this lack of experimental evidence. First, we have mentioned previously that further multiphoton excitation can occur through the doubly excited states of H_2 which are first accessed by the B state. This would have the tendency to deplete the B state. Second, we must point out that no account of the ponderomotive energy of the electron has been included in description of the ionization of H_2. This energy acquired by the electron in the field, is defined by the quantity $eE^2/4m\omega^2$, and can be estimated (see Chapter 1) to be 10^{-1} a.u. (~ 3 eV for an intensity of 10^{14} W/cm^2 at $\lambda = 532$ nm). As a first approximation, this implies that all H_2^+ states are shifted up in energy with respect to H_2 states in Figure 3.4, by the ponderomotive energy. Since Rydberg states are expected to undergo similar ponderomotive energy shifts as a free electron [72], these shifts therefore could reduce radiative and nondiabatic transitions from valence states (B and F) into the Rydberg states (F and G) which are the doorway states for the H_2^+ photodissociation. This remains an open problem. In the next section we describe a coupled equations approach to calculate the transition amplitudes for the various open channels illustrated in Figure 3.4, where ponderomotive and autoionization effects have been neglected.

3.6 COUPLED EQUATIONS AND LASER-INDUCED AVOIDED CROSSINGS

We shall consider photodissociation of a molecule so that one is dealing with bound to continuum state transitions. The most convenient approach is to use a scattering theory method in the field-molecule representation which can lead to a proper and accurate description of dressed molecular states from weak to strong field intensities including continua. The Hamiltonian for the field-molecule system may be partitioned into four components,

$$\hat{\mathbf{H}} = \hat{\mathbf{H}}_R + \hat{\mathbf{H}}_M + \hat{\mathbf{H}}_{MR} + \hat{\mathbf{H}}_{nd} \tag{3.67}$$

where the zeroth-order field, $\hat{\mathbf{H}}_R$ and molecular, $\hat{\mathbf{H}}_M$, Hamiltonian have been discussed in previous sections of this chapter. The field-molecule radiative $\hat{\mathbf{H}}_{MR}$ interactions can be expressed as the Rabi frequency (3.1) in terms of the classical maximum field amplitude

$$\mathbf{E} = \left(\frac{I \cdot 8\pi}{c} \right)^{1/2}$$

or the quantum equivalent (Equation 3.50). We furthermore include the nondiabatic couplings $\hat{\mathbf{H}}_{nd}$ which occur at the crossings of diabatic electronic states such as the GF and EK states. Thus both the radiative $\hat{\mathbf{H}}_{MR}$ and nonradiative $\hat{\mathbf{H}}_{nd}$ interactions are of potential type (i.e., they are functions of the nuclear coordinate R). The total Hamiltonian is therefore completely *diabatic* with respect to photon and electron states simultaneously. We see that the effect of electronic nondiabacity can be treated on equal footing with the radiative interaction, and has been shown previously to play an important role in the multiphoton infrared ATD of ionic molecules [23].

A measure of the various interstate couplings involved will help in understanding the dynamics. Radiative couplings can be estimated from the Rabi frequency expression ω_R (Equations 3.1 or 3.50). For a dipole transition moment d ~ 1 a.u. and an intensity $I = 10^{12}$ W/cm^2, one obtains an electronic radiative interaction (matrix element) of ~ 1000 cm^{-1} (~ 0.125 e.v.). This is to be compared with the nonradiative nondiabatic interaction $\hat{\mathbf{H}}_{nd} = \langle GF|\hat{\mathbf{H}}_{el}|EK\rangle \simeq 3000$ cm^{-1}, obtained by deperturbing the adiabatic *ab initio* calculations [3, 23, 74], and the vibrational frequency of H_2^+, $\omega_v \simeq$ 2000 cm^{-1}. It is clear that at the intensities, $I \geq 10^{12}$ W/cm^2, radiative interactions are nonperturbative and will compete with the nonradiative interactions, hence influencing considerably the photodissociation branching ratios of the various product atomic states in the open channels (channels below zero-energy line; Figure 3.4).

A. Artificial Channel Method

We will endeavor to show in this section that the field molecule state approach leads to the determination of the dressed or field-molecule eigenstates as solutions to coupled differential Equations that describe the nuclear motion in the presence of the laser field. Thus bound-bound, bound-continuum, continuum-continuum, radiative and nonradiative transitions can all be treated simultaneously for any coupling strength, thus allowing us to go beyond the usual perturbative strengths. Since we shall be dealing with bound states as initial conditions, the presence of dissociative (continuum) nuclear states presents a problem which is circumvented through the use of a scattering matrix, **S**-matrix, formalism. Thus, by introducing the technique of artificial channels for entrance channels [75] and generalized to include exit channels [20–23], one can simultaneously treat bound and continuum states. It is thus possible by the present method to calculate rigorously transition amplitudes for any radiative or nonradiative interaction strength in the presence of bound and continuum states, thus covering both perturbative (Fermi-Golden rule) and nonperturbative regimes.

We rewrite the total Hamiltonian (Equation 3.67) as a nonperturbed, $\hat{\mathbf{H}}_o$ and interaction part, $\hat{\mathbf{V}}$,

$$\mathbf{H} = \hat{\mathbf{H}}_o + \hat{\mathbf{V}}, \quad \hat{\mathbf{H}}_o = \hat{\mathbf{H}}_R + \hat{\mathbf{H}}_M, \quad \hat{\mathbf{V}} = \hat{\mathbf{H}}_{MR} + \hat{\mathbf{H}}_{nd} \tag{3.68}$$

where $\hat{\mathbf{H}}_o$ is the zeroth-order field-molecule Hamiltonian and $\hat{\mathbf{V}}$ is the total interaction, radiative and nonradiative. We therefore define the field-electronic states

$$|e,n\rangle = |e\rangle\,|n\rangle \tag{3.69}$$

where e is a collective quantum number (symmetry, spin, etc. [3]) for the electronic states and n is the photon number. We now look for solutions of the total Schroedinger Equation $\hat{\mathbf{H}}|\Psi_E\rangle = E|\Psi_E\rangle$ with the total wavefunction expanded in terms of the field-electronic states defined above,

$$|\Psi_E\rangle = \frac{1}{R}\sum_{e,n} \mathbf{F}_{en}(R)\, |e,n\rangle \tag{3.70}$$

The $\mathbf{F}_{en}(R)$'s are appropriate nuclear radial functions propagating on the potential surfaces which are the R-dependent eigenvalues $\mathbf{V}_{en}(R) = \mathbf{V}_e(R) + n\hbar\omega$ of $\hat{\mathbf{H}}_o$ with eigenfunctions $|e,n\rangle$. By substituting the expansion of Equation 3.70 into the Schrödinger Equation with total Hamiltonian $\hat{\mathbf{H}}$, and premultiplying by a particular $|e,n\rangle$ state, one obtains the set of one-dimensional second-order differential equations for $\mathbf{F}_{en}(R)$ (for a diatomic molecule),

$$\left\{\frac{d^2}{dR^2} + \frac{2M}{\hbar^2}\left[E - \mathbf{V}_e(R) - n\hbar\omega\right]\right\} \mathbf{F}_{en}(R) = \frac{2M}{\hbar^2}\sum_{e',n'} \mathbf{V}^{(R)}_{en,e'n'}\, \mathbf{F}_{e',n'}(R) \tag{3.71}$$

M is the reduced mass of the molecule, whereas $\mathbf{V}_e(R)$ is the field-free diabatic electronic potential. We treat here rotationless molecules although in principle both rotational quantum numbers (J,M) can be included rigorously [20, 40, 41, 56].

Equation 3.71 for the field-molecule problem can be more succinctly expressed in matrix form as

$$\mathbf{F}''(R) + \mathbf{W}(R)\,\mathbf{F} = 0 \tag{3.72}$$

where the diagonal energy matrix elements are

$$\mathbf{W}_{en,en}(R) = \frac{2M}{\hbar^2}\,[E - \mathbf{V}_e(R) - n\hbar\omega] \tag{3.73}$$

The nondiagonal elements that describe the couplings,

$$\mathbf{W}_{en,e'n'}(R) = \frac{2M}{\hbar^2}\left[\mathbf{V}^{nd}_{en,e'n} + \mathbf{V}^{R}_{en,e'n\pm1}\right] \tag{3.74}$$

are of two types: nonradiative ($\mathbf{V}^{nd} = \hat{\mathbf{H}}_{nd}$) and radiative ($\mathbf{V}^{R} = \hat{\mathbf{H}}_{MR}$). Since each electronic potential $\mathbf{V}_e(R)$ has its own bound (vibrational) and continuous (dissociative) spectrum of nuclear eigenstates, the numerical solution of Equation 3.72 sums automatically contributions from all these nuclear states. Thus only electronic |e> and photon |n> states need be specified explicitly in any numerical calculation (i.e., Figure 3.4). We note that contrary to the adiabatic representation treated earlier (Equations 3.22–3.24), where adiabatic states are mixed field-molecule states (Equation 3.25), in the present diabatic representation, nonradiative (nondiabatic) couplings $\mathbf{V}^{nd}$ remain diagonal in the photon number n since these are field independent. Both representations are of course equivalent provided the basis expansions are complete.

All numerical calculations are performed using a Fox-Goodwin integrator method which has proved to be very accurate for molecular problems [20, 70]. The asymptotic numerical radial functions are projected onto asymptotic field-molecular states |en> and are expressed as

$$\mathbf{F}_{en}(R) = \sum_{e'n'} \mathbf{F}_{en}^{e'n'}(R)$$

$$\mathbf{F}_{en}^{e'n'}(R) = k_{en}^{-1/2}\left\{\delta_{ee'}\delta_{nn'} \exp\left[-i(k_{en}R + \delta_{en})\right] - S_{en,e'n'} \exp\left[i(k_{e'n'}R + \delta_{e'n'})\right]\right\}$$

$$k_{en}^2 = \frac{2M}{\hbar^2}\left[E - \mathbf{V}_e(R_\infty) - n\hbar\omega\right] \tag{3.75}$$

δ is an elastic scattering phase factor which is zero for neutral dissociating products but needs to be modified for charged products [23]. The coefficients $S_{en,e'n'}$ are defined as the scattering, S-matrix, elements which can be reexpressed in terms of the transition matrix **T** as

$$\mathbf{S} = 1 - 2\pi i\mathbf{T} \tag{3.76}$$

from which one can obtain transition amplitudes of physical interest.

In the molecular problem discussed here, invariably the initial state is a bound state, so one encounters new boundary value problems in calculating bound-bound or bound-continuum transition amplitudes. In one method, the complex-coordinate method [76], one calculates linewidths Γ directly from the imaginary part of the energy, and these are related to the photodissociation amplitudes T_{vc} by the relation

$$\Gamma_v = 2\frac{\pi}{\hbar} |T_{vc}|^2$$

for some initial bound state |v> dissociating to some continuum state |c>. We have shown previously [21, 22, 77], that it is possible to obtain transition amplitudes involving bound states as initial and (or) final state boundary conditions directly from the coupled Equation 3.72. This is achieved by transforming all transition amplitude calculations, including bound-bound transitions, into a scattering problem by introducing additional artificial channels (continua) as entrance and exit channels. The introduction of such artificial channels (first suggested by Shapiro for calculating Franck-Condon factors [75]) into the coupled Equation 3.72 permits one to exploit the various relations between transition matrices in order to extract the relevant photophysical amplitudes. Thus using the following relations between the total Green's functions $\hat{\mathbf{G}}$ and the transition operator $\hat{\mathbf{T}}$ [69, 78], (see also Chapter 1),

$$\hat{\mathbf{T}} = \hat{\mathrm{V}} + \hat{\mathrm{V}}\hat{\mathbf{G}}_o\hat{\mathbf{T}} = \hat{\mathrm{V}} + \hat{\mathbf{T}}\hat{\mathbf{G}}_o\hat{\mathrm{V}} \tag{3.77}$$

$$\hat{\mathbf{G}} = \hat{\mathbf{G}}_o + \hat{\mathbf{G}}_o\hat{\mathbf{T}}\hat{\mathbf{G}}_o, \quad \hat{\mathbf{G}} = (\mathrm{E} - \hat{\mathbf{H}})^{-1}, \quad \hat{\mathbf{G}}_o = (\mathrm{E} - \hat{\mathbf{H}}_o)^{-1} \tag{3.78}$$

then one can obtain an exact expression for the transition amplitude $\mathbf{T}_{cl,c}$ between an (artificial) entrance channel |c1> and a physical continuum (dissociative) channel |c>,

$$\mathbf{T}_{cl,c} = \exp(i\eta_1)\, \mathrm{V}_{c1,o}\, \mathbf{G}_o^o\, \mathbf{T}_{oc} \tag{3.79}$$

where

$$\mathbf{G}_o^o = (\mathrm{E} - \mathrm{E}_o + i\Gamma_{o,c1})^{-1}, \quad \Gamma_{o,c1} = 2\frac{\pi}{\hbar} |\mathrm{V}_{o,c1}|^2 \tag{3.80}$$

is the zeroth-order field-molecule Green's function of the initial bound state |o,n> coupled to the artificial channel |c1> via an arbitrary (chosen to be small) perturbation $\mathrm{V}_{c1,o}$, thus inducing a width $\Gamma_{o,c1}$. η_1 is the elastic phase shift for scattering on the arbitrary extraneous continuum potential of |c1>. The numerical solutions of the coupled Equation 3.72 including the artificial channel |c1> coupled to the initial state |0,n> permits one to extract each photodissociation amplitude for a given final dissociative channel |c>, the open channels in Figure 3.4, from the final equation

$$\mathbf{T}_{oc} = \mathbf{T}_{c1,c} \exp(-i\eta_1)(V_{c1,o}\mathbf{G}_o^o)^{-1} \tag{3.81}$$

All quantities on the right hand side of Equation 3.81 can be calculated exactly numerically. Thus $\Gamma_{o,c1}$ and $V_{c1,o}$ (Equation 3.80) can be obtained from a separate two-channel, **S**-matrix calculation involving the bound state |o,n> and the continuum |c1> coupled by $V_{c1,o}$, as a resonant pole in $\mathbf{T}_{o,c1}$. This then gives $\mathbf{G}_0^o$ (Equation 3.80). Numerical calculation of the full **T** matrix (involving all channels, Figure 3.4 including the artificial channel |c1>) gives the matrix elements $\mathbf{T}_{c1,c}$ from which $\mathbf{T}_{oc}$ is obtained via Equation 3.81 [20–22].

Equation 3.81 implies that the initial state |o,n> is only weakly perturbed during the multiphoton process (i.e., no energy shifts and widths from these physical processes appear in Equation 3.80). This will be the case if the initial state is coupled nonresonantly to some doorway state. This is precisely the case in Figures 3.3–3.4 where the initial five-photon transition from the initial $X^1\Sigma_g^+$ state to the doorway state $B^1\Sigma_u^+$ is nonresonant and therefore will be very weak, hence producing negligible perturbations on the initial state of H_2. All multiphoton resonant processes and nonadiabatic interactions from the doorway state, B, are calculated exactly in $\mathbf{T}_{o,c}$, thereby allowing us to join the weak, perturbative regime ($I < 10^{10}$ W/cm^2) to the strong nonperturbative regime ($I > 10^{10}$ W/cm^2). We conclude by saying that in practice one can generalize the artificial channel to include strong perturbations of the initial state [21, 77], however this necessitates more elaborate calculations [22, 23]. The simpler artificial channel version used here, based on Equation 3.80, was found to be most convenient since the initial state, the $|X^1\Sigma_g^+, v = 0\rangle$ state of H_2 is nonresonantly and therefore only weakly coupled to the doorway, $B^1\Sigma_u^+$, states which are themselves strongly radiatively coupled to the higher excited states (see Figure 3.3).

B. Numerical Results

As discussed previously, scattering (collision) theory allows one to obtain (using an arbitrary, extraneous entrance channel |c1>), transition amplitudes from initial bound states to final bound or continuum states. In the present case, we shall be calculating transition amplitudes from the initial v = 0 vibrational state of the ground $X^1\Sigma_g^+$ electronic state of H_2 to the various channels that are open according to the field-molecule diagram (i.e., all the channels which are below the zero energy line which corresponds to the initial energy; Figure 3.4).

The input into the coupled Equations 3.82 are the *ab-initio* potentials of H_2 [74] and H_2^+ [1, 3]. In the case of EF and GK states, since these were calculated adiabatically (i.e., for fixed nuclei) [74], these well-known non-crossing double well potentials were deperturbed using 2×2 unitary matrices [3, 20], in order to produce the crossing diabatic potentials GF and EK. A Gaussian nondiabatic interaction V^{nd} of the form $V^{nd}(R) = 3023\ (cm^{-1})\ \exp[-38.2(R - 0.29)^2]$ was found to give the adiabatic potentials, EF and GK, when the nondiabatic, field-free matrix is diagonalized (Equation 3.56).

As to the radiative couplings, two equivalent gauges are possible: the EF or Coulomb gauge. The radiative coupling for the first is the Rabi frequency (Equations 3.1 or 3.50) whereas in the latter, the coupling is

$$\frac{e\vec{A}}{mc} \cdot \hat{p}$$

(the A^2 term can be eliminated in the dipole approximation by a unitary transformation for all levels). Both gauges will give identical results if complete sets of states (i.e., electronic and photon) are used. The use of one gauge or another thus depends on its convenience. In the present problem, as stated in the previous section, the $B \rightarrow F$ and $X^+ \rightarrow A^+$ transitions are both $1s\sigma_g \rightarrow 2p\sigma_u$ molecular orbital transitions with transition moment given by Equation 3.44 (i.e., R/2), in the limit of nonoverlapping atomic orbitals. Clearly, in the asymptotic atomic limit, these moments become infinite, implying very strong coupling of the molecule to the electromagnetic field upon dissociation. Unfortunately this creates divergent radiative couplings in the EF gauge. Using the adiabatic EF field-molecule representation (based upon Equations 3.59, 3.63–3.66), then new adiabatic EF gauge radiative couplings are obtained which are no longer divergent [66, 79]. These in fact vanish at $R \rightarrow \infty$ since the adiabatic levels themselves now diverge as $\pm R/2$ [80] (see also Chapter 1).

Since we are working in a diabatic representation, the divergence of the diabatic EF transition moment, R/2 can be circumvented by adopting the diabatic Coulomb gauge. Thus in view of the commutation relation

$$\frac{\hat{\mathbf{p}}}{m} = \frac{i}{\hbar}[\hat{\mathbf{H}}_o, \hat{\mathbf{r}}]$$

then one obtains the fully convergent radiative coupling

$$\frac{\mathbf{A}}{\mathbf{mc}} \cdot \mathbf{p}_{ij} = \frac{V_i(R) - V_j(R)}{\hbar\omega} \frac{R}{2} \cdot \mathbf{E} \tag{3.82}$$

where V_i and V_j are field-free electronic potentials and we have used the relation

$$\mathbf{E} = \frac{-1}{c} \frac{\partial \mathbf{A}}{\partial t}$$

Thus for the $\sigma_g \rightarrow \sigma_u$ transitions, due to the asymptotic degeneracy of electronic resonance states, then the expression (Equation 3.82) converges to zero at $R = \infty$ and also at $R = 0$. Furthermore, on resonance, corresponding to crossings of field-molecule states (Figure 3.4), then $\hbar\omega = V_i(R_c) - V_j(R_c)$ where R_c is the crossing point and both gauge radiative couplings are identical. This gauge has previously been used in H_2^+ photodissociation calculations [24, 25, 65, 76], and is being used here for all the transition moments. The X–B five-photon transition in H_2 was simulated by using an effective one-photon transition, of weak intensity, in accord with the nonresonant nature of this process. Finally the E,G to X^+ transitions, corresponding to Rydberg H_2 to $H_2^+ + e^-$ photoionization, being unknown was given an arbitrary transition moment $d = 1\ cm^{-1}$. This gives relative transition amplitudes and not absolute results. The transition moments B $\rightarrow$ E,G were obtained from the literature [81].

Calculations were performed for the transition amplitude $T_{o,c}$ (Equation 3.81) for the open channels below the initial zero energy, corresponding to the $v = 0$ level of $X^1\Sigma_g^+$ of H_2 (Figure 3.4). Since the seventh $\lambda = 532$ nm photon falls in energy just above the $v = 3$ level of $X^+\Sigma_g^+$ of H_2^+, then the $v = 0$ to 3 vibrational levels of the $X^+(n-2)$ dressed H_2^+ potential all lie below this zero energy line. It is to be emphasized that this figure corresponds to zero kinetic energy of the electron. In actual fact, in the photoionization process $H_2 \rightarrow H_2^+ + e^-$, the electron acquires considerable kinetic energy which when analyzed exhibits ATI peaks [12, 13]. Since we are interested in the dressing of the molecular ion H_2^+, we can obtain the energies and photodissociation widths of the vibrational levels of H_2^+ in the presence of the laser field by examining the resonance structure of $\mathbf{T}_{0c}$. In fact these resonances were found to show up in the numerical calculations of all transition amplitudes when one scans as a function of energy the levels of H_2^+ in any one open channel. One thus takes into account the electronic kinetic energy by displacing up all the H_2^+ channels, X^+, and A^+ by the electron energy. As one calculates $\mathbf{T}_{0c}$ as a function of the displacements of the H_2^+ channels, one finds resonances to appear corresponding to v = 3, 2, 1, 0 successively with increasing displacement (i.e., increasing electron energy). This is as expected since at low electron energy, the molecule remains in high vibrational states and conversely for high electron energy, as a result of conservation of total energy. The same resonances were found to occur in all the open channels, thus confirming that their energy shifts and widths are consistently treated by the present numerical procedure.

The effect of laser-induced avoided crossings on proton yields is summarized in Table 3.1 where we tabulate the peak intensities of the laser-induced resonances. These should appear as peaks in the kinetic energy distribution of the protons produced in the open channels. At intensities below $I = 10^{11}$ W/cm^2, the relative peak heights for different laser intensities was found to follow the expected perturbative I^n law, where n is the number of photons involved in an nth-order process. Above $I = 10^{11}$ W/cm^2, such a law is no longer obeyed. As an example, the $B(n) \rightarrow A^+(n-5)$ transition should scale as I^5. One observes in fact a maximum in the resonance peaks at $I = 10^{12}$ W/cm^2 and a decrease at 10^{13} W/cm^2 for the final channel, $A^+(n-5)$. The $X^+(n-4)$ channel exhibits an unusual increase in yield with respect to the $A^+(n-5)$ channel between $I = 10^{12}$ and 10^{13} W/cm^2. The most logical explanation for this reversal in relative yield of $X^+(n-4)$ compared to $A^+(n-5)$ is the appearance of a laser-induced avoided crossing (see Figure 3.3) between these states. Thus since the transition moment $\langle X^+|r|A^+\rangle \simeq R/2 \simeq 2.5$ a.u. at the crossing, the radiative interaction (Rabi frequency) at $I = 10^{12}$ W/cm^2 is approximately $\omega_R = 3000$ cm^{-1}, inducing a gap of $2\omega_R = 6000$ cm^{-1} (0.75 ev) between the adiabatic surfaces $\mathbf{W}_+$ and $\mathbf{W}_-$ (Equation 3.62). This avoided crossing will therefore favor dissociation on the $X^+(n-4)$ surface over the $A^+(n-5)$ products. This can be predicted from the Landau-Zener theory of nonadiabatic transitions (next section) from which it follows that increasing avoided crossings render these crossings

Table 3.1 Peak Intensities of Laser-Induced Resonances (v) for Different Open Channels (Figure 3.4) as a Function of Intensity I (W/cm^2). Numbers are Relative Intensities. λ = 532 nm.

I	v	$A^+(n-3)$	$B(n-2)$	$X^+(n-4)$	$EK(n-3)$	$A^+(n-5)$
Photons absorbed:		8	7	9	8	10
10^{11}	3	1×10^{-5}	4×10^{-3}	10^{-6}	2×10^{-4}	2×10^{-5}
	2	9×10^{-6}	4×10^{-3}	8×10^{-4}	10^{-4}	9×10^{-3}
	1	–	5×10^{-3}	4×10^{-1}	2×10^{-4}	4
	0	–	4×10^{-3}	4×10^{-6}	10^{-4}	4×10^{-5}
10^{12}	4	2	9×10^{-3}	$4\times10^{+2}$	10^{-2}	$3\times10^{+2}$
	3	10^{-5}	8×10^{-3}	4	3×10^{-3}	2.5
	2	4×10^{-8}	8×10^{-3}	2×10^{-1}	3×10^{-3}	2×10^{-1}
	1	–	9×10^{-3}	10^{+3}	3×10^{-3}	$7\times10^{+2}$
	0	–	9×10^{-3}	15	3×10^{-3}	10^{+1}
10^{13}	7	$3\times10^{+2}$	7×10^{-1}	10^{+2}	5×10^{-1}	4×10^{-1}
	6	3×10^{-2}	6×10^{-1}	$3\times10^{+1}$	5×10^{-1}	10^{-1}
	5	7×10^{-2}	6×10^{-1}	3	5×10^{-1}	10^{-2}
	4	–	7×10^{-1}	$2\times10^{+2}$	5×10^{-1}	4×10^{-2}

more adiabatic [3, 32]. This reversal of proton yield at higher laser intensities (i.e., increase in lower kinetic energy protons with increasing laser intensity) has been observed by Bucksbaum et al. [13, 14] (see also Chapter 2). This important experimental observation confirms that the laser-induced avoided crossing mechanism becomes operative at laser intensities above 10^{12} W/cm^2.

Table 3.1 shows other interesting results. As discussed in Section 3.5, the lowest kinetic energy protons are produced in the $A^+(n-3)$ channel by the tunneling of the bound states created in the lower adiabatic well, $\mathbf{W}_-(R)$ from the laser-induced avoided crossing between the ground $X^+(n-2)$ state of H_2^+ and its $A^+(n-3)$ repulsive state. This avoided crossing, see Figure 3.4, at intensities above 10^{13} W/cm^2 pushes $\mathbf{W}_-(R)$ below the initial energy so that eventually it becomes a new physical open channel. This explains the fact observed in Table 3.1 that the proton yield from this channel dominates all other channels at the higher intensities. Secondly, ATD is observed also for the two open channels, B(n − 2) and X^+(n − 4). In both cases, at low fields one is dealing with two-photon transitions from bound states of each electronic state. At higher intensities, laser-induced avoided crossings with resonant partners, GF(n − 3) in the first and A^+(n − 5) in the second, enhance ATD. In fact, above 10^{13} W/cm^2, the B → F radiative coupling becomes larger than the nondiabatic coupling of the F (valence) state to the Rydberg states, so that ATD in the B state is enhanced. Clearly, radiative transitions out of the B state become important at these high intensities. Since we have neglected transitions from this state into the higher doubly excited states of H_2 which can autoionize to H_2^+ (Chapter 5), this ionization pathway needs to be considered in a more complete treatment.

We have thus seen that *laser-induced avoided crossings* are operative at intensities such that the radiative Rabi frequency ω_R is of the order of vibrational frequencies, thus implying strong nonperturbative radiative effects. At the experimental wavelength λ = 532 m, resonances have been seen to appear as vibronic structure in the ATI [12] and proton yield peaks [13]. From Table 3.1, four of these resonances can be assigned to v = 0–3 of H_2^+ in low fields (I < 10^{11} W/cm^2). At intensities I > 10^{12} W/cm^2, new resonances occur as the radiative coupling increases and makes the lower adiabatic well $\mathbf{W}_-(R)$ deeper. As a result, new resonances will occur at the higher intensities (Table 3.1). Unfortunately, the Coulomb gauge makes the radiative coupling (Equation 3.81) larger at the equilibrium distance of H_2^+ than the EF coupling (Equations 3.1 and 3.50). Hence this effect is exaggerated in the Coulomb gauge since the H_2^+ problem has been reduced here to a two-electronic state model, Σ_g^+ and Σ_u^+, whereas equivalence between the two gauges requires a complete set of electronic states [82]. We note nevertheless that at the laser-induced crossing points $\mathbf{R}_c$, the radiative matrix elements for both gauges are equal. Furthermore, it is clear from Figure 3.3 that new *adiabatic* levels will also occur above the crossing point (i.e., the new vibrational states trapped by the laser-induced potential $\mathbf{W}_+(R)$). These were first predicted from a semiclassical theory of predissociation [32], in view of the analogy between molecular predissociation and laser-induced photodissociation [28–31]. There is evidence for such adiabatic states from ATI measurements [15] and new proton yield results in H_2^+ [68]. We turn next to the stability theory of these new laser-induced quasibound states or resonances and their properties when created by short, time-dependent pulses.

3.7 LASER-INDUCED RESONANCES AND SEMICLASSICAL PREDISSOCIATION

In the field-molecule representation, multiphoton direct photodissociation is determined by time-independent coupled equations with the potential matrix (Equation 3.54) in the diabatic representation, which is made up of 2×2 submatrices or Floquet blocks, $\mathbf{W}^n(R)$ (Equation 3.56). These submatrices are connected to other submatrices of different photon number by virtual transition due to the nonresonant part of the electric field operator (Equation 3.51). These transitions are illustrated in Figure 3.1 by the dotted arrows and are clearly seen to be $\pm 2\hbar\omega$ photon energies away from the resonant transition. Thus if the electronic Rabi frequency ω_R is less than the photon energy (Equation 3.52), then RWA applies and the single 2×2 matrix (Equation 3.56) suffices to consider the dressed states. In such a case one is dealing with a two-channel problem where the bound states of some initial diabatic potential $\mathbf{W}_1(R) = \mathbf{V}_1(R) + n\hbar\omega$ are coupled to continuum potentials $\mathbf{W}_2(R) = \mathbf{V}_2(R) + (n-1)\hbar\omega$ by some radiative coupling $\mathbf{V}_{12}(R) = \overline{\mathbf{V}}_{gu}(R)$ (Equation 3.50). This is precisely analogous to the molecular *predissociation* problem as shown first in our earlier work [28, 30, 31]. Thus one can apply the technique of semiclassical collision theory to treat the resonances of such a system as a function of coupling strength and number of channels [32, 33, 83].

The semiclassical theory of quantum systems is very useful since it is an approximation exact to all of orders of the perturbation but only to order of $\hbar$ with respect to this last parameter [84]. Even so it gives exact answers for many potentials [85] including field-theory models [86]. In the present case we shall be applying this method to the solution of the two coupled differential equations for the two-state problem defined above. This will give in principle highly accurate results for any coupling strengths. However we first examine the Fermi-Golden rule expression for transition amplitudes $\mathbf{T}_{vc}$ between same initial bound state $|v,n\rangle$ and a final continuum state $|c,n-1\rangle$ in the two limits of interaction, diabatic (i.e., weak interaction) and adiabatic (i.e., strong interaction) limit.

A. Diabatic Resonances

The photodissociation width for weak radiative coupling

$$\frac{\hbar\omega_R}{2} = V_{12} < \hbar\omega_{v_d}$$

is given by the standard perturbation formula (Fermi-Golden rule) [87].

$$\Gamma_{v_d}(\text{cm}^{-1}) = 2\pi\gamma^2 \left| \int_0^\infty \chi_{v_d}(R)\, d_{12}(R)\chi_c(R)\, dR \right|^2 \tag{3.83}$$

where γ is the radiative coupling parameter $5.85 \times 10^{-4}\, I(W/cm^2)^{1/2}/a.u.$ (Equation 3.50) and $d_{12}(R)$ is the electronic transition moment in atomic units (a.u.). $\chi_{v_d}(R)$ is an initial bound state vibrational wave function and $\chi_c(R)$ is a final dissociative nuclear function, all parametrically dependent on the internuclear distance R. We now replace these functions by semi-classical (WKB) functions,

$$\chi(R) = C[k(R)]^{-1/2} \sin\left[\varphi(R) + \frac{\pi}{4}\right] \tag{3.84}$$

$$\varphi(R) = \int_a^R k(R)\, dR; \quad k(R) = \left\{\frac{2M}{\hbar^2}[E - V(R)]\right\}^{1/2} \tag{3.85}$$

where a is the classical turning point (k(a) = 0); for bound states

$$C_{v_d} = \left[\frac{2M\omega_{v_d}}{\pi\hbar}\right]^{1/2}$$

and for continuum states

$$C_c = \left(\frac{2M}{\pi\hbar^2}\right)^{1/2}$$

M is the reduced mass of the diatom and

$$\omega_{v_d} = \frac{\partial E_{v_d}}{\partial v_d}$$

is the local diabatic vibrational level density. The continuum wavefunctions are normalized to a delta function of energy [87]. Reexpressing the Franck-Condon integral (Equation 3.83) in terms of exponentials for the semiclassical functions and assuming monotonic variation of the dipole moment, one can approximate the integral by the stationary phase method [87], thus obtaining the stationary phase condition for the phase difference $\Delta\varphi(R) = \varphi_c(R) - \varphi_{v_d}(R)$,

$$\frac{d\Delta\varphi}{dR} = k_c(R) - k_{v_d}(R) = 0 \tag{3.86}$$

or equivalently,

$$E_{v_d} - W_1(R) = E_c - W_2(R) \tag{3.87}$$

This condition is therefore satisfied at the crossing point R_c (Figure 3.5) when $E_o = E_c$ and therefore $\mathbf{W}_1(R_c) = \mathbf{W}_2(R_c)$.

Expanding the phase difference to second order about R_c so that

$$\Delta\varphi(R) = \Delta\varphi(R_c) + \frac{1}{2}\Delta\varphi''(R - R_c)^2 + \ldots,$$

performing the Gaussian integration in Equation 3.83 finally gives for the diabatic photodissociation linewidth of the initial level v_d

$$\Gamma_{v_d} = \frac{4\pi\omega_{v_d}}{v_c \Delta F_c}[\gamma\, d_{12}(R_c)]^2 \sin^2\left[\Delta\varphi(R_c) + \frac{\pi}{4}\right] \tag{3.88}$$

v_c is the particle velocity at the crossing point R_c, ΔF_c is the difference in slope between the dressed diabatic potentials,

$$\Delta F_c = \left|\frac{dW_1}{dR} - \frac{dW_2}{dR}\right| R_c$$

and $\Delta\varphi(R_c)$ is a composite phase integral

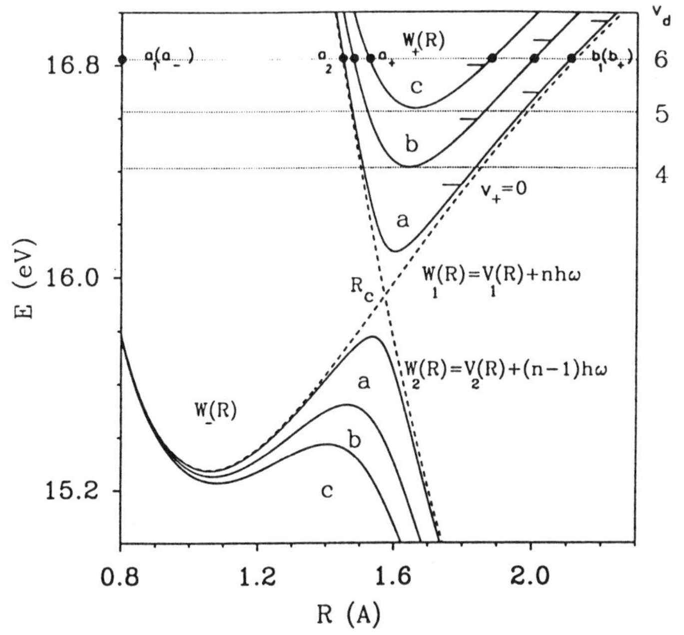

Figure 3.5 Laser-induced avoided crossing region for $H_2^+\ {}^2\Sigma_g^+ \rightarrow {}^2\Sigma_u^+$ transition at $\lambda = 212$ nm. v_d are diabatic levels (dot ~cdot~ cdot) of ${}^2\Sigma_g^+$, v_+ are adiabatic levels (—) induced by laser field at intensities: a) 3.2×10^{12}; b) 2.6×10^{13}; c) 5.2×10^{13} W/cm^2. $\mathbf{W}_1(R) = V_1({}^2\Sigma_g^+) + n\hbar\omega$, $\mathbf{W}_2(R) = V_2({}^2\Sigma_u^+) + (n-1)\hbar\omega$; $\mathbf{W}_\pm(R)$ are adiabatic potentials (Equation 3.62). $a_1(a_-)$ is left turning point of $\mathbf{W}_1(\mathbf{W}_-)$; $a_2(a_+)$ is left turning point of $\mathbf{W}_2(\mathbf{W}_+)$ and $b_1(b_+)$ is right turning point of $\mathbf{W}_1(\mathbf{W}_+)$. R_c is crossing point where $\mathbf{W}_1(R_c) = \mathbf{W}_2(R_c)$.

$$\Delta\varphi(R_c) = -\int_{a_1}^{R_c} k_1(R)dR + \int_{a_2}^{R_c} k_2(R)dR = \int_{a_2}^{R_c} k_2(R)dR + \int_{R_c}^{b_1} k_1(R)dR \tag{3.89}$$

In Equation 3.89 we have used the semiclassical quantization conditions [87],

$$\int_{a_1}^{b_1} k_1(R)dR = \left(v_d + \frac{1}{2}\right)\pi$$

where b_1 is the right hand turning point of the bound state potential $W_1(R)$ for the vibrational level v_d (Figure 3.5). We see therefore in Equation 3.89 the appearance of a new state $\chi_+(R)$ *trapped* in the potential $W_+(R) = W_2(R)$ for $a_2 \geq R \geq R_c$ and $W_1(R)$ for $R_c \geq R \geq b_1$. Furthermore, we see that the width Γ_{v_d} or photodissociation rate is zero when

$$\left[\Delta\varphi(R_c) + \frac{\pi}{4}\right] = m\pi$$

where m is some integer. In this case, *stability* of the diabatic resonance (i.e., $\Gamma_{v_d} = 0$) occurs at these phase integral conditions (i.e., when the new state $\chi_+(R)$ is quasidegenerate with the initial state). The degeneracy is not exact due to the factor of $\pi/4$ in the phase integral; Equation 3.89. We conclude by emphasizing that the diabatic photodissociation probability is linear with intensity I through the parameter γ^2 (Equation 3.83).

B. Adiabatic Resonances

We now turn to the strong coupling case where a real avoided crossing is now to be expected. Thus the new trapped state $\chi_+(R)$ discussed in the previous section on diabatic resonances will now evolve into a true adiabatic state of the upper adiabatic potential $W_+(R)$ (Equation 3.62) but will remain coupled to the states of the lower laser-induced adiabatic potential $W_-(R)$ by the nonadiabatic couplings $Q(R)d/dR$ (Equation 64). In fact writing the adiabatic electronic states $\Psi_\pm(R)$ in terms of the diabatic states $\Psi_{1,2}$ (Equation 3.66), one obtains [32]

$$\begin{bmatrix} \Psi_+(R) \\ \Psi_-(R) \end{bmatrix} = \begin{bmatrix} \cos\theta & \sin\theta \\ -\sin\theta & \cos\theta \end{bmatrix} \begin{bmatrix} \Psi_1 \\ \Psi_2 \end{bmatrix} \tag{3.90}$$

$$\theta = \frac{1}{2}\tan^{-1}\left[2\frac{V_{12}(R)}{\mathbf{W}_1(R) - \mathbf{W}_2(R)}\right] \tag{3.91}$$

$$V_{\pm}(R) = -V_{\pm} = -\frac{\hbar^2}{2M}\frac{d\theta}{dR}\frac{d}{dR} \tag{3.92}$$

where now $V_{\pm}(R)$ is the nonadiabatic coupling $\mathbf{Q}(R)$ d/dR of Equation 3.63.

The adiabatic level with vibrational number v_+ is defined by the semi-classical quantization rule

$$\varphi_+ = \int_{a_+}^{b_+} k_+(R)dR = \left(v_+ + \frac{1}{2}\right)\pi \tag{3.93}$$

where a_+, b_+ are the turning points on $\mathbf{W}_+(R)$ (see Figure 3.5) and

$$k_+(R) = \left[\frac{2M}{\hbar^2}(E - \mathbf{W}_+(R))\right]^{1/2} \tag{3.94}$$

The adiabatic Fermi-Golden rule expression for the width Γ_{v_+} now becomes

$$\Gamma_{v_+} = 2\pi|\langle\chi_+(R)|V_{\pm}|\chi_-(R)\rangle|^2 \tag{3.95}$$

Introducing the semiclassical approximation (Equation 3.84) for the two nuclear adiabatic states: bound $\chi_+(R)$ and continuum $\chi_-(R)$, assuming V_{12} is constant and $\mathbf{W}_1(R) - \mathbf{W}_2(R) = \Delta F(R - R_c) + \ldots$ around the crossing point, the integral can be performed giving as a result [32],

$$\Gamma_{v_+} = \frac{\pi}{4}\hbar\omega_{v_+}\exp\left[-\frac{\pi V_{12}^2}{\hbar v_c \Delta F_c}\right]\cos^2\beta, \quad V_{12} = \gamma d_{12}(R_c) \tag{3.96}$$

$$\beta = \int_{a_-}^{R_c} k_-(R)dR + \int_{R_c}^{b_+} k_+(R)dR, \quad \omega_{v_+} = \frac{\partial E_{v_+}}{\partial v_+} \tag{3.97}$$

a_- and b_+ are the left and right turning points on the potentials $\mathbf{W}_{-(+)}(R)$. The width of the adiabatic resonance is now a function of an adiabatic phase integral β which defines a new state nearly equal to the original diabatic state. In fact from the exact coupled Equations treatment, β is essentially the phase integral for the original unperturbed diabatic level [31–33]. Finally the preexponential factor is the well-known Landau Zener formula for non-adiabatic transitions [7, 32]. We note that *stability* of the adiabatic states, $\Gamma_{v_+} = 0$, occurs when $\beta = (m + 1/2)\pi$ (i.e., when there is coincidence (degeneracy) between the diabatic and adiabatic levels). Furthermore the adiabatic photodissociation probability is exponentially decreasing with intensity I contrary to the diabatic probability which increases linearly with I (Equation 3.83).

A more complete theory using a semiclassical approximation to the S-matrix has been obtained for the two coupled differential equations (3.48–3.49), neglecting the nonresonant terms (i.e., using RWA). This has been applied to define dressed states for the two-state problem [30] and multistate problems [31, 33]. In each case one can define weak field, perturbative, diabatic resonances in terms of adiabatic phase integrals and strong field, nonperturbative, adiabatic resonances in terms of diabatic phase integrals. The S-matrix result is very general, allowing us to go continuously from the diabatic to adiabatic limit by varying the radiative interaction $V_{12} = d_{12}(R)\,E_0$ from weak to high fields. This is due to the fact that as we have mentioned above, the semiclassical approximation treats perturbations exactly. The only approximation is in the nuclear motion which is treated to order $\hbar$ only. This is adequate for nuclear motions since the nuclear mass M is 2×10^3 times greater than the electron mass m in H_2^+.

From Equations 3.88 and 3.96 and the exact semiclassical results one obtains the following general rule: *stability* of dressed states in photodissociation (i.e., vanishing photodissociation rates) will be obtained whenever diabatic states are quasiresonant with adiabatic states. In the strong field (i.e., adiabatic limit), the stability occurs for exact resonance between the two states. In the intermediate coupling case (i.e., neither diabatic or adiabatic), it can be shown that the actual stable dressed state is a linear combination of the diabatic and adiabatic states [30–33]. This result is very economical in description as it implies that only two states are needed to describe the true dressed state. However one must realize that adiabatic states are not orthogonal to diabatic states and vice-versa (see Equations 3.66 and 3.88). Thus, dressed states are linear combinations of a complete (infinite) set of diabatic or adiabatic states. Highly stable dressed states can be expressed as a linear combination of two nonorthogonal states, coincident diabatic and adiabatic states. Physically this means that the stable dressed state is *trapped* simultaneously by the diabatic ($\mathbf{W}_1(R)$) and adiabatic ($\mathbf{W}_+(R)$) potentials at a particular energy E due to the presence of the turning points $a_1(a_-)$, a_2, and $bb_1(b_2)$ (Figure 3.5). Trapping of atomic Rydberg states in intense laser fields has been considered in the literature [88, 89]. The trapping and therefore stabilization of molecular vibrational states in high field photodissociation was first predicted by Bandrauk et al. [28, 30, 31], where it was shown that stabilization by the laser-induced avoided crossing mechanism can be expressed in terms of molecular parameters in analogy with the theory of nonadiabatic predissociation [32].

3.8 MOLECULAR STABILIZATION BY LASER-INDUCED AVOIDED CROSSINGS

One of the most useful concepts emanating from the dressed state description of molecular multiphoton transition is the idea developed in previous sections on laser-induced avoided crossings between different resonant field-molecule surfaces. This is illustrated in Figure 3.5 in detail for the H_2^+ photodissociation at the wavelength $\lambda = 212$ nm from an initial bound nuclear lead v_d in the ground $^2\Sigma_g^+(1\sigma_g)$ electronic surface to a continuum dissociate nuclear surface, the repulsive $^2\Sigma_u^+(2p\sigma_u)$ electronic state. In the dressed picture the field-molecular surface $\mathbf{W}_1 = \mathbf{V}_1(^2\Sigma_g^+) + n\hbar\omega$ crosses the $\mathbf{W}_2 = \mathbf{V}_2(^2\Sigma_u^+) + (n-1)\hbar\omega$ surface as a consequence of conservation of total energy after absorption of one photon. The radiative interaction (Equations 3.1 and 3.50) is operative between the two diabatic (unperturbed) states of $\mathbf{W}_1$ and $\mathbf{W}_2$. Figure 3.5 shows that one can describe the molecular states either in the original unperturbed (crossing) diabatic representation or the new field-induced adiabatic states obtained from the diagonalized potentials $\mathbf{W}_\pm(R)$ (Equation 3.62). The upper adiabatic surface, $\mathbf{W}_+(R)$ supports new nuclear bound states,

$\chi_+(R)$, which are quasibound as they remain coupled to the continuum nuclear states $\chi_-(R)$ via nonadiabatic couplings (Equations 3.68 or 3.92). Thus in either representation, bound levels have a finite lifetime or energy width which were defined in the Fermi-Golden rule limit in the previous section.

A. Time-Independent Dressed States

The linewidth of the exact dressed quasibound states can be found by calculating the S-matrix for the two-channel, RWA, coupled equations or the many-channel coupled equations beyond RWA (Equation 3.71). Poles of the S-matrix occur at complex energies

$$E = E_r - i\Gamma_r \tag{3.98}$$

where Γ_r is the photodissociation width related to the photodissociation rate by the relation

$$\frac{\Gamma}{\hbar} = \frac{2\pi}{\hbar} |\mathbf{T}_{vc}|^2$$

and $\mathbf{T}_{vc}$ is the nonperturbative photodissociation amplitude. Alternatively the asymptotic analysis of the nuclear wavefunction gives an energy dependent S-matrix which can be expressed in terms of a phase [69, 78, 84],

$$\det [\mathbf{S}(E)] = \exp [2i\delta(E)] \tag{3.99}$$

A reduced S-matrix, defined as

$$S_r(E) = S_o^{-1/2}\, S(E)\, S_o^{-1/2} \tag{3.100}$$

where S_o are the elastic uncoupled open channel scattering matrices, enables one to remove background elastic phases δ_o and thus isolate the resonance phase

$$\det [S_r(E)] = \exp \{2i\, [\delta(E) - \delta_o(E)]\} \tag{3.101}$$

A jump in π of the phase difference $\delta(E) - \delta_o(E)$ is therefore indicative of an isolated resonance. This was the general case found at the intensities

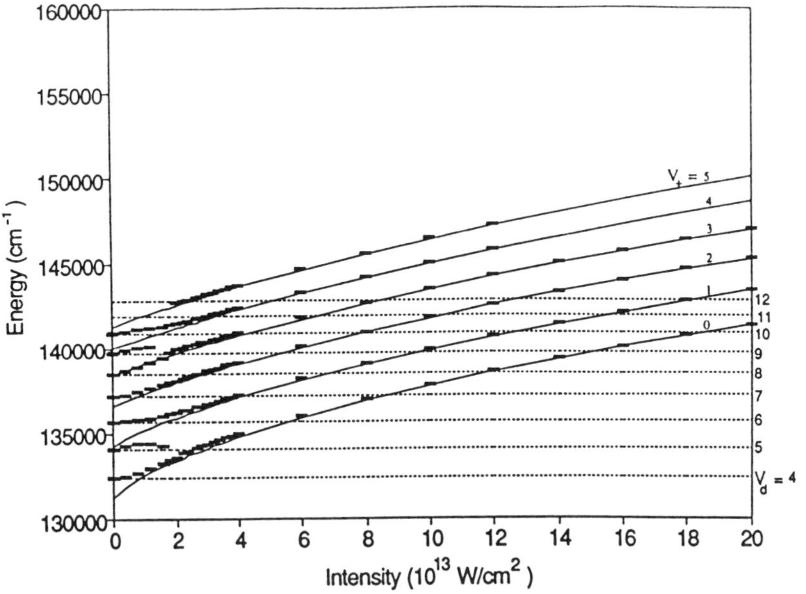

Figure 3.6 *Diabatic* vibrational levels v_d of $V_1(^2\Sigma_g^+) + n\hbar\omega$ and *adiabatic* vibrational levels v_+ of $\mathbf{W}_+(R)$ showing crossings of these as a function of laser intensity I. Energy scale is with respect to $v_d = 0$ of $X^1\Sigma_g^+$ of H_2. Broken (short) lines correspond to exact two-channel laser-induced resonances starting at weak fields with $v_d = 4, 5, 6$ etc.

investigated here. One difficulty was the divergent $\sigma_g \rightarrow \sigma_u$ transition moment, R/2, in the EF gauge, which was the gauge used in all the time-independent calculations reported here. This creates no problem for the time-dependent calculations to be presented below. In the case of the time-independent **S**-matrix calculations, the EF radiative interaction, was truncated smoothly around 20 Å, thus giving stable resonance positions (this was checked by extending the truncation to larger distances).

First, we illustrate in Figure 3.6 the phenomenon of crossing of diabatic and adiabatic levels in RWA (i.e., two-channel calculation as the intensity is increased). This is also confirmed by the position of the levels in the adiabatic potentials (Figure 3.5). Thus a crossing occurs for the level v_d = 4 with $v_+ = 0$ at 3.2×10^{12} W/cm^2, $v_d = 6$ with $v_+ = 1$ at 2.6×10^{13} W/cm^2,

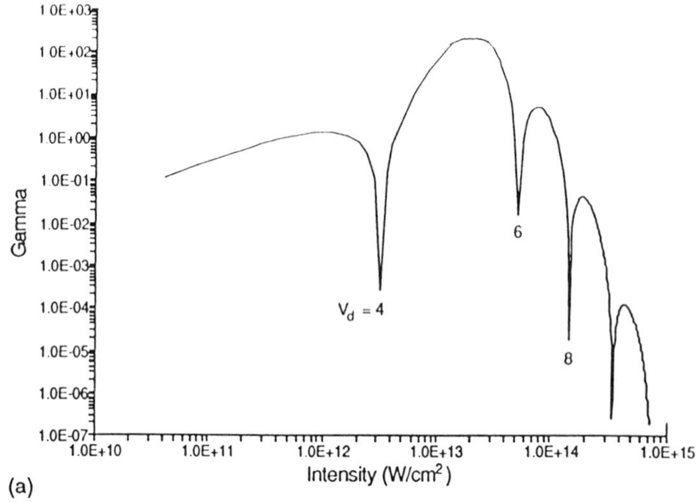

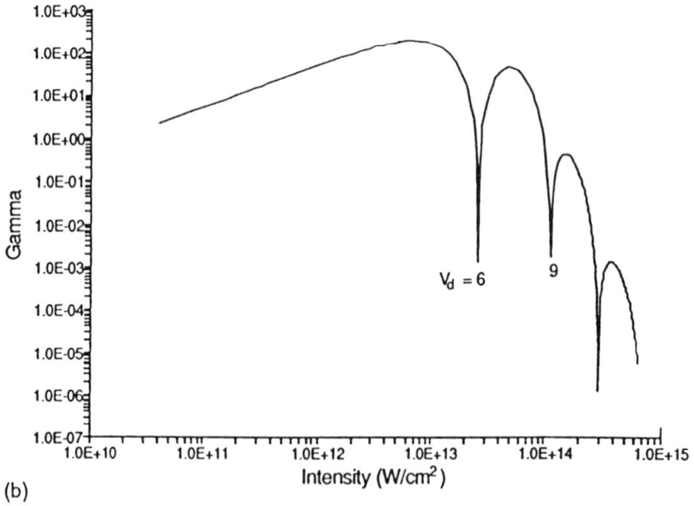

Figure 3.7 Linewidths Γ of two-channel laser-induced resonances illustrated in Figure 3.6, which correlate with adiabatic levels, a) $v_+ = 0$; b) $v_+ = 1$. Minima correspond to intensities where diabatic (v_d) cross the adiabatic level (Figure 3.6).

$v_d = 6$ with $v_+ = 0$ at 6×10^{13} W/cm^2, and so on. This creates deep minima, in the linewidths of the two-channel dressed states as illustrated in Figure 3.7. In this last figure, at weak fields, the levels correspond to the initial diabatic levels $v_d = 4$ and 6, which are just above the adiabatic levels v_+ of $\mathbf{W}_+(R)$. Increasing the field-intensity induces coincidence between diabatic and adiabatic levels at intensities where maximum stability (i.e., vanishing widths Γ) occur. Of note is that in Figure 3.7, the widths initially follow a linear increase with intensity, corresponding to the diabatic Fermi-Golden rule photodissociation probability (Equation 3.88) and attains a maximum (saturates) in order to obey the adiabatic exponentially decreasing law (Equation 3.96).

Therefore the minima in Figure 3.7 correspond to the *coincidence* of diabatic and adiabatic levels. This is a confirmation of the rule derived from semiclassical predissociation theory, discussed previously, which predicts stability of laser-induced resonances (i.e., of dressed states at such intensities). We now investigate next the effect of time-dependent pulses on the photodissociation process.

B. Time-Dependent Photodissociation

The time-dependent Schrödinger Equation for the time-dependent nuclear states $\chi_1(R,t)$ and $\chi_2(R,t) = [\chi_1, \chi_2]$

$$\left(i\hbar\frac{\partial}{\partial t} + \frac{\hbar^2 \nabla^2}{2M}\right)\chi(R,t) = \begin{bmatrix} V_1(R) & V_{12}(R,t) \\ V_{12}(R,t) & V_2(R) \end{bmatrix} \chi(R,t) \tag{3.102}$$

was solved numerically using an efficient, high-order, split exponential operator algorithm [90]. The radiative interaction for such calculations consisted of two types,

$$\text{exact:} \qquad V_{12}(R,t) = \frac{eR}{2} E_o(t) \cos \omega t \tag{3.103}$$

$$\text{RWA:} \qquad V_{12}(R,t) = \frac{eR}{4} E_o(t)\, e^{i\omega t} \tag{3.104}$$

Equation 3.104 corresponds to the two-state dressed picture (Figure 3.5) since only the resonant process is included (see also Figure 3.2). Corrections to RWA will occur when virtual processes from the neglected term $e^{-i\omega t}$ become

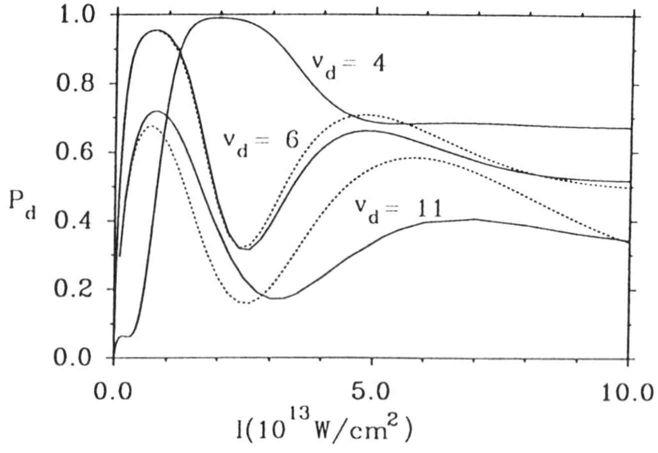

Figure 3.8 Photodissociation probability P_d as a function of intensity I (W/cm^2) for initial diabatic levels v_d = 4, 6, 11 from exact (solid line) and RWA (dashed line) time-dependent calculations. Pulses are 100 femtosecond long with 1 femtosecond rise and fall.

important. This is expected to occur when the electronic Rabi frequency ω_R (Equation 3.1), approaches the photon frequency ω. For the present calculation, $\hbar\omega$ ($\lambda = 212$ nm) $\simeq$ 6 e.v., $R_c \simeq 4$ a.u., then from Equation 3.44 $\hbar\omega_R = \hbar\omega$ at an intensity $I \simeq 6 \times 10^{14}$ W/cm^2. Clearly RWA will only be useful below intensities of 10^{14} W/cm^2.

Taking as initial condition $\chi_v(R,t = 0)$, one propagates in time (Equation 3.102) the nuclear wavefunctions discretized in space, allowing us to obtain the ground $\chi_1(R,t)$ and excited state $\chi_2(R,t)$ nuclear functions until they propagate freely on the diabatic unperturbed surfaces $V_1(R)$ and $V_2(R)$ after the end of the pulse. One then integrates the density $|\chi_2(R,t)|^2$ from a point R outside the right turning point b_1 of the upper $W_+(R)$ surface (Figure 3.5) in order to obtain the dissociation probabilities P_d (Figure 3.8). The momentum Fourier transform of $\chi_2(R,t)$ gives the kinetic energy distribution of the photodissociating fragments through the relation

$$P(E,t) = \frac{m}{p} |\chi_2(p,t)|^2$$

Table 3.2 Kinetic Energy Peaks (eV) in Photodissociation of H_2^+ at $\lambda = 212$ nm and Corresponding Adiabatic Levels v_+ ($E_{ad} = E_{RWA}$) at Intensity $I = 10^{12}$ W/cm^2 for v_d = 4, 6, 11 as Initial States (1200 fs. pulse).

exact:	4.77*	5.00*	5.20*	5.36*	5.50	5.62	5.71	5.77
RWA:	4.85*	5.11*	5.34*	5.55*	5.74	5.91	6.07	6.22
v_+:	0	1	2	3	4	5	6	7

* present in v_d = 4, 6 only.

These results are reported in Table 3.2.

We now discuss the numerical results. Figure 3.8 shows an initial increase with intensity I of the dissociation probability P_d for the initial vibrational levels v_d = 4, 6, 11. The linear rise of P_d with I is consistent with the diabatic Fermi-Golden rule (Equation 3.88). Then with increasing intensity, saturation (i.e., a maximum) occurs followed by a deep minimum for each initial level v_d. The minima in P_d is a clear manifestation of *molecular stabilization* by a laser-induced avoided crossing mechanism, as we will show next. Finally a constant plateau up to I = 1014 W/cm^2 is reached. For v_d = 4, an initial plateau in P_d occurs already at the lower intensities, between $I = 1–4 \times 10^{12}$ W/cm^2. These are also manifestations of stabilization by the laser field.

In Figure 3.8 we in fact present exact (solid) and RWA (dashed) results. We see that for the lower levels, v_d = 4 and 6, RWA calculations agree fairly well with the exact results. Deviations begin to occur for v_d = 11 at 10^{13} W/cm^2. Thus for higher vibrational levels, due to their larger right turning points b_1 on the adiabatic surface $W_+(R)$, these undergo larger radiative interactions from the divergent transition moment. In summary, Figure 3.8 illustrates clearly the existence of dissociation probability maxima approaching unity (~ 100% yield) and minima (10–20% yield) and plateaus (50% yield) at higher intensities.

This manifestation of *stabilization* of molecular levels in high intensity photodissociation can be correlated with the laser-induced resonances created by the laser-induced avoided crossings during the pulse duration. Further

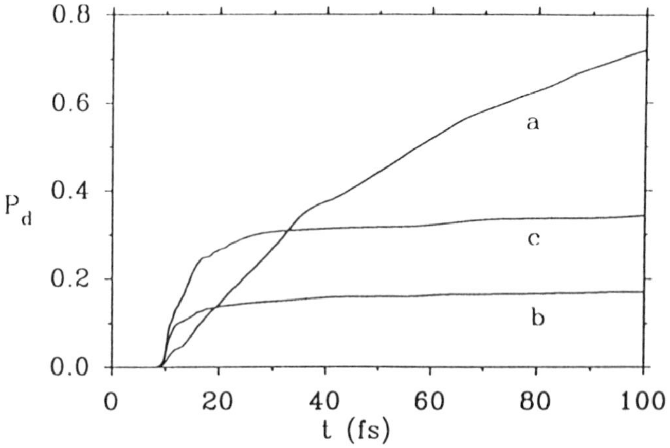

Figure 3.9 Photodissociation probability P_d as a function of time for $v_d = 11$ level in Figure 3.8 as initial state with I = a) 8×10^{12} W/cm^2; maximum P_d (~ 100%); b) 3.2×10^{12} W/cm^2; minimum P_d (~ 20%); c) 10^{14} W/cm^2; P_d (~ 30%). b) and c) show the laser-induced molecular stabilization at the higher intensities.

proof of stabilization is first shown in Figure 3.9 where we illustrate the time-dependent dissociation probability during the pulse. As described above, since in the time-dependent calculations one integrates everything to the right of the turning point $b_1(b_1)$ of the adiabatic level (see Figure 3.5), then Figure 3.9 is a measure of the presence of the nuclear wavepacket well past the laser-induced interaction region around the crossing point R_c. Figure 3.9 emphasizes the different behavior of $v_d = 11$ at its maximum dissociation ($I = 8 \times 10^{12}$ W/cm^2) versus minimum dissociation ($I = 3.2 \times 10^{13}$ W/cm^2). Thus for the lower intensity, the dissociation rises linearly continuously with time during the pulse suggesting a constant dissociation rate or width (lifetime). In contrast to this, at the two higher intensities, rapid dissociation occurs in the first 10–20 femtoseconds, then there is stabilization of the initial molecular state.

We can now correlate the minima in the time-dependent dissociation probabilities (Figures 3.8–3.9) with the time-independent dressed state energies and widths (Figures 3.6–3.7). Thus the minima in the linewidths of

$v_+ = 0, 1$ were found to occur at intensities of 3.2×10^{12} and 2.6×10^{13} W/cm^2 (Figure 3.7). These minima correspond to coincidences of diabatic and adiabatic levels (Figure 3.6). Minima in the time-dependence photodissociation probabilities P_d occur precisely at the same intensities and for the same levels as in the dressed-state time-independent results. Inspection of Figure 3.5 also illustrates these coincidences. Thus $v_+ = 0$ is quasidegenerate with $v_d = 4$ at $I = 3.2 \times 10^{12}$ W/cm^2, $v_+ = 1$ with $v_d = 6$ at $I \simeq 2.6 \times 10^{13}$ W/cm^2 and finally $v_+ = 0$ with $v_d = 6$ at $I \simeq 6 \times 10^{13}$ W/cm^2. The latter coincidence correlates with the onset of a constant plateau beyond that intensity.

In the Fano language of configuration interactions between bound and continuum states [91], the laser-induced resonances are either linear combinations of all the diabatic (field-free) nuclear states $\chi_d(R)$ or all the adiabatic states $\chi_\pm(R)$ [i.e., the eigenstates of the new laser-induced adiabatic potentials $W_\pm(R)$ (Equation 3.62) when RWA applies]. The adiabatic states are not orthogonal to the diabatic states. The semiclassical theory of laser-induced resonances [28, 30–32] shows that at quasidegeneracies of diabatic and adiabatic levels, the true resonance, which becomes stable, is a simple combination of these two nonorthogonal quasidegenerate states. The composition is determined by a nonadiabacity parameter u, which we express here in terms of the Landau-Zener parameter ε (see Equation 3.96),

$$u = \exp(\varepsilon) - 1, \quad \varepsilon = \frac{2\pi |V_{12}|^2}{\hbar v_c |F_1 - F_2|} \tag{3.105}$$

V_{12} is the radiative coupling $\hbar\omega_R/2$ (Equations 3.1 and 3.96), v_c is the nuclear velocity at the crossing point R_c and the F's are the slopes of the diabatic potentials at the crossing point. The parameter ε is tabulated in Table 3.3 for the intensities illustrated in Figure 3.9. $u \sim 1$ corresponds to intermediate coupling ($\varepsilon \sim 1$) (i.e. about equal mixing of diabatic and adiabatic levels), whereas $u >> 1$ implies adiabatic behavior ($\varepsilon > 1$) (i.e., there is a predominant adiabatic component in the exact dressed quasibound nuclear wavefunction). We see immediately that at $I = 3.2 \times 10^{13}$ and 10^{14} W/cm^2, the states are more adiabatic than diabatic so that a considerable amount of the resonant states at these critical intensities should be trapped in the new laser-induced

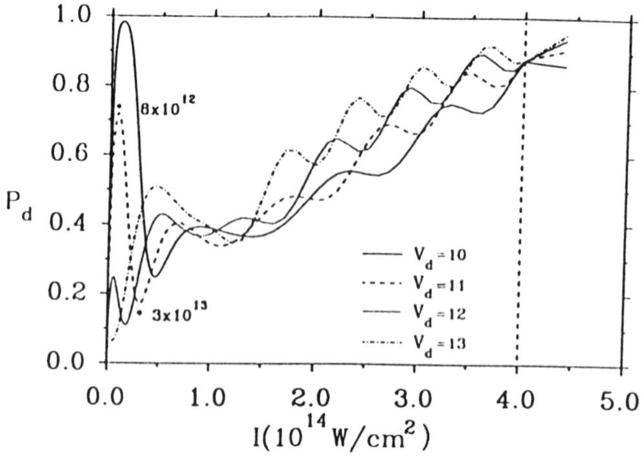

Figure 3.10 Photodissociation probability P_d for v_d = 10–13 as a function of intensity I (W/cm^2) showing laser-induced molecular stabilization at high intensities (i.e., incomplete photodissociation). I = 4 × 10^{14} W/cm^2 is intensity where adiabatic potential $\mathbf{W}_+(R)$ is above dissociation limit of H_2^+.

potential $\mathbf{W}_+(R)$. We have previously emphasized the importance of such trapped, and therefore stable, adiabatic states in causing highly anomalous photodissociation angular distributions [40, 41, 64].

The asymptotic limit of the dissociation probability P_d to values less than unity (~ 50% yield) at higher intensities up to 10^{14} W/cm^2 (see Figures 3.8–3.9) is further manifestation of *stabilization* by *trapping* by the adiabatic potential $\mathbf{W}_+(R)$ [92, 93]. This is dramatically demonstrated for the higher levels of H_2^+ in Figure 3.10. All adiabatic levels are above the dissociation limit at I = 4 × 10^{14} W/cm^2 for λ = 212 nm excitation. Figure 3.10 clearly demonstrates that dissociation is never complete due to trapping by the adiabatic potential $\mathbf{W}_+(R)$ since above 10^{14} W/cm^2 one is clearly in the adiabatic regime (see also Table 3.3). Calculations by Mies and Guisti-Suzor [94] at intensities above 10^{14} W/cm^2 also confirm this adiabatic trapping phenomenon, first predicted by Bandrauk et al. [28, 30, 31, 40, 41]. However at these higher intensities, since Rabi frequencies exceed rotational energies and approach the ionization energy of H_2^+, destabilization of these resonances will occur by these last two excitation processes [56, 95, 96].

Table 3.3 Landau-Zener Parameter ε (Equation 3.105); λ = 212 nm.

		ε		
v_+	I(W/cm²):	8×10^{12}	3.2×10^{13}	10^{14}
0		0.74	2.5	6.4
1		0.57	2.2	5.8
2		0.52	2.0	5.4

These stabilizations are expected to manifest themselves for one-photon resonant transitions or three-photon transitions at around intensities of 10^{13} W/cm^2 (see Chapter 2).

A final confirmation of the adiabatic behavior of the laser-induced resonances is presented in Table 3.2 in which we report the kinetic energy distribution of the photofragments. We report there the exact time-dependent peak positions, the time-dependent RWA energies and the corresponding levels of the two-state coupled Equation time-independent adiabatic resonances. The time-independent resonances have the same energies as the time-dependent RWA peaks. Thus at the highest intensity reported here, and for long pulses (τ = 1200 fs) well defined energy peaks occur for the initial levels v_d = 4, 6, 11. The peaks for v_d = 11 as initial state are identical to the high energy peaks of v_d = 4 and 6. All the peaks appearing in the RWA calculations are at somewhat higher energies than the exact results, indicative of the breakdown of RWA at 10^{14} W/cm^2. Thus the true dressed resonances generally appear at somewhat lower energies than the RWA resonances (i.e., corrections to RWA have a much larger effect on energies than on dissociation rates). We emphasize that all time-dependent RWA peaks fall at the same energies as the time-independent RWA (two-level) dressed states, so that a direct correlation can be made between the energy peaks of the photodissociation fragments and the laser-induced resonances [92].

A multitude of kinetic energy peaks occurs for subpicosecond (τ = 100 fs) excitation and is evidence for considerable nonadiabatic excitations due to the pulse envelope itself. These effects only occur for short pulses and disappear for longer pulses (e.g., 1 psec) [92]. Careful examination of the time-dependent nuclear functions $\chi(R,t)$ show, in agreement with Figure 3.9, that the nonadiabatic pulse shape dependent excitation occurs within the first 20 fs and stabilization occurs afterwards, with considerable amplitude of the nuclear function $\chi(R,t)$ oscillating between a_2, the left turning point of the upper diabatic potential $\mathbf{W}_2(R)$ and the right turning point b_1 of the lower diabatic potential $\mathbf{W}_1(R)$. This is consistent with the adiabatic viewpoint that the wavepacket oscillates in the upper adiabatic laser-induced potential, $\mathbf{W}_+(R)$ (Equation 3.62 and Figure 3.5).

The time-independent dressed state approach allows therefore one to establish *two mechanisms for stabilization*, vibrational trapping and consequently suppression of photodissociation at high field intensities: (a) degeneracies of diabatic (unperturbed) and adiabatic (laser-induced) vibrational levels; (b) decoupling of the adiabatic levels of the laser-induced potentials $\mathbf{W}_+(R)$ at high intensities (see Equation 3.88). These two mechanisms have their common origin in the laser-induced avoided crossings. The presence of such laser-induced avoided crossings have now been inferred from molecular ATI experiments [12–16] and the anomalous kinetic energy distribution of photofragments [13, 14]. The latter experiments have also shown the presence of ATD, which can be also explained in terms of the dressed states (see Section 3.5). The stabilization or equivalently suppression of photodissociation persists also in the time-dependent, short pulse simulations, and are consistent with the time-independent dressed state laser-induced avoided crossing mechanism. Renormalization of the spectrum into new adiabatic states and multistate coherent superpositions giving rise to these new states by the above mechanisms find their analogue in atomic stabilization and the resulting suppression of ionization [38, 39, 62, 63, 88, 89, 97, 98].

An open problem is the role of higher excitations. Calculations on the ionization rate of H_2^+ indicate that rates approach 10^{13} s^{-1} at intensities $I = 10^{14}$ W/cm^2 [96]. This implies that the stable laser-induced resonances predicted by the time-independent dressed state method might still appear as resonances in the photoionization [68]. Rotational excitation is another factor to be considered. Previous time-independent coupled Equations calculations have shown that rotational excitations can stabilize resonances by inducing degeneracy between nonresonant diabatic and adiabatic levels. Alternatively, stable laser-induced resonances can become destabilized because of the destruction of the required degeneracy by rotational excita-

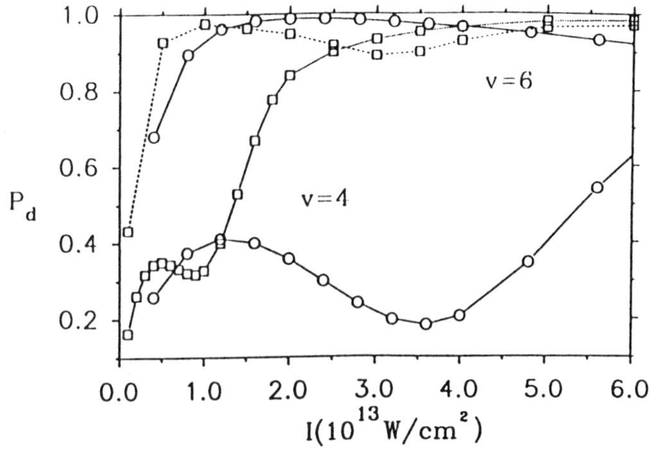

Figure 3.11 Photodissociation probability P_d for $v_d = 4, 6$ as a function of intensity I (W/cm^2) for initial rotational quantum numbers J = 5.5, M_J = 0.5. Solid line: two-channel calculation, $\Delta J = +1$, $\Delta M_J = 0$; dotted line: complete calculation (all ΔJ), $\Delta M_J = 0$. Stabilization intensity I is displaced to lower values upon rotational excitation.

tions [40, 41, 56, 64]. The same effect, rotational induced stabilization or destabilization occurs in the time-dependent coupled Equations simulations. This is illustrated in Figure 3.11 for a J = 5.5, M_J = 0.5 level. This particular rotational sublevel orients the molecule parallel to the laser linear polarization, so that for a $\Sigma_g \rightarrow \Sigma_u$ transition, this is the strongest radiatively coupled sublevel. The two-channel (no rotation), $v_d = 4$, minimum in the photodissociation P_d which occurs at an intensity around 3.5×10^{13} W/cm^2 is now shifted to lower intensities, 1×10^{13} W/cm^2. The stability region is much smaller as rotational excitation introduces more dissociation pathways, while at the same time lining up the molecule with the field. For $v_d = 6$, slight stabilization occurs as a result of rotational excitation. In all cases, the plateaux illustrated in Figures 3.9–3.10 at the high intensities 10^{14} W/cm^2, increased to 100% dissociation at and beyond these intensities. This is consistent with the general finding that high rotational excitation and ionization generally occurs in molecules at intensities beyond 10^{14} W/cm^2 [14–18] (see also Chapter 4).

ACKNOWLEDGMENTS

We thank the National Science and Engineering Research Council of Canada and the National Center of Excellence in Molecular Dynamics for supporting this research. Finally, one of us (ADB) wishes to express gratitude to Prof. H. Nakamura of the Institute of Molecular Science, Okazaki, Japan for his hospitality and continuing support of this project.

REFERENCES

1. J. C. Slater, *Quantum Theory of Molecules and Solids*, McGraw-Hill, New York, (1963), vol. I.
2. N. Mataga, T. Kubota, *Molecular Interactions and Electronic Spectra*, M. Dekker Inc., New York, (1970).
3. H. Lefebvre-Brion, R. W. Field, *Perturbations in Spectra of Diatomic Molecules*, Academic Press, Orlando FL, (1986).
4. J. Gallagher, C. E. Brion, J. A. R. Samson, P. W. Langhoff, *J. Phys. Chem. Ref. Data*, **17**, 9 (1988).
5. A. D. Bawagan, C. E. Brion, *Chem. Phys. Lett.*, **137**, 573 (1987).
6. C. B. Duke, N. D. Lipari, L. Pietronero, *J. Chem. Phys.*, **65**, 1165 (1976).
7. M. H. Mittleman, *Introduction to the Theory of Laser-Atom Interactions*, Plenum Press, New York, (1982).
8. F. H .M. Faisal, *Theory of Multiphoton Processes*, Plenum Press, New York, (1988).
9. A. D. Bandrauk, (Ed.), *Atomic and Molecular Processes with Short Intense Laser Pulses*, Plenum Press, New York, (1988), NATO ASI, vol. 171B.
10. A. D. Bandrauk, S. C. Wallace (Eds.), *Coherence Phenomena in Atoms and Molecules in Laser Fields*, Plenum Press, New York, (1992), NATO ASI, vol. 287B.
11. C. Cornaggia, D. Normand, J. Morellec, G. Mainfray, C. Manus, *Phys. Rev.*, **A34**, 207 (1986).
12. J. W. J. Verschuur, L. D. Noordam, H. B. van Linden van den Heuvell, *Phys. Rev.*, **A40**, 4383 (1989).
13. P. H. Bucksbaum, A. Zavriyev, H. G. Muller, D. W. Schumacker, *Phys. Rev. Lett.*, **64**, 1883 (1990).
14. A. Zavriyev, P. H. Bucksbaum, H. G. Muller, D. W. Schumacker, *Phys. Rev.*, **A42**, 5500 (1990).
15. S. W. Allendorf, A. Szoke, *Phys. Rev.*, **A44**, 518 (1991).
16. H. P. Helm, M. J. Dyer, H. Bissantz, *Phys. Rev. Lett.*, **67**, 1234 (1991).
17. D. T. Strickland, Y. Beaudoin, P. Dietrich, P. B. Corkum, *Phys. Rev. Lett.*, **68**, 2755 (1992).
18. P. Dietrich, P. B. Corkum. *J. Chem. Phys.*, **97**, 3187 (1992).

19. C. Cohen-Tannoudji, S. Reynaud, in *Multiphoton Processes*, (J. H. Eberly, P. Lambropoulos, Eds.), J. Wiley, New York, (1978).
20. A. D. Bandrauk, N. Gélinas, *J. Comput. Chem.*, **8**, 313 (1987), A. D. Bandrauk, N. Gélinas, *J. Chem. Phys.* 86, 5257 (1987).
21. A. D. Bandrauk, O. Atabek, in *Advances in Chem. Phys.*, (J. O. Hirschfelder, R. Wyatt, R. D. Coalson, Eds.), J. Wiley, New York, (1988), vol. 73, chap. 19.
22. S. Miret-Artes, O. Atabek, A. D. Bandrauk, *Phys. Rev.*, **A45**, 8056 (1992).
23. A. D. Bandrauk, J. M. Gauthier, *J. Opt. Soc. Amer.*, **B7**, 1422 (1990).
24. X. He, O. Atabek, A. Giusti-Suzor, F. H. Mies, *Phys. Rev. Lett.*, **64**, 515 (1990).
25. A .D. Bandrauk, E. Constant, J. M. Gauthier, *J. Phys. II* (France), **1**, 1033 (1991).
26. A. I. Voronin, A. A. Samokhin, *Sov. Phys. JETP*, **43**, 4 (1976).
27. A. M. Lau, C. K. Rhodes, *Phys. Rev.*, **A16**, 2392 (1977).
28. A. D. Bandrauk, M. L. Sink, *Chem. Phys. Lett.*, **57**, 569 (1978).
29. J. M. Yuan, T. F. George, *J. Chem. Phys.*, **68**, 3040 (1978).
30. A. D. Bandrauk, M. L. Sink, *J. Chem. Phys.*, **74**, 1110 (1981).
31. A. D. Bandrauk, J. F. McCann, *Comments Atom. Molec. Phys.*, **22**, 325 (1989).
32. A. D. Bandrauk, M. S. Child, *Molec. Phys.*, **19**, 95 (1970).
33. A. D. Bandrauk, O. Atabek, *J. Phys. Chem.*, **91**, 6469 (1987).
34. R.S. Mulliken, *J. Chem. Phys.*, **7**, 20 (1939).
35. J. P. Connerade, J. M. Esteva, R. C. Karnatak (Eds.), *Giant Resonances in Atoms, Molecules and Solids*, Plenum Press, New York, (1987), NATO ASI, vol. 151B.
36. J. W. Negele, H. Orland, *Quantum Many-Particle Systems*, Addison-Wesley, New York, (1988).
37. S. Chelkowski, A. D. Bandrauk, *J. Chem. Phys.*, **89**, 3618 (1988).
38. Q. Su, J. H. Eberly, J. Javanainen, *Phys. Rev. Lett.*, **64**, 862 (1990).
39. J. Parker, C. R. Stroud, *Phys. Rev.*, **A41**, 1602 (1990).
40. J. F. McCann, A. D. Bandrauk, *Phys. Rev.*, **A42**, 2806 (1990).
41. J. F. McCann, A. D. Bandrauk, *J. Chem. Phys.*, **96**, 903 (1992).
42. S. Mukamel, J. Jortner, *J. Chem. Phys.*, **65**, 5204 (1976).
43. P. B. Corkum, *IEEE J. Quant. Electron.*, **21**, 216 (1985).
44. S. Chelkowski, A. D. Bandrauk, P. B. Corkum, *Phys. Rev. Lett.*, **65**, 2355 (1990).
45. F.T. Smith, *Phys. Rev.*, **179**, 111 (1969).
46. T. T. Nguyen-Dang, A. D. Bandrauk, *J. Chem. Phys.*, **79**, 3256 (1983), *J. Chem. Phys.*, **80**, 4926 (1984).
47. A. D. Bandrauk, T. T. Nguyen-Dang, *J. Chem. Phys.*, **83**, 2840 (1985).
48. A. D. Bandrauk, O. F. Kalman, T. T. Nguyen-Dang, *J. Chem. Phys.*, **84**, 6761 (1986).
49. F. Bloch, A. Nordsieck, *Phys. Rev.*, **52**, 54 (1937).
50. W. Pauli, M. Fierz, *Nuovo Cimento*, **15**, 167 (1933).
51. T. A. Welton, *Phys. Rev.*, **74**, 1157 (1948).
52. J. Bergou, S. Varro, M. V. Fedorov, *J. Phys.*, **A14**, 2305 (1981).
53. J. Bergou, S. Varro, *J. Phys.*, **A14**, 1469 (1981).

54. T. F. George, I. H. Zimmerman, J. M. Yuan, J. R. Laing, P. L. Devries, *Acc. Chem. Res.*, **10**, 449 (1977).
55. T. F. George, *J. Phys. Chem.*, **86**, 10 (1982).
56. A. D. Bandrauk, G. Turcotte, *J. Chem. Phys.*, **77**, 3867 (1982).
57. A. D. Bandrauk, *Internat. Rev. Phys. Chem,* Jan, 1994 (in press).
58. C. A. S. Lima, L. C. M. Miranda, *J. Chem. Phys.*, **78**, 6102 (1983).
59. N. V. Cohan, H. F. Hameka, *J. Chem. Phys.*, **45**, 4392 (1966).
60. L. A. Nafie, J. B. Freedman, *J. Chem. Phys.*, **78**, 7108 (1983).
61. G. E. O. Giacaglia, *Perturbation Methods in Non-Linear Systems*, Springer, New York, (1972), vol. 8.
62. M. Pont, N. R. Walet, M. Gavrila, C. W. McCurdy, *Phys. Rev. Lett.*, **61**, 939 (1988).
63. J. vande Ree, J. Z. Kaminski, M. Gavrila, *Phys. Rev.*, **37A**, 4536 (1988).
64. A. D. Bandrauk, G. Turcotte, *J. Phys. Chem.*, **87**, 5098 (1983).
65. S. I. Chu, *J. Chem. Phys.*, **94**, 7901 (1991).
66. T. T. Nguyen-Dang, S. Manoli, *Phys. Rev.*, **44A**, 5841 (1991).
67. R. Loudon, *The Quantum Theory of Light*, Oxford University Press, London, (1973).
68. A. Zavriyev, P. H. Bucksbaum, J. Squier, F. Salin, *Phys. Rev. Lett.*, **70**, 1077 (1993).
69. R. D. Levine, *Quantum Mechanics of Molecular Rate Processes*, Oxford Press, London, (1969), Sections 3.1-3.4.
70. T. T. Nguyen-Dang, S. Durocher, O. Atabek, *Chem. Phys.*, **129**, 451 (1989).
71. H. Feshbach, *Ann. Phys.*, **19**, 287 (1962).
72. H. P. Helm, M. J. Dyer, H. Bissantz, D. L. Huestis, in *Coherence Phenomena in Atoms and Molecules in Laser Fields*, (A. D. Bandrauk, S. C. Wallace, Eds.), Plenum Press, New York, (1992), NATO ASI, vol. 287B, p. 109-123.
73. L. F. di Mauro, B. Yang, M. Saed, in *Coherence Phenomena in Atoms and Molecules in Laser Fields*, (A. D. Bandrauk, S.C. Wallace, Eds.), Plenum Press, New York, (1992), NATO ASI, vol. 287B, p. 75-88.
74. L. Wolniewicz, K. Dressler, *J. Molec. Spectr.*, **77**, 286 (1979); L. Wolniewicz, K. Dressler, *J. Chem. Phys.*, **82**, 3292 (1985).
75. M. Shapiro, *J. Chem. Phys.*, **56**, 2582 (1972).
76. X. He, O. Atabek, A. Guisti-Suzor, *Phys. Rev.*, **A38**, 5586 (1988).
77. A. D. Bandrauk, G. Turcotte, *J. Phys. Chem.*, **89**, 3039 (1985).
78. K. M. Watson, J. Nuttal, *Topics in Several Particle Dynamics*, Holden Day Publishers, San Francisco, (1967).
79. T. T. Nguyen-Dang, S. Manoli, *Phys. Rev.*, **A44**, 5841 (1991).
80. J. R. Hiskes, *Phys. Rev.*, **122**, 1207 (1961).
81. L. Wolniewicz, *J. Chem. Phys.*, **51**, 5002 (1969).
82. H. G. Muller, in *Coherence Phenomena in Atoms and Molecules in Laser Fields*, (A. D. Bandrauk, S. C. Wallace, Eds.), Plenum Press, New York, (1992), NATO ASI, vol. 287B, p. 89-98.
83. M. L. Sink, A. D. Bandrauk, *J. Chem. Phys.*, **66**, 5313 (1977).

84. M. S. Child, *Molecular Collision Theory*, Academic Press, London, (1974).
85. A. Comtet, A. D. Bandrauk, D. K. Campbell, *Phys. Lett.*, **1508**, 159 (1985).
86. R. Rajaraman, *Solitons and Instantons*, North Holland, Amsterdam (1982)).
87. L. D. Landau, E. M. Lifshitz, *Quantum Mechanics*, Pergamon Press, London, (1965), 2nd edition.
88. M. Fedorov, A. Movsevian, *J. Opt. Soc. Amer.*, **B6**, 928 (1989).
89. M. Fedorov, M. Ivanov, A. Movsesian, *J. Phys.*, **B23**, 2245 (1990).
90. A. D. Bandrauk, H. Shen, *Chem. Phys. Lett.*, **176**, 428 (1991); A. D. Bandrauk, H. Shen, *J. Chem. Phys.*, **99**, 1185 (1993).
91. U. Fano, *Phys. Rev.*, **124**, 1866 (1961).
92. E. Aubanel, A. D. Bandrauk, P. Rancourt, *Chem. Phys. Lett.*, **197**, 419 (1992).
93. A. D. Bandrauk, E. Aubanel, J. M. Gauthier, *Laser Physics*, **3**, 381 (1993).
94. F. Mies, A. Giusti-Suzor, *Phys. Rev. Lett.*, **68**, 3869 (1992).
95. E. Aubanel, A. D. Bandrauk, J. M. Gauthier, *Phys. Rev. A*, **48**, 2145 (1993).
96. S. Chelkowski, T. Zuo, A. D. Bandrauk, *Phys. Rev. A*, **46**, 5342 (1992).
97. Y. Dubrovskii, M. Ivanov, M. Federov, *Sov. Phys. JETP*, **72**, 228 (1991).
98. L. D. Noordham, H. Stapelfeldt, D. I. Duncan, T. F. Gallagher, *Phys. Rev. Lett.*, **68**, 1496 (1992).

4

Molecular Ions in Intense Laser Fields

Peter Dietrich and P. B. Corkum

Steacie Institute for Molecular Sciences, National Research Council of Canada, Ottawa, Ontario, Canada

Donna T. Strickland

Princeton University, Princeton, New Jersey

Michel Laberge

CREO Products Inc., Burnaby, British Columbia, Canada

4.1 INTRODUCTION

High-power femtosecond laser pulses are now so short [1] that even for highly charged molecules the internuclear motion can be frozen on the time scale of the pulse. This means that rather unusual species, such as inertially confined molecules, can be studied optically.

Because of the model role of diatomic molecules in the understanding of molecule-light interactions, this chapter deals only with diatomic molecules and molecular ions. We introduce the broad outline of the molecular science appropriate for inertially confined diatomic ions. To do this, we use knowledge of the hydrogen molecule H_2 and the molecular ion H_2^+ to identify the most important states [2] which are then used to calculate Stark shifts and polarizability in the strong laser field. The coupling of these states (charge transfer in the case of even charged molecules, charge resonance for odd charged molecules) corresponds to the motion of electrons on the internuclear scale. The transition dipole moments between these states are very large so that these couplings dominate nonresonant interactions. In the high-intensity limit, this analysis leads to the same results as the classical description of the molecular ion in the field. The concepts that we introduce can be generalized to other diatomic molecules [3] although in this chapter we consider only homonuclear molecules. They are also important for neutral molecules in strong laser fields (e.g., neutral molecular hydrogen). In our view, nonresonant couplings will effect many high-intensity molecular experiments.

Our theoretical analysis is supported by experimental observations of inertially confined iodine molecules [4] produced with ultrashort 625 nm pulses. The use of an ultrashort pulse together with a heavy molecule allows a field-free Coulomb explosion of the fragments. This is essential to the interpretation of the results because the coupling of electronic states that is characteristic of strong-field interactions with molecules [5, 6] is unlikely to allow a clear signature of the electronic state in the kinetic energy spectrum of the fragments if the pulse duration is comparable to, or larger than the lifetime of the molecular ion. Therefore we use pulses of 80 fs duration. Iodine is chosen because it remains essentially at rest on this time scale. Its ground state vibrational period is 155 fs and the rotational period is 446 ps/J where J is the rotational quantum number.

This chapter combines work published in three different papers [7–9]. The work is significant not only because of the intrinsic interest of optical experiments on this unusual species, but it is also important for high-field molecular science in general. For example, it can be used to establish the maximum laser intensity that a molecule can withstand without ionization [3]. Furthermore, any new development in multiphoton ionization has implications for the physics of plasmas produced by ultrashort laser pulses [10]. It is clear that plasmas can be formed of inertially confined ions; such plasmas are far from equilibrium. Understanding them, particularly their nonlinear properties [11], will be an important frontier of plasma research.

The outline of the chapter is as follows. In Section 4.2, we discuss the charge-transfer and charge-resonance states in H_2 and H_2^+. From this we derive the Stark shift together with the associated induced dipole moment of a molecule in a strong laser field. We show how the classical limit emerges from the quantum mechanical description. Section 4.3 is devoted to the modifications that are required to atomic concepts of multiphoton ionization to obtain an accurate description of molecular ionization. Both the Stark shifts and the induced dipole moments will be important. Section 4.4 contains a brief description of the experimental setup. In Section 4.5 we discuss the experimental results. We start with the basic features of the ion kinetic energy spectrum in Section 4.5.A. Coulomb explosions impress a characteristic kinetic energy on the fragment ions that allows us to measure the ionization yields as a function of intensity. The results and a comparison with model calculations are shown in Section 4.5.B. Section 4.5.C deals with the observation of anisotropic angular distributions of the ion fragments. We explain this in terms of deflection of the fragment trajectory due to the interaction of the polarizable molecule with the strong field. In Section 4.6 we discuss consequences of this work.

4.2 MOLECULAR IONS IN STRONG FIELDS

In this section we consider the interaction of a diatomic molecule or molecular ion in a strong electric field. This problem is, in general, a rather complex one that involves many electronic states, each with its own rovibrational structure. In a strong field, many of these states might be coupled, leading to complicated nuclear and electronic motions. Our aim is to develop ideas which are simple but capture the essential features. To achieve this, we first identify the important electronic states which have to be considered because they are most strongly coupled. These states are those coupled by charge-resonance or charge-transfer transitions. They were first introduced by Mulliken to explain the observed intensities in molecular electronic spectra [2]. Concentrating on these states, we have only two or three states to consider for a given molecule (or molecular ion) with known transition moments between these states. We discuss these states in a strong static field to demonstrate some of the important features, for example the classical behavior of two ions in a strong field. Specifically, we calculate Stark shifts and induced dipole moments which are important for ionization and dissociation of the molecule [3].

We start by briefly introducing charge-resonance and charge-transfer couplings (for a detailed discussion see [2]). Both transitions involve moving electrons on the internuclear scale and have no analog in atoms. A charge-resonance transition couples two states which we will call *charge-resonance states* and which lead asymptotically to the same product states with differently charged fragments. A simple example is H_2^+ where the ground state $X\,^2\Sigma_g^+$ and the repulsive state $A\,^2\Sigma_u$ both yield asymptotically $H + H^+$. In general, one would have two states leading to $A^{k+} + A^{l+}$ $(k \neq l)$. From the atomic wave functions of the atomic ions, one can construct the molecular wave functions (to first-order) of a symmetric (gerade) and antisymmetric (ungerade) state similar to H_2^+. It is easy to show [2] that the transition dipole moment μ_{12} between these two states is approximately given by:

$$\mu_{12} = \delta\, e\, \frac{R}{2} \tag{4.1}$$

where $\delta = k - l$ and R = internuclear separation. Here we neglected the overlap integral which is justified provided the internuclear separation is not too small. One can rationalize this dipole moment by arguing that δ electrons are moved coherently over a distance $R/2$ because the transition couples the gerade state where the electrons are in the middle of the molecule to the ungerade state where they are close to the ends. This dipole moment increases linearly with R leading to very strong coupling between the two charge-resonance states. (This does not violate the sum rule because the energy gap decreases exponentially with increasing internuclear separation so that the intensity of the transition also goes to zero exponentially.)

A charge-transfer transition is similar to a charge-resonance transition in that it also involves moving electrons on the internuclear scale. However, the states coupled by a charge-transfer transition lead asymptotically to different product states with the same total charge. An example is the transition between the ground state $X\,^1\Sigma_g^+$ of H_2 and the $B\,^1\Sigma_u^+$ state which would lead to $H^+ + H^-$ if there were no avoided crossings with states of the same symmetry coming from H (n = 1) + H (n = 2, 3). In general, we have a charge-transfer transition between a state leading to $A^{k+} + A^{l+}$ and a state leading to $A^{(k+1)+} + A^{(l-1)+}$. In practice, $k = l$ is the most important case because the symmetrically charged state is the lowest electronic state for a given molecular ion (in the framework of the Born-Oppenheimer approximation). The transition dipole moment for the coupling between these *charge-transfer states* is given by [2]:

$$\mu_{12} = \sqrt{\delta}\, e\, \frac{R}{2} \tag{4.2}$$

where δ is the number of electrons which can be excited in the transition (e.g., $\delta = 2$ for H_2 and I_2). The square root enters because the δ electrons are excited incoherently. Note, however, that this form holds only for $R \approx R_e$, the equilibrium internuclear separation. For larger distances μ_{12} falls exponentially with the overlap integral. Due to limitations imposed by the sum rule, the actual transition moment might be smaller than given by Equation 4.2 even at the equilibrium distance because, for large molecules (like I_2), R_e is large and the energy gap between the two states is typically several eV. In this chapter we will neglect charge-transfer couplings and concentrate on charge-resonance transitions although the experimental results show the importance of the former (see Section 4.5.A). However, our main goal is to introduce the important concepts which can be done with charge-resonance coupling only.

After having determined the important electronic states and the couplings between them, it remains to calculate the perturbed states in the presence of the strong electric field. For a time-varying field, one could use a dressed-state or Floquet approach [6, 12, 13]. However, this is difficult for such high-order processes as we consider here. We will discuss the simpler case of a static electric field and then assume that the field is slowly varied. In this case, one can calculate the new states by simply diagonalizing the new Hamiltonian. For two states (e.g., for charge-resonance states) especially simple formulas can be derived. This quasistatic approach [14] is well founded in the case of 10 μm ionization of atoms [10] or molecules [3]. It has also been shown to adequately describe shorter wavelength ionization of rare gas atoms as well [15, 16].

When coupling two states with field-free energies E_1^o, E_2^o, $(E_1^o > E_2^o)$ with a coupling V_{12}, the perturbed energies are given by [3]:

$$E_{1(2)} = E_{1(2)}^o + (-) V_{12} \tan\beta \tag{4.3}$$

where the coupling angle β is given by

$$\tan 2\beta = \frac{2V_{12}}{E_1^o - E_2^o}$$

The induced dipole moments are:

$$\mu_{1(2)} = (-)\,\mu_{12} \sin 2\beta \tag{4.4}$$

In the high-intensity limit where $V_{12} >> E_1^o - E_2^o$, one has $\beta = \pm\,\pi/4$ depending on the sign of V_{12}. Then the energies and dipole moment become:

$$E_{1(2)} = E_{1(2)}^o + (-)|V_{12}| \tag{4.5}$$

$$|\mu_{12}| = \mu_{12} \tag{4.6}$$

which is the linear Stark effect. For dipole coupling, $V_{12} = -\mu_{12}\,\xi \cos \Psi$ where ξ is the electric field and Ψ is the angle between the molecular axis and the direction of the electric field vector. As we will show now, this is just the classical expression for the energy of the molecular ion in the electric field.

Consider two point like ions A^{k+} and A^{l+} held together by a field-free intramolecular potential $V_{mol}(R)$. In an external electric field, the total potential energy of this molecular ion in the center-of-mass frame is given by:

$$E_{pot} = V_{mol}(R) - \frac{1}{2}(k - l)\, e\, R\, \xi \cos \Psi \tag{4.7}$$

where the electric field is assumed to be constant over the range of the molecule. In this classical description the Stark shift depends on the charge asymmetry $\delta = k - l$. For coupling charge-resonance states, we have $\mu_{12} = \delta\, eR/2$ according to Equation 4.1. Thus the classical expression (Equation 4.7) is identical to the high-intensity limit of the quantum mechanical expression (Equation 4.5). We will use the classical expression in much of the following analysis because we do not know the spectroscopy of highly charged particles very well. The classical expression gives the high-field limit and, since we are concerned with strong laser fields, this is the appropriate limit.

In the next section we apply the ideas developed here to the ionization of diatomic molecules where we generalize concepts first given by Dietrich and Corkum [3]. The connection between Stark shifts and strong-field dissociation has been discussed by Hiskes [17] and Dietrich and Corkum [3].

4.3 MOLECULAR MULTIPHOTON IONIZATION

In this section we discuss in some detail the modifications to atomic ionization models necessary to describe ionization of diatomic molecules, both charged and neutral. These considerations are supported by ionization experiments of I_2^{n+} to be described in the following two sections. As has been discussed briefly by Dietrich and Corkum [3], the basic idea is to use an intensity-dependent ionization potential to account for the effects of the interaction of the molecule with the strong laser field. There are two effects to be considered: the Stark shifts of the initial and final molecular states and the multipole moments of the aspherical charge distribution of the final state.

We want to apply concepts of atomic ionization models to molecular ionization. Specifically, we consider the tunnel ionization model [18, 19]. Our approach is also valid for barrier-suppression ionization [16]. These models basically treat ionization as the process of removing an electron from an ion core whose Coulomb potential is modified by the external electric field. The essential parameters are the (negative) energy of the electron and the height of the barrier created by the superposition of the Coulomb potential and the external field. It has been shown for atoms that these quasistatic models successfully describe multiphoton ionization [10], even for visible ionization of rare gases where the quasistatic approximation might not be expected to be valid [16]. It has also been shown that atomic models can be generalized to accurately describe ionization of neutral diatomic molecules in the infrared [3].

In atoms, Stark shifts of the bound electron are small. As we have seen above, this is no longer the case in molecules. Furthermore, the potential of the molecular ion core (after ionization) is not spherical. Because of the ellipsoidal electron distribution, there will be quadrupole moments and, for an applied electric field, also induced dipole moments (for homonuclear molecules treated here the permanent dipole moments vanish for symmetry reasons). Due to these higher-order multipole moments, the height of the barrier now depends on the orientation of the molecule with respect to the electric field [3]. A change in the barrier height is equivalent to a change in the electron energy in the models mentioned previously because the important quantity is the difference between barrier energy and electron energy. Thus dipole and quadrupole moments lead to a change of the ionization potential which together with the molecular Stark shifts yield an effective ionization potential for molecular ionization.

We consider the ionization of a molecule A_2:

$$A_2^{(Z-1)} \rightarrow A_2^{Z+} + e^- \tag{4.8}$$

where the initial ion (or molecule) is a state corresponding asymptotically to product ions:

$$A_2^{(Z-1)+} \rightarrow A^{k'+} + A^{l'+}, \quad (Z-1 = k' + l') \tag{4.9}$$

and the final ion is a state:

$$A_2^{Z+} \rightarrow A^{k+} + A^{l+}, \quad (Z = k + l) \tag{4.10}$$

We denote the charge asymmetry with δ:

$$\delta = k - 1 \text{ and } \delta' = k' - l' \tag{4.11}$$

and the change of the asymmetry with Δ:

$$\Delta = \delta - \delta' \tag{4.12}$$

According to Section 4.2, an ion with charge asymmetry δ shows a classical energy shift:

$$\Delta E = -\frac{1}{2} e\, \delta\, R\, \xi_o \cos \Psi \tag{4.13}$$

In the following discussion we will use only this classical expression because it does not require knowledge of the spectroscopy of the molecular ions and because we are interested in the high-intensity behavior. The effect on the ionization potential of the Stark shifts is given by the difference of the shifts in the upper and the lower charged state:

$$\Delta E_{mol} = -\frac{1}{2}\,e\,\Delta R\,\xi_o \cos\Psi \tag{4.14}$$

To obtain the change of the barrier height due to the aspherical charge distribution, we start with the total classical energy of the electron which is being removed during ionization in the field of the residual ion core. We take only the lowest three multipole moments into account because the influence of the higher-order moments is small. Thus we have:

$$E_{\text{tot}} = \frac{1}{4\pi\varepsilon_o} e^2 \left[\frac{Z}{r} + \frac{p\cos\Psi}{r^2} + \frac{Q}{r^3}\frac{1}{2}(3\cos^2\Psi - 1) \right] + e\,\xi\,\boldsymbol{r} \tag{4.15}$$

where $\boldsymbol{r}$ is the position of the electron, $r = |\boldsymbol{r}|$, Z is the total charge (monopole moment), p is the dipole moment, Q is the quadrupole moment, and ξ is the electric field vector. Calculating the barrier position and barrier energy from the total electron energy is straightforward.

For $p = Q = 0$, one has:

$$r_0 = \left[\frac{1}{4\pi\varepsilon_0} \frac{Z\,e}{\xi} \right]^{1/2} \tag{4.16a}$$

$$E_0 = -\left[\frac{1}{4\pi\varepsilon_0} 4\,Z\,e^3\,\xi \right]^{1/2} \tag{4.16b}$$

Setting $E_0 = -E_s^0$ where E_s^0 is the field-free ionization potential of the atom or ion being ionized, gives a threshold value for the electric field ξ which is used in the barrier suppression ionization model [16]. For $p \neq 0$ and $Q \equiv 0$, we assume a small dipole moment in the sense that it does not change r_0 too much. Then we calculate the barrier position up to first-order corrections and energy up to second-order corrections:

$$r_1 = r_0 - p\,\frac{\cos\Psi}{Z} \tag{4.17a}$$

$$E_1 = E_o + \xi \frac{p}{Z} \cos \Psi + \xi \frac{p^2}{eZr_o} \cos^2 \Psi$$

$$= E_o + \Delta E_{p1} + \Delta E_{p2} \tag{4.17b}$$

For $p \equiv 0$ and $Q \neq 0$, we again assume a small quadrupole contribution and calculate up to first-order corrections:

$$r_2 = \left[r_o^2 + \frac{3/2\, Q\, (3 \cos^2 \Psi - 1)}{Z} \right]^{1/2} \tag{4.18a}$$

$$E_2 = E_o + \frac{Q \cdot \xi}{Z \cdot r_o} \frac{1}{2} (3 \cos^2 \Psi - 1)$$

$$= E_o + \Delta E_{q1} \tag{4.18b}$$

In the general case of $p \neq 0$ and $Q \neq 0$, one can add the changes of the barrier energy in the first-order. The change of the barrier energy is equivalent to a change of the electron energy so that we obtain for the effective ionization potential:

$$E_s = E_s^0 + \Delta E_{mol} + \Delta E_{p1} + \Delta E_{p2} + \Delta E_{q1} \tag{4.19}$$

Obviously, one could refine this approach by including higher multipole moments or calculate the exact barrier energy numerically.

However, for R not too large (i.e., less than or on the order of 0.2 nm), it is sufficient to include only the first-order dipole term ΔE_{p1}, as has been done by Dietrich and Corkum [3]. In the classical case and using the center-of-molecule frame, we have $p = \delta\, \mathrm{e}\, R/2$, so that the effective ionization potential in this approximation is given by:

$$E_s = E_s^0 + \frac{1}{2} \mathrm{e}\, R \left(\frac{\delta}{Z} - \Delta \right) \mathrm{E} \cos \Psi \tag{4.20}$$

For the first ionization step ($A_2 \rightarrow A_2^+$) the second term vanishes ($\delta' = 0$, $\delta = \Delta = 1, Z = 1$), so that the atomic model using only the field-free ionization potential is sufficient as has been shown in infrared multiphoton ionization of HCl [3]. For higher ionization steps this cancellation of Stark shifts and induced dipole moments does not hold, except for $\delta = Z$.

For a larger R, one should include the next terms, that is, the second-order dipole term and the first-order quadrupole term which reflects the ellipsoidal shape of the molecular ion. Both have a similar angular dependence and different signs, thus partly canceling each other. We will neglect these terms in this chapter and use the effective ionization potential given in Equation 4.20 to calculate ionization rates with the tunnel ionization model [18, 19] (see also Appendix A).

It is well known that the higher-order moments are not unique but depend on the coordinate system. However, the (partial) cancellation of the dipole contribution and the Stark shifts in Equation 4.20 is independent from the choice of the origin. We used the center-of-molecule frame in the formulas given above. This coordinate system has the disadvantage that the special case $\delta = Z$ (all charge on one end) does not reduce to the atomic behavior when including the quadrupole term because $Q = Z\,e\,R^2/4 \neq 0$. (An exception is $\Psi = 0$ where the second-order dipole term ΔE_{p2} cancels exactly the first-order quadrupole term ΔE_{q1}.) Obviously, one would have to include the higher-order moments. A much more natural choice for the origin is the center of charge of the final ion. Now we have $p \equiv 0$ and $Q = k\,l\,e\,R^2/Z$. The special case $\delta = Z$ now gives $Q = 0$ so that Equation 4.14 reduces correctly to the atomic result.

4.4 EXPERIMENTAL SETUP

A CPM dye laser [1] is used to produce 70 fs pulses at a wavelength of 630 nm. The pulses are amplified using a four-stage dye amplifier [20], pumped by a Q-switched Nd:YAG laser operating at a 5 Hz repetition rate. A single pass through a double grating compressor is used at the output of the amplifier chain to compensate for the dispersion of the amplifier. After the compressor the pulses have energies of 600 μJ and durations of 80 fs. A good spatial profile is obtained by using a spatial filter which selects only the Airy disk of the transmitted radiation.

The laser pulses are focused inside a vacuum chamber by an on-axis f/2 parabolic mirror. The aperture is limited to f/20 focusing in order to have a large number of molecules in the interaction region at a low density. To keep the intensity at the window of the chamber below the limit where nonlinearities may distort the laser focus, the energy is kept below 150 μJ, giving a maximum intensity of 10^{15} W/cm^2. The relative intensity of the laser pulses is determined by measuring the pulse energy. Absolute intensities are calculated from these relative intensities by measuring Xe ionization and fitting the experimental results using the atomic tunneling model. As we want to compare the iodine results with a similar molecular tunneling model, this selfconsistent procedure seems most appropriate. The background pressure is 10^{-8} mbar. Iodine is leaked into the chamber up to a maximum pressure of 4×10^{-6} mbar. The pressure is typically below 10^{-6} mbar to avoid space-charge effects.

To perform the double-pulse experiments, the laser pulses are split in a Michelson interferometer. In one arm a λ/4 wave plate rotates the linear polarization by 90°. It is crucial that both pulses overlap inside the vacuum chamber. Therefore we use an undersized pinhole in a vacuum spatial filter. Due to the two Brewster windows on the spatial filter, the pulses with polarization parallel to the time-of-flight axis are weaker by a factor of 2 than the perpendicular pulses. Both pulses are focused inside the vacuum chamber.

The ions are measured in a time-of-flight mass spectrometer (Figure 4.1). In order to have the exploding ions situated symmetrically around the zero kinetic energy time, the accelerating and field-free regions of the TOF spectrometer have approximately equal distance of 30 and 32 mm, respectively. The focal spot is in the center of the accelerating plates. The ions pass through a 2 mm and 4 mm aperture at the output of the accelerating and field-free regions. The apertures define a solid angle for detection which varies for different fragmentation channels [17]. At the end of the time-of-flight spectrometer, the ions are detected by a microchannel plate detector.

The time-of-flight spectra are recorded using a digital oscilloscope with a sampling rate of 400 MS/s. Ion yields are measured with boxcar integrators. The results shown here are corrected for the different size of the integration window. No attempt has been made to determine the sensitivity of the microchannel plate as a function of the energy (and charge) of the detected ions. Although the microchannel plate is operated at high gains near saturation, the detection efficiency may vary. Therefore direct comparison of the yields for different ions should take the uncertainty of the detection efficiency and the acceptance angle into account. All results shown here are an average over 1000 shots.

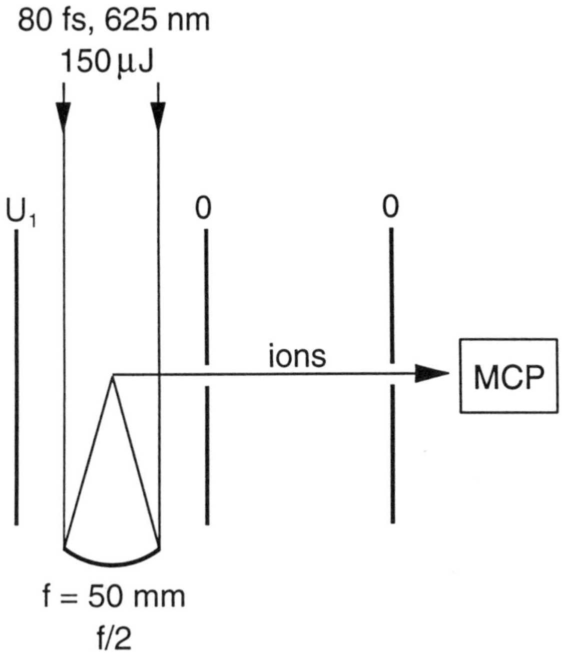

Figure 4.1 Scheme of the time-of-flight mass spectrometer. The extraction plates are separated by 30 mm and kept on a voltage of 500–1000 V. The field-free drift length is 32 mm.

4.5 RESULTS AND DISCUSSION

In this section we discuss the experimental results. First, we show that the structure in the observed time-of-flight spectra is a clear signature of the molecular ion state which fragments to different product ions. This information is used to determine the ion yields as a function of the laser intensity for a number of different channels. Finally, we discuss the observed angular distributions of the ion fragments.

A. Time-Of-Flight Spectra

We now discuss the time-of-flight spectra shown in Figures 4.2 and 4.3. The spectrum has been measured at high peak intensities ($I \approx 10^{15}$ W/cm^2) with linearly polarized light whose electric field vector points along the time-of-flight axis (i.e., parallel polarization). The results with perpendicular polarization will be discussed in Section 4.5.C.

We observe a large number of different ion signals (Figure 4.2). Except for water signals and some very small signals from nitrogen, all signals result from molecular or atomic iodine. I_2^+, which has a stable ground state, is detected, whereas the metastable I_2^{2+} is not observed; however the latter could be hidden in the strong I^+ peaks. Atomic iodine ions are observed up to I^{4+}. The structure of the I^{n+} peaks is reproducible. To understand the significance of the structure, we first consider the geometry of the time-of-flight spectrometer. Then we discuss the fragmentation process.

Apertures in the time-of-flight mass spectrometer along the drift path select ions with small transverse velocity. The limit on the transverse velocity depends on the charge over mass ratio, the extraction voltage, and the geometry of the time-of-flight mass spectrometer. This effect yields pairs of peaks which, for the time-of-flight mass spectrometer used here, are symmetric around the position of the ion signal with zero initial kinetic energy. The peaks have slightly different strength because the transverse velocity limit for the forward peak (i.e., ions flying directly onto the detector and arriving earlier than ions initially at rest) is higher than for the backward peak (i.e., ions flying initially away from the detector and arriving later than ions initially at rest). The forward peak is consequently always larger then the backward peak as can be seen in Figures 4.2 and 4.3.

The structure is therefore a signature of the molecular ion state which leads to the observed product ion. Calculating the kinetic energy for the ion

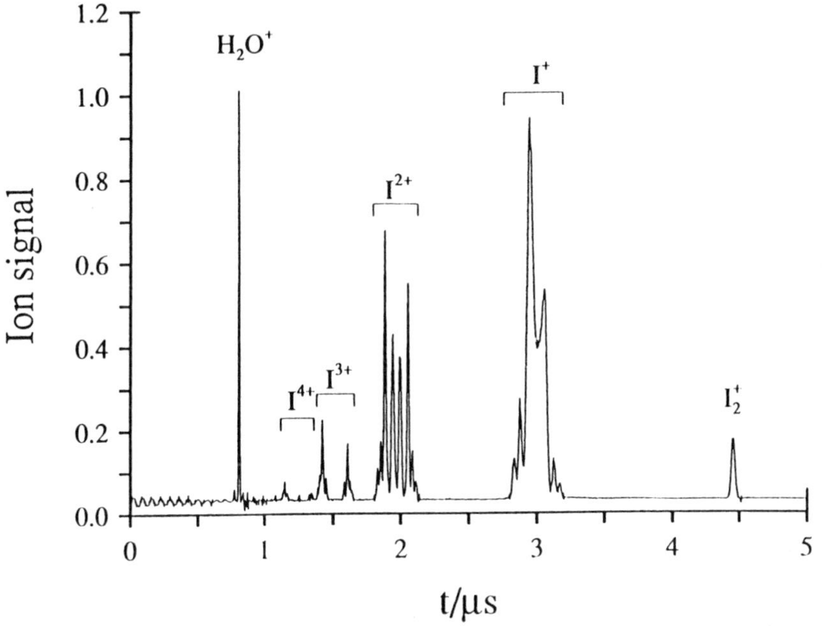

Figure 4.2 Time-of-flight spectrum for dissociative ionization of iodine. The intensity is approximately 10^{15} W/cm^2. The laser pulse is linearly polarized with the electric field vector pointing along the time-of-flight axis (parallel polarization).

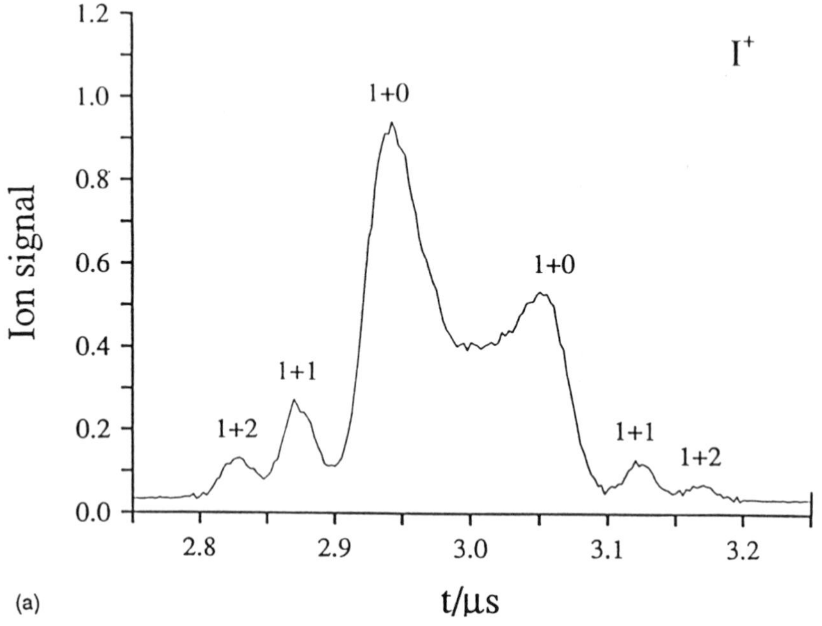

Figure 4.3 Expanded details from Figure 4.2. Shown are the parts of the time-of-flight spectrum corresponding to a) I^+, b) I^{2+}, and c) I^{3+}. The assignment of the peaks is based on the comparison of the observed kinetic energy with the Coulomb energy of two ions and is explained in the text.

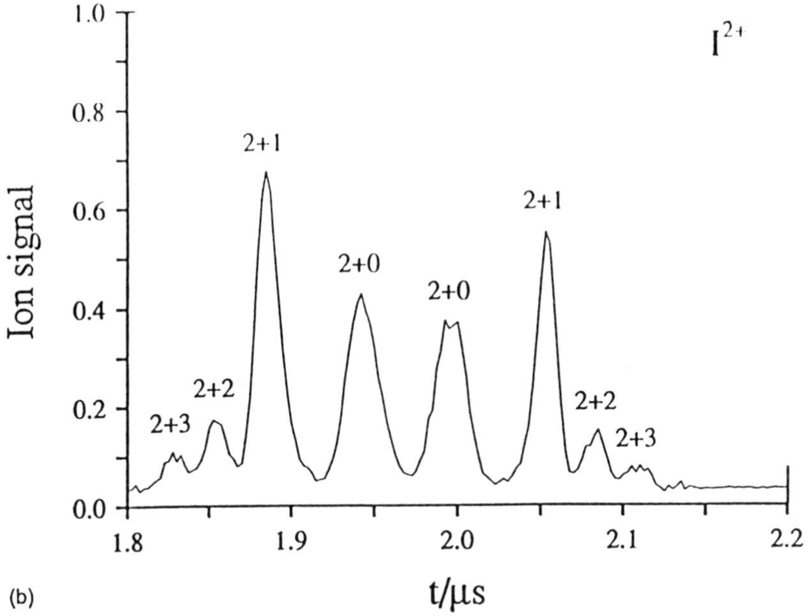
1.0
0.8
0.6
0.4
0.2
0.0
Ion signal
I^{2+}
2+1
2+0
2+0
2+1
2+2
2+3
2+2
2+3
1.8
1.9
2.0
2.1
2.2
t/μs

(b)

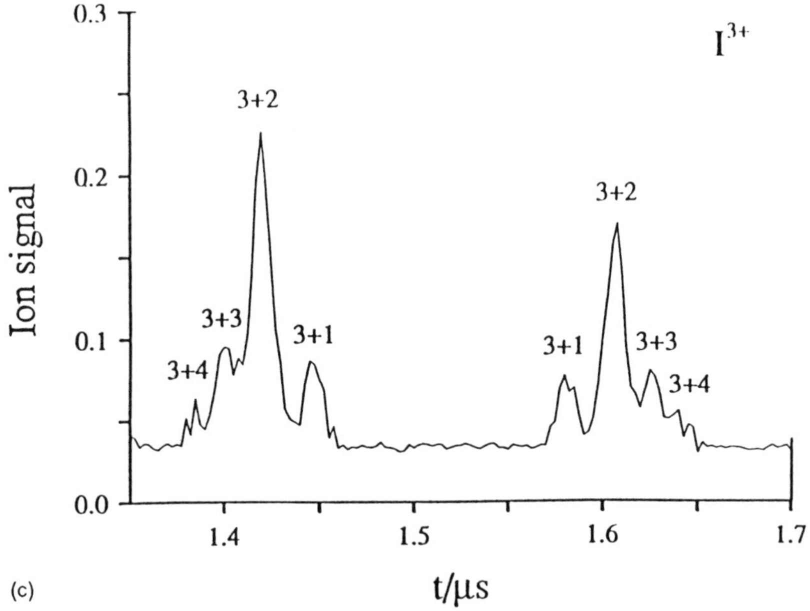
0.3
0.2
0.1
0.0
Ion signal
I^{3+}
3+2
3+3
3+4
3+1
3+1
3+2
3+3
3+4
1.4
1.5
1.6
1.7
t/μs

(c)

peaks from the measured time of flight one finds, in many cases, that the energy corresponds to the Coulomb potential between two positive ions at an internuclear separation R_e = 0.2666 nm, the equilibrium distance for neutral iodine molecules. Therefore we assign these ion peaks to molecular ion states according to:

$$I_2^{(k+l)+} \rightarrow I^{k+} + I^{l+} \tag{4.21}$$

(k+l) is used in Figure 4.3 to label the peaks. This assignment has been confirmed by covariance mapping experiments [9] where the correlation between the charged fragments are measured. We note that the observed kinetic energy is always slightly less than the Coulomb energy:

$$E_{\text{coul}} = \frac{1}{4\pi\varepsilon_0} \frac{k \cdot l \cdot e^2}{R_e} \tag{4.22}$$

which may be due to some binding contributions in the molecular potential [21] or to an increase of the internuclear distance during the laser pulse which leads to a decrease of the energy gap between two ion states [22].

Exceptions (i.e., peaks that cannot be explained by Coulomb energies) are the peaks labeled (1+0) and (2+0). They should result from dissociation of I_2^+ and I_2^{2+}. Products from these fragmentation channels should show no Coulomb energy. The observed kinetic energy is probably due to dissociation on one or a set of repulsive potential curves of I_2^+ and I_2^{2+}.

The fact that we observe ions in the (2+0) and (3+1) channels means that electronically excited molecular ion states are populated during the laser pulse either by direct ionization into these states or by coupling of these states with the charge symmetric ground state of the molecular ion. To be able to observe these channels it is absolutely necessary to have inertially confined molecules [8]. If the molecular ion fragments during the laser pulse (as is the case for N_2, even with 20 fs pulses) the neutral or weakly ionized fragment will immediately ionize because it has, in general, a lower ionization potential than the molecular ion. For example, the ionization potential for $I_2^+ \rightarrow I_2^{2+}$ is roughly 16.5 eV, as compared to 9.3 eV for $I \rightarrow I^+$. The same kind of relation holds for I_2^{4+} compared to I^+. Long-pulse experiments (defined by the fact that the molecule dissociates during the laser pulse) are therefore unable to observe asymmetric charge state even if they are produced. This issue is discussed in detailed by Dietrich et al. [8].

We conclude that we observe ion peaks resulting from dissociation of well-defined molecular ion states. We can use these peaks to measure the creation of such molecular states directly. This will be done in the next subsection.

B. Ionization Yields

Figure 4.4 shows the ionization yields for a number of different fragmentation channels as a function of the peak intensity in the laser pulse. The intensity scale has been obtained by fitting the Xe^+ yields (not shown) using the atomic tunneling model. The data in Figure 4.4 have not been corrected for the different acceptance angle for different channels or possible differences of the detection sensitivity.

Modeling the data, as has been done for the infrared multiphoton ionization of HCl [3], is difficult to do here partly because of these experimental uncertainties and partly because the exact values of the field-free ionization potentials and the spectroscopy of the molecular ions are unknown.

Even for a less ambitious problem of an approximate calculation, a serious problem arises from the observation of excited (i.e., charge asymmetric) molecular ion states. It is not clear whether these states are reached in a direct ionization process from a lower ion state (i.e., whether the ionization process leaves the molecular ion in an excited state) or whether first the ground state, which is always charge symmetric, is reached and then in a second step the excited state is coupled to the ground state by the strong external field. The discussion in Section 4.3 implicitly assumes the first possibility. However, the Rabi oscillations which couple the charge transfer states have a frequency larger than the energy separation for intensities required to produce I_2^{4+}. Without a better understanding of the coupling of electronic states in the strong field this question remains open.

We present in Figure 4.5 a comparison between experimental appearance intensities and calculated threshold intensities. (The important equations are summarized in the Appendix A.) These values are less sensitive to details of the ionization potentials. The experimental values have been taken as the

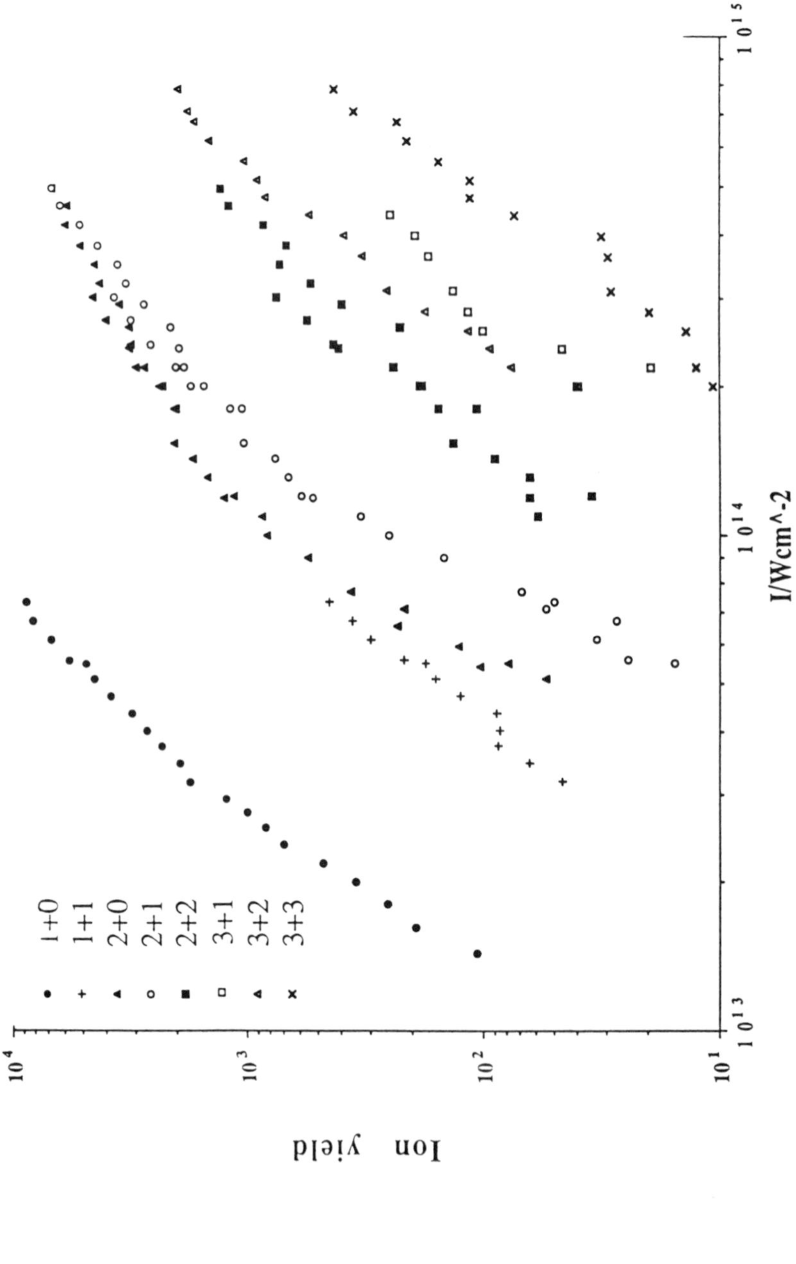

Figure 4.4 Ion yields as a function of the peak intensity in the laser pulse for dissociative multiphoton ionization of I_2. The relative intensity scale is obtained by measuring the pulse energy. Absolute values are obtained by fitting the yields for Xe ionization (not shown) using the atomic tunnel ionization model. The yields are not corrected for varying detection efficiency and acceptance angle.

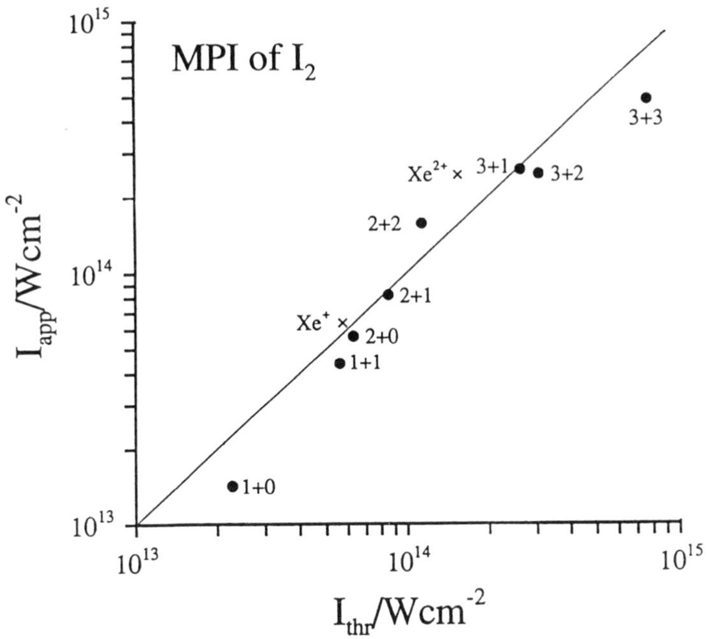

Figure 4.5 Experimental appearance intensities versus calculated threshold intensities. The appearance intensity is read from Figure 4.4 as the intensity for an ion yield S = 100. The calculated threshold intensity is the intensity for an ionization rate of 10^{12} s^{-1}.

intensity for a yield S = 100 (all curves have been determined to this low yield). Theoretical values are obtained from the molecular tunneling model which includes both Stark shifts and induced dipole moments together with the assumption of direct ionization into asymmetric states (i.e., only charge-resonance couplings are considered). We considered all possible routes to a given charge state (i.e., with either sign of the dipole moment) and determined the most likely. Shown in Figure 4.5 are the intensities which give an ionization rate of 10^{12} s^{-1} for the rate determining step. The predicted threshold intensities agree very well with the experimental values given in Figure 4.5.

C. Angular Distribution

Not only do we observe inertially confined ions and study their formation, but we can also study their dissociation dynamics. Most dramatically, we observe an anisotropic angular distribution of the ion fragments. The angular distribution is strongly peaked around the direction of the polarization of the laser field. Although this phenomenon is not new [4, 24, 25], there has been no quantitative explanation and its observation in inertially confined ions places a strong constraint on possible models to explain the observations. After presenting the experimental data, we discuss two different explanations. Experiment and theory are used to distinguish between these. Then we present model calculations which reproduce the essential features of the experimental observations. The model is based on the anisotropy of the molecular polarizibility.

i. Experimental Results

The time-of-flight spectra (shown in Figures 4.2 and 4.3) are obtained using parallel polarization (electric field vector along the time-of-flight axis). For perpendicular polarization under otherwise unchanged conditions the spectrum is very different (Figure 4.6). All ions from higher charged states disappear, only I_2^+, I^+ and I^{2+} are observed. This can be seen more quantitatively by measuring the ion yields as a function of the angle between the polarization vector and the time-of-flight axis. In these experiments, the linear laser polarization was rotated from -90° to $+90^o$ with respect to the time-of-flight axis using a $\lambda/2$ plate. A typical result is shown in Figure 4.7. The distribution of the I^{2+} ions resulting from fragmentation of $I_2^{3+} \rightarrow I^{2+} + I^+$ is clearly peaked around the direction of the laser polarization. The intensity of the laser pulses just exceeds that required to produce I_2^{3+}.

Angular distribution of the ion fragments have been measured for a number of different fragmentation channels, some of which are listed in Table 4.1. With increasing degree of ionization, the anisotropy of the angular distribution becomes more pronounced. All distributions listed in Table 4.1 have been fit using the empirical formula:

$$S(\theta) = a \cos^n \theta \tag{4.23}$$

Both a and n are used as fitting parameters. The exponent n depends on the fragmentation channel. The values range from $n = 3$ to $n = 7$ and increase with increasing degree of ionization. The values for n seem to be inde-

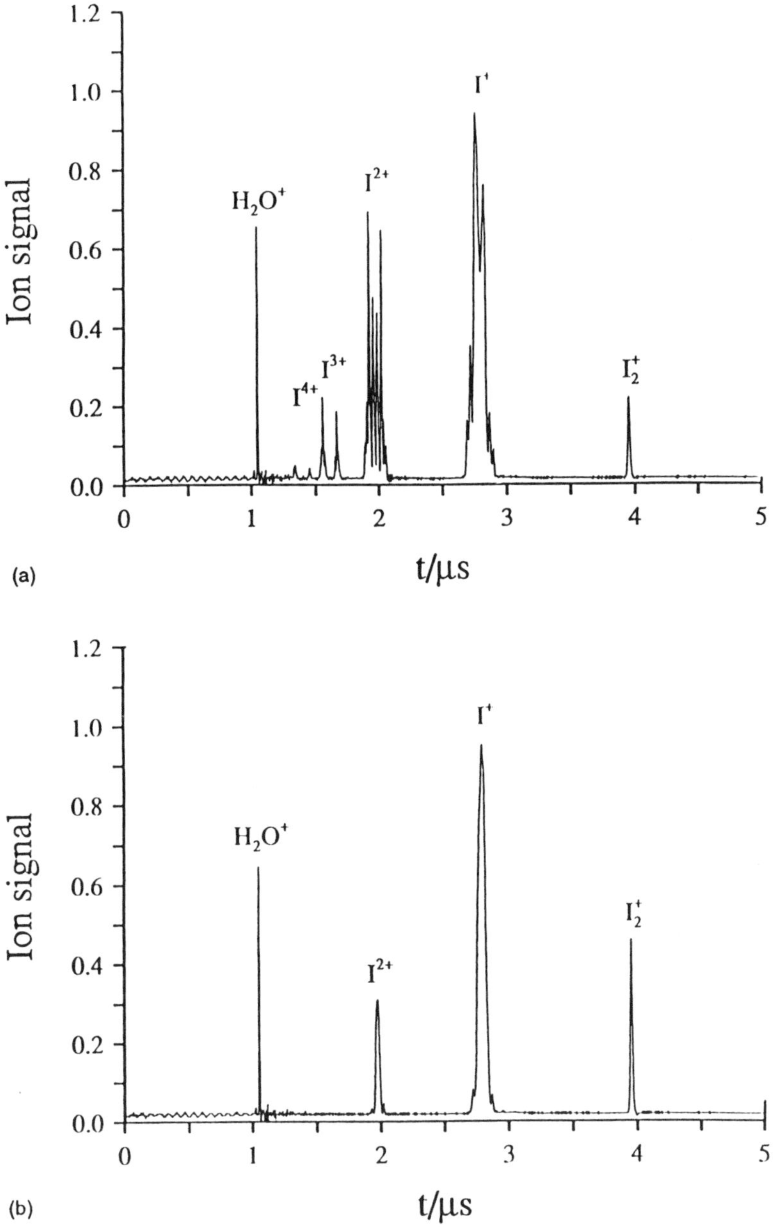

Figure 4.6 Time-of-flight spectrum for dissociative ionization of iodine. Conditions similar to Figure 4.2. a) parallel polarization, b) perpendicular polarization.

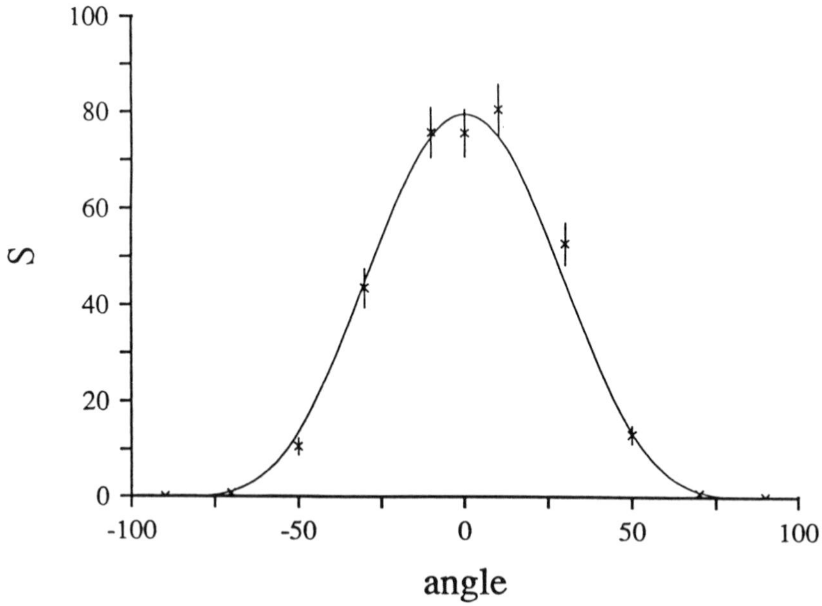

Figure 4.7 Angular distribution for dissociative multiphoton ionization of I_2 using linearly polarized laser light. Shown is the ion yield of the fragmentation channel $I_2^{3+} \rightarrow I^{2+} + I^+$ as a function of the angle between the laser polarization and the time-of-flight axis. The peak intensity is 1.8×10^{14} W/cm^2, the acceptance angle is estimated to be $\pm$ 29°. The solid line is a fit with the empirical function $S(\theta) = a \cos^n \theta$ where $n = 4$ in this case.

Table 4.1 List of Angular Distributions

Molecular ion	Ion fragments	I_o/W cm^{-2}	n	$\Delta\theta$
I_2^{2+}	$I^+ + I^+$	1.0×10^{14}	3.10 ± 0.23	28
		1.5×10^{14}	3.00 ± 0.24	28
I_2^{3+}	$I^{2+} + I^+$	1.8×10^{14}	4.28 ± 0.25	28
I_2^{4+}	$I^{2+} + I^{2+}$	3.7×10^{14}	6.11 ± 0.46	19
I_2^{4+}	$I^{3+} + I^+$	3.7×10^{14}	7.11 ± 0.36	28

Listed are the of fragmentation channels for which angular distributions have been measured and fitted using $S(\theta) = a \cos^n \theta$. The fragment which has been measured is listed first. Also given are the average peak intensity and the fitting parameter n with its standard deviation. The acceptance angle $\Delta\theta$ has been calculated for a point-like source. Only ions with an angle smaller than $\pm\, \Delta\theta$ will be detected.

pendent of the peak intensity for a given fragmentation channel. This is consistent with the fact that a certain fragmentation channel is produced at a certain intensity. Changing the peak intensity only changes the volume of the shell in which this channel is produced. We note that both Equation 4.23 and n are purely empirical and do not imply theoretical foundations. We also measured channels other than listed in Table 4.1. As expected, we observe no angular dependence for I_2^+. The distribution for the channel $I_2^{2+} \rightarrow I_{2+} + I$ shows a much less pronounced anisotropy. The ion yields drop to about 50% at 90° relative to 0°.

The measured distributions are broadened due to the geometry of the time-of-flight mass spectrometer which accepts ions with a finite transverse velocity with respect to the time-of-flight axis. The acceptance angle is determined by the extraction voltage and the size and the positions of the apertures along the drift path. For the given experimental conditions (extraction voltage 600 V, radius of the first aperture R = 1 mm), we estimate it assuming a point-like ion source. Only ions within $\pm\, \Delta\theta$ of the time-of-flight axis will be detected. The resulting values are listed in Table 4.1 and show that there is significant broadening of the observed distributions.

Angular fragment distributions similar to the one shown in Figure 4.7 have been observed previously [24] and explained by a strong angular dependence of the ionization rates. In other words, only molecules parallel with the electric field vector are ionized. Such a strong angular dependence does not agree with our analysis of molecular ionization in Section 4.3. In the next section, we discuss possible explanations of the observed fragment distributions is more detail and show conclusive experimental results.

ii. Possible Explanations

For highly charged molecular ions, there are only two possible explanations for the observed angular distributions. The first argues that molecules perpendicular to the laser field are not ionized (or at least have a much smaller probability of ionization). The second assumes that the ionization is (more or less) independent of the angle between molecule and field but that the fragments are deflected towards the direction of the electric field vector during dissociation. There are no other explanations possible because once a highly charged molecular ion is produced, it will dissociate. So either the molecular ion is not produced at 90^0 or the fragments are not observed.

Let us have closer look at the first explanation. This approach assumes that molecules perpendicular to the polarization of the laser field are not ionized [24]. That is, from the random distribution of molecular orientations in the gas phase, only molecules more or less parallel with the electric field vector are ionized. Subsequently, the fragments come apart in the initial direction of the molecule. This explanation can be ruled out for both theoretical and experimental reasons.

We have discussed in Section C that, indeed, there is a angular dependence of the ionization rates. This is a result of the angular dependence of the Stark shifts and the induced dipole moments. Furthermore, the quadrupole moments will also suppress ionization for perpendicular molecules compared to parallel molecules. However, these effects do not change the ionization potential significantly. A typical change of 15% can easily be compensated by increasing the intensity by a factor of 2. Therefore, including these shifts into the calculation of the ionization yields would shift the calculated yields to higher intensities. But many experiments are performed at intensities in the saturation regime. Even under these conditions, there are hardly any ions observed at perpendicular polarization. This is not consistent with the molecular ionization model which is otherwise quite successful.

In order to confirm this conclusion experimentally, we have performed double-pulse experiments with parallel and perpendicular pulses (Figure 4.8).

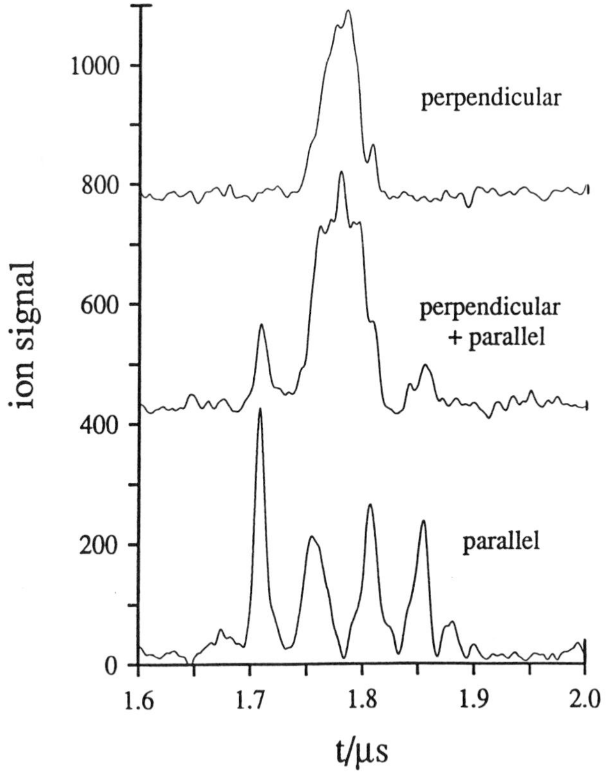

Figure 4.8 Results of the double-pulse experiment with two pulses linearly polarized perpendicular to each other. Shown is the I^{2+} region of the time-of-flight spectrum. Upper trace: perpendicular pulse only (polarization perpendicular to time-of-flight axis); middle trace: perpendicular pulse followed by a parallel pulse after 5 ps; bottom trace: parallel pulse only (polarization parallel with time-of-flight axis). The perpendicular pulse is a factor of 2 more intense than the parallel pulse.

Both pulses have a similar energy with a peak intensity of 10^{14} W/cm^2, the perpendicular being more intense by a factor of 2. Using either of the two pulses (upper and lower trace in Figure 4.8), we obtain the characteristic spectra shown in Figure 4.6. The middle trace shows the result when a perpendicular pulse precedes the parallel pulse by 5 ps. It is clearly seen that the perpendicular pulse strongly decreases the signals due to the parallel pulse. Similar results are observed for a delay of 1 ps. This proves that both polarizations interact with the same class of molecules. It is important to recall that the classical period of rotation for I_2 is 446 ps/J. Therefore J = 100 is required to allow the molecules to reorient in 5 ps; even higher values of J are required for shorter delays. Thus we conclude that a variation of the ionization rates with angle between molecule and field cannot explain the observed angular fragment distribution.

The other explanation argues that due to the anisotropy of the molecular polarizibility, the fragments are deflected during dissociation. The interaction between the external field and the induced dipole moments results in a torque towards alignment with the field. This will lead to the observed fragment distribution peaked around the direction of the field. In the next subsection we will discuss this in more detail.

iii. Classical Trajectory Calculations

We have performed classical trajectory calculations to obtain a better understanding of the influence of the molecular polarizability. The model molecule bears some similarities with iodine but leaves out many of the not well-known details of I_2^{n+}. It is important to include the polarizability of the neutral molecule because it will lead to an increase of the angular momentum before ionization. Once the molecule is ionized to a repulsive state, it immediately starts dissociating. The initial angular momentum at the start of the fragmentation is important for the calculated angular distribution. For the same reason, it is also important to use pulses with a Gaussian temporal profile instead of pulses with constant intensity. A square pulse leads to earlier ionization and less initial angular momentum than a Gaussian pulse where the molecule experiences a slowly increasing field and can increase angular momentum before being ionized. The neutral molecule is modeled by taking a ground state potential curve $X\,^1\Sigma_g^+$ and an ionic state ($B\,^1\Sigma_u^+$ from I_2) coupled by charge transfer. The transition dipole moment $\mu_{12} = \sqrt{2}\,e\,R/2$ leads to a parallel polarizibility $\alpha_{\parallel}$ = 14.7 Å^3 for low intensities in good agreement with the experimental value [26]. To take the perpendicular polarizibility into account, we reduce the transition dipole moment by a factor $\sqrt{2}$ to obtain the value γ = 7.2 Å^3 for the anisotropic polarizibility [26].

We concentrate on the dissociation dynamics of the first purely repulsive charge state I_2^{3+} decaying into $I^{2+} + I^+$. From the fragment ion states, one can construct the two states coupled by charge resonance. We assume that the potential curve of the upper state is given by a simple Coulomb potential. The potential curve of the lower state is the same Coulomb potential but superimposed is the attractive potential of the isoelectronic molecules Te_2^+ [27]. In doing this, we include approximately the attractive forces between the two ions due to the overlap of the electron shells (see [21]). The gap between the two curves at the equilibrium distance of the neutral molecule is 3.4 eV, the binding energy of Te_2^+. The charge-resonance coupling leads to a transition dipole moment $\mu = e\, R/2$. The absolute energy of the two charged states are adjusted so that the ionization potential equals 26.5 eV which is the value for the ionization of $I_2^{2+} \rightarrow I_2^{3+}$ assuming a Coulomb potential for the upper charge state. The energy of the lower state is derived from the estimated ionization potential of 16.4 eV for $I_2^+ \rightarrow I_2^{2+}$ [28].

We neglect the intermediate charge states (I_2^+ and I_2^{2+}) for several reasons. These intermediate states (especially I_2^{2+}) are not well known, but the main parameter in our approach, the molecular polarizibility, should not be very different from that of neutral iodine. The most important contribution in all charge species results from charge-transfer or charge-resonance couplings with very similar transition moments. Furthermore, the dynamics of the molecule should be similar as long as the species has a bound state (I_2^{3+} is the first purely repulsive state). Thus, we treat a problem that should resemble the real iodine molecule.

The classical dynamics on one potential energy curve has been calculated by direct integration of the equation of motions (see Appendix B). The ionization rates are calculated from the tunneling model (see Appendix A). To simulate the random nature of the ionization, several trajectories with the same initial conditions are calculated where the exact time of ionization depends on the ionization rate at the momentary intensity and a random number. The initial conditions used in the results presented here are those of a molecule at rest in the equilibrium configuration.

Figure 4.9 shows the normalized angular distribution for a peak intensity $I_0 = 3 \times 10^{14}$ W/cm^2. This intensity gives population of I_2^{3+} without significant further ionization. The spatial intensity distribution has not been taken into account. The distribution is clearly peaked around 0° which is the direction of the electric field vector. The width is comparable to the measured ones. We do not expect perfect agreement as our model neglects a number of details of the real I_2 problem; nevertheless, the agreement is impressive.

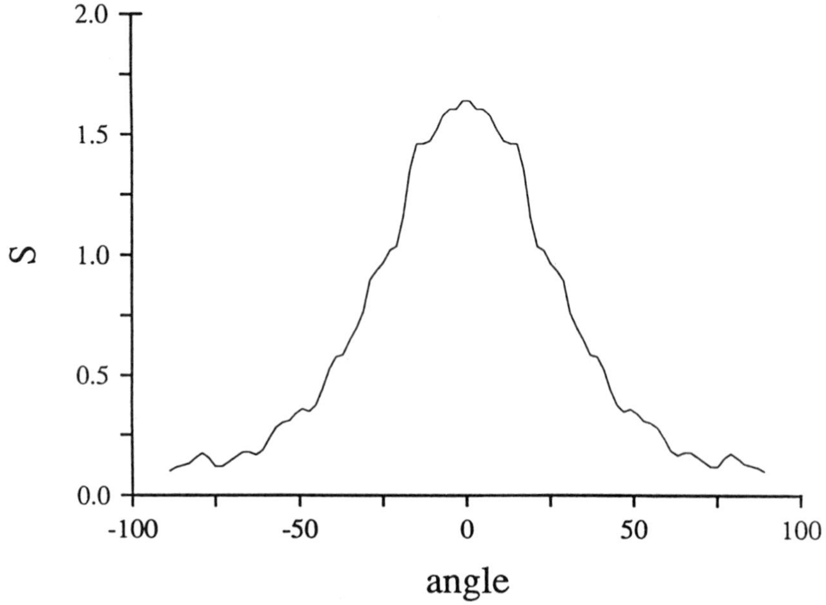

Figure 4.9 Calculated angular distribution for dissociative multiphoton ionization of I_2. The calculation uses a simplified model for I_2. A pulse with Gaussian temporal profile with a full width at half maximum of 70 fs and a peak intensity of 3×10^{14} W/cm^2 is used. The spatial distribution has not been taken into account.

It clearly demonstrates that deflection of the ion fragments due to an increase of angular momentum can explain the observed angular distributions. With higher intensity, the torque on the molecule becomes larger so that higher charged molecules should show a more pronounced deflection, as indeed is observed.

An initial angular momentum of the molecule has little effect on the calculated angular distribution. Calculations with $J = \pm 50$, the maximum angular momentum at room temperature, yield identical distributions that are shifted by $\pm 5^{\circ}$. Superimposing them leads to the same distribution as $J = 0$.

Our model can be improved in a number of ways. First, this problem should be addressed rigorously using a quantum mechanical approach. We have used classical trajectory calculations because there seems to be no simple theoretical treatment of a rotating molecule in a strong time-dependent field available [29]. Second, the energy gap between the charge-resonance states (the states most strongly coupled) is on the order of the photon energy. Therefore single and multiphoton resonances can occur and will clearly play a role in the dynamics of the molecular ion (though not so much in the neutral molecule). Instead of the static polarizability, one should use the dynamic one which again is not known in the high-intensity limit. However, the error induced by using the static instead of the dynamic polarizability is unlikely to change the angular distribution qualitatively, because the angular distribution is more sensitive to the angular momentum absorbed at small internuclear separation where the laser frequency is well below the resonance frequency. It will, however, influence the kinetic energy of the ion fragments, which will increase as long as the laser pulse is on. Third, we have not included an angular dependence of the ionization rates which is straightforward to do. This might enhance the directionality because ionization will be more likely for parallel than for perpendicular molecules. But because this effect alone cannot explain the observations, as has been discussed previously, we have concentrated on the influence of the polarizability. Note also that the present calculation does not include the spatial integration over the focal volume.

We thus think that the rather simple model shown here gives the basic explanation of the observed anisotropic angular distributions. Classically speaking, the polarizable molecule is rotated in the presence of the strong field which deflects the fragments into the direction of the electric field.

4.6 CONCLUSIONS

By monitoring the kinetic energy of the dissociation fragments we have shown that it is possible to optically produce inertially confined ions. Although inertially confined ions have been produced in beam-foil experiments, optical production offers many advantages. We should be able to produce inertially confined molecular ions in large quantities and in high densities. We also should be able to study them using the techniques of ultrafast spectroscopy. This would mean that the potential energy curves of a whole new class of molecules can be studied for the first time.

We also introduced experimental techniques for studying ionization of highly charged molecular ions. For measuring ionization rates we use an approach very similar to that used in atomic experiments. In atomic experiments we measure the production of a given charge state by observing the ion. For molecular multiphoton ionization experiments we observe the fragment ions with the appropriate Coulomb energy as kinetic energy.

We have also introduced the first quantitative model of molecular ionization. We compare the experimental ionization yields with calculated results. Our approach is to modify well-developed atomic models of multiphoton ionization by including the most important effects of the interaction of the molecule with the ionizing field. At the present stage, the theory predicts appearance intensities of molecular ions as accurately as the atomic theory (on which it is based) describes atomic appearance intensities.

Our analysis has been extended to describe the angular dependence of the fragments produced by dissociative multiphoton ionization. Experiment and theory are in good agreement; both indicate that high rotational states are excited by the presence of the strong laser field.

We have based all of our calculations on the assumption that only strongly coupled states need to be considered to describe the response of a molecule to an intense short-pulse laser field. We have identified a general class of states that must be considered. These states are those coupled by charge-resonance or charge-transfer transitions. This electronic coupling determines the details of the multiphoton ionization of the molecule. In an earlier paper [3] we have shown that, irrespective of resonances, it can also dominate the vibrational excitation of molecular ions. In an almost direct analogy, we show in this chapter that it dominates the rotational coupling as well.

The resultant simplicity of this first-order approach should help us to identify new phenomena in strong field laser-molecule interactions. Our results foreshadow one new area of investigation. In Section 4.2, we showed the field dependence of the induced dipole moment. Inertially confined molecular ions promise to be a new and very nonlinear medium of considerable interest for nonlinear optics [11, 30].

ACKNOWLEDGEMENTS

M. Laberge acknowledges support from the Centers of Excellence for Molecular and Interfacial Dynamics.

APPENDIX A

We briefly summarize the equations used to calculate the ionization rates. The basic formula is the tunneling rate of the electron in a static electric field [18, 19]:

$$W_{DC} = \omega_s \, |C_{n^*l^*}|^2 \, G_{lm} \left(\frac{4\omega_s}{\omega_t} \right)^{2n^*-m-1} \exp\left(\frac{-4\omega_s}{3\omega_t} \right) \tag{4.A1}$$

with:

$$\omega_s = \frac{E_s^o}{\hbar}$$

$$\omega_t = e\xi \, (2m_e E_s^o)^{-1/2}$$

$$n^* = Z \left(\frac{E_s^h}{E_s^o} \right)^{1/2}$$

$$G_{lm} = \frac{(2l+1)\,(l+|m|)!}{|m|!(e-|m|)!} \, 2^{-|m|}$$

$$|C_{n^*l^*}|^2 = 2^{2n^*} \, [n^* \Gamma(n^* + l^* + 1) \Gamma(n^* - l^*)]^{-1}$$

Here E_s^o is the ionization potential of the species of interest and E_s^h is the ionization potential of atomic hydrogen. The initial atom is characterized by azimuthal and magnetic quantum numbers l and m, and an effective principal quantum number n^*. The residual ion has the charge Z. ξ is the electric field. $C_{n^*l^*}$ is determined by the asymptotic expansion of the wave function. The effective quantum number l^* is given by $l^* = 0$ for $l << n$ and $l^* = n^* - 1$ otherwise [19]. Using the quasistatic assumption, the rate can easily be averaged over one optical cycle for the case of an alternating field. One obtains the following approximate formula [18]:

$$W_{AC} = \left(\frac{3\omega_t}{2\pi\omega_s} \right)^{1/2} W_{DC} \tag{4.A2}$$

We used $l = m = 0$ in all our calculations. The intensity-dependent ionization potential is used in Equation 4.A1 but not in the prefactor in Equation 4.A2 because this gives the best approximation to the exact average over one optical cycle as has been tested by direct integration of Equation 4.A1 over one period.

APPENDIX B

We briefly summarize the important equations used in the classical trajectory calculations in Section 4.5.C. We start with the classical Hamilton function. For simplicity, we assume that the direction of the electric field is in the plane of rotation. Then the Hamilton function for the homonuclear molecule in the center-of-mass frame is given by:

$$\mathrm{H}(r,p_r,\theta,L,t) = \frac{p_r^2}{2\mu} + \frac{L^2}{2I} + V(r) + V_{int} \tag{4.B1}$$

The first term on the right hand side is the radial kinetic energy with the reduced mass μ and the radial momentum p_r. The second term is the rotational energy with the angular momentum L and the moment of inertia $I = \mu \cdot r^2$ where r is the internuclear separation. $V(r)$ is the field-free potential of the states where the dynamic is calculated. The last term results from the coupling of the two states and includes the effect of the external field on the molecule. The classical energy of a polarizable system in an electric field is $V_{\mathrm{int}} = -\gamma \mathrm{E}^2 \cos^2 \theta/2$ where γ is the anisotropic polarizibility and θ the angle between the molecule and the electric field. This form has a serious fault of having the wrong high-intensity behavior. For very high fields the induced dipole moments saturate. Therefore we use the quantum mechanical expression where the polarizability results from coupling of electronic states. Coupling two states, one has [3]:

$$V_{int} = -V_{12} \tan \beta \tag{4.B2}$$

where $V_{12} = -\mu_{12}\, \xi(t) \cos\theta$ with the transition moment μ_{12}, the time-dependent electric field $\xi(\mathrm{t})$. β is given by $\tan 2\beta = 2V_{12}/\Delta E$ where ΔE is the energy gap between the two coupled states. Applying the Hamilton equations to Equation 4.B1, we obtain the equations of motions:

$$\dot{r} = \frac{\partial H}{\partial p} = \frac{p}{\mu} \tag{4.B3}$$

$$\dot{p} = \frac{-\partial H}{\partial r} = \frac{V_{12} \sin 2\beta \, \mu_{12}'}{\mu_{12}} + \frac{L^2}{rI} - V'(r) - \Delta E' \sin^2 \beta \tag{4.B4}$$

$$\dot{\theta} = \frac{\partial H}{\partial L} = \frac{L}{I} \tag{4.B5}$$

$$\dot{L} = \frac{-\partial H}{\partial \theta} = -V_{12} \tan \theta \sin 2\beta \tag{4.B6}$$

where the dot indicates the time derivative and the prime the spatial derivative. The integration was done using the Bulirsch-Stoer method [31]. Most of the calculations are done by averaging the right-hand side of Equations 4.B3–4.B6 over one optical cycle. This leads to elliptic integrals and is justified because the molecule does not move on the time scale of the laser period (2 fs).

REFERENCES

1. C. Rolland and P. B. Corkum, *J. Opt. Soc. Amer.* **B5**, 641 (1988).
2. R. S. Mulliken, *J. Chem. Phys.* **7**, 20 (1939).
3. P. Dietrich and P. B. Corkum, *J. Chem. Phys.* **97**, 3187 (1992).
4. D. T. Strickland, Y. Beaudoin, P. Dietrich and P. B. Corkum, *Phys. Rev. Lett.* **68**, 2755 (1992).
5. A. Zavriyev, P. H. Bucksbaum, H. G. Muller and D. W. Schumacher, *Phys. Rev. A*, **42**, 5500 (1990).
6. A. D. Bandrauk and M. L. Sink, *J. Chem. Phys.* **74**, 1110 (1981).
7. P. Dietrich, D. T. Strickland, M. Laberge and P. B. Corkum, *Phys. Rev.*, **A47**, 2305 (1993)
8. P. Dietrich, D. T. Strickland and P. B. Corkum, *J. Phys.* **B26** *(in press)*.
9. M. Laberge, P. Dietrich and P. B. Corkum, to be submitted to *Phys. Rev. A*.
10. P.B. Corkum, N. H. Burnett and F. Brunel, *Phys. Rev. Lett.* **62**, 1259 (1989).
11. M. Ivanov and P. B. Corkum, *Phys. Rev.*, **A48**, 580 (1993)
12. A. Giusti-Suzor, X. He, O. Atabek and F. H. Mies, *Phys. Rev. Lett.* **64**, 515 (1990).
13. T. T. Nguyen-Dang and S. Manoli, *Phys. Rev. A* **44**, 5841 (1991).
14. P. B. Corkum and P. Dietrich, *Comments At. Mol. Phys.* **28**, 357 (1993).

15. S. Augst, D. Strickland, D. D. Meyerhofer, S. L. Chin and J.H. Eberly, *Phys. Rev. Lett.* **63**, 2212 (1989).
16. S. Augst, D. D. Meyerhofer, D. Strickland, and S. L. Chin, *J. Opt. Soc. Amer.* **B8**, 858 (1991).
17. J. R. Hiskes, *Phys. Rev.* **122**, 1207 (1961).
18. A. M. Perelomov, V. S. Popov, and M. V. Terent'ev, *Sov. Phys. JETP* **23**, 924 (1966).
19. M. V. Ammosov, N. B. Delone, and V. P. Kravinov, *Sov. Phys. JETP* **64**, 1191 (1986).
20. C. Rolland and P. B. Corkum, *Opt. Comm.* **59**, 64 (1986).
21. J. Senekowitsch and S. O'Neil, *J. Chem. Phys.* **95**, 1847 (1991).
22. K. Codling, L. J. Frasinski and P. A. Hatherly, *J. Phys. B* **22**, L321 (1989).
23. J. Li and K. Balasubramanian, *J. Mol. Spectrosc.* **138**, 162 (1989).
24. L. J. Frasinski, K. Codling, P. Hatherly, J. Barr, I. N. Ross and W. T. Toner, *Phys. Rev. Lett.* **58**, 2424 (1987).
25. P. A. Hatherly, L. J. Frasinski, K. Codling, A. J. Langley and W. Shaikh, *J. Phys. B: At. Mol. Opt. Phys.* **23**, L291 (1990).
26. D. W. Callahan, A. Yokozeki and J. S. Muenter, *J. Chem. Phys.* **72**, 4791 (1980).
27. K. P. Huber and G. Herzberg, *Molecular Spectra and Molecular Structure. IV. Constants of Diatomic Molecules*, Van Nostrand Reinhold Co., New York, 1979.
28. A. C. Hurley and V. W. Maslen, *J. Chem. Phys.* **34**, 1919 (1961).
29. A. D. Bandrauk and J. McCann, *Comments At. Mol. Phys.*, **22**, 325 (1989).
30. M. Ivanov, P. B. Corkum and P. Dietrich, *Laser Physics*, **3**, 375 (1993)
31. W. H. Press, B. P. Flannery, S. A. Teukolsky and W. T. Vetterling, *Numerical Recipes*, Cambridge University Press, Cambridge, 1986.

5

Characteristics and Dynamics of Doubly Excited States of Molecules

Sungyul Lee

Kyunghee University, Yongin-Kun, Kyungki-Do, Korea

Masahiro Iwai* and Hiroki Nakamura

Institute for Molecular Science, Myodaiji, Okazaki, Japan

5.1 INTRODUCTION

The physics and chemistry of electronically excited molecules have been attracting much attention and their importance in various dynamic processes is now well recognized [1–5]. Among those states, the so-called superexcited states (SES) have been paid a particular attention because of their peculiar properties and decaying mechanisms which are different from the ordinary excited states. A variety of dynamic processes can occur in this high energy region and thus the dynamics of SES are expected to open a new field of science. Thanks to the rapid progress of laser technology, the SES are becoming rather easily accessible experimentally and basic theoretical information on them is strongly desired. Close cooperation among quantum

---

* *Deceased*

chemistry, dynamic study, and experiment would thus be very important to reveal this world of science. Out of the two kinds of SES [doubly or inner-shell-excited states (the first kind of SES) and rovibrationally excited Rydberg states (the second kind of SES), here we focus our attention mainly on the doubly excited SES. As mentioned above, quantitative information on these states is required in order to understand the dynamic processes involving these states in a strong laser field. The results of the quantum chemical calculations based on the Feshbach projection operator technique are reported here for the potential (resonance) energies $E_r(R)$ and the autoionization widths $\Gamma(R)$ of the doubly excited states of H_2 corresponding to the $n = 2$ intrashell and ($n = 2$, $n' = 3$)-intershell doubly excited states of He in the united atom limit. The data are provided from small to intermediate internuclear distances ($0.2a_0 \leq R \leq 2.0a_0$).

In addition to this quantitative information, another important subject is to acquire a qualitative understanding of electron correlation or collective electron motion in these states. In the atomic doubly excited states electron correlation is well studied by the various methods such as group theoretical method and the hyperspherical coordinate approach [6–8]. The rovibrator model holds well and the supermultiplets are well classified by new quantum numbers. Furthermore, it was found that the rovibrator model can actually be given a quantum mechanical foundation [9].

Next, natural questions are how the electron correlation patterns observed in the atomic doubly excited states survive or disappear as the two nuclei split apart from the united atom limit and how the autoionization mechanisms change accordingly. Based on the conditional electron density maps and the discrete-to-continuum transition density maps, these problems are analyzed for the aforementioned doubly excited states of H_2 [10]. Some new interesting features of electron correlation are found for the molecular states such as a drastic change in correlation pattern by avoided curve crossing and an effect of axial symmetry.

This chapter is organized as follows: a variety of studies on the atomic doubly excited states are briefly summarized in Section 5.2. In Section 5.3 the quantitative information on the potential energies $E_r(R)$ and the autoionization widths $\Gamma(R)$ are presented for the doubly excited states of H_2. Electron correlation and autoionization mechanisms in these molecular doubly excited states are analyzed in Section 5.4 in comparison with those of the corresponding atomic states. Section 5.5 briefly reviews the dynamic processes involving SES studied so far. The basic and powerful theoretical tool, that is, the multichannel quantum defect theory (MQDT) is also outlined there.

5.2 ELECTRON CORRELATION IN ATOMIC DOUBLY EXCITED STATES

The independent particle model has long been used to describe the motion of electrons in many-electron atoms. According to the model, electrons in atoms move independently of each other and the motion of each electron is determined by an average potential due to the nucleus and the other electrons. Deviations from this approximation has been successfully treated by many different methods, such as many-body perturbation theory or the configuration-interaction method, for atomic ground states or low-lying excited states.

While low-lying states are amenable to the perturbative treatments based on the independent particle approximation, extensive mixing between degenerate and near degenerate configurations occurs in doubly excited states, rendering the concept of dominant single configuration meaningless. Hence a new way of classification scheme or new quantum numbers are needed to describe these atomic doubly excited states.

Since the observation of some peculiar features in the absorption spectra of doubly excited states of He by Madden and Codling [11, 12], Fano et al. [13] have noticed that completely new concepts must be employed to describe these highly excited states. The observation of only a single prominent Rydberg series converging to the $n = 2$ limit of He in the photon energy of 60–65 eV has led Fano and coworkers to suggest that the extensive mixing of ${}^1P^0$ Rydberg series results in a collective motion of the electrons. Thus, the concept of collective motion of particles, so familiar in rotations and vibrations of nuclei in molecules, has found its analog in atomic physics.

Electron correlation has profound effects on the dynamic processes such as photoexcitation and inelastic collision of a low-energy electron with an atom or ion. The $nsn'p$ or $ns\varepsilon p$ excited state of the alkaline earth atoms are extensively mixed with such doubly excited states as npn's, npεs, npn'd, npεd, $(n-1)$dn'p and $(n-1)$dεp configurations, due to the very strong electron correlation. Thus, the photoexcitation spectra of these atoms show strong perturbations and broad resonance features, especially above the resonance ionization threshold [14]. Electron correlation also affects the inelastic scattering cross-sections between electron and atoms. The excitation cross-section of open-shell atoms by slow electrons is usually much larger than that of closed-shell atoms, since the incident electron can be incorporated easily into the open shell, resulting in large electron correlation. Electrons incident on the closed shell atoms are, on the other hand, excluded from the target by the Pauli exclusion principle. These qualitative difference clearly indicates the importance of electron correlations in electron-atom scattering processes. Hence, it is expected that electron correlation may have significant influence in many dynamic processes of doubly excited states of atoms as well as molecules. Our review of this important topic will not be exhaustive, of course, but we will concentrate on some important findings that are relevant to our discussions on molecular doubly excited states in later Sections.

A. Group Theoretical Method

Classification of the doubly excited states in terms of electron configuration (or orbital) sometime dose not result in clear picture of the atomic states. For $N = 3$ intrashell doubly excited states (N is the principal quantum number), for example, there are six configurations, corresponding to $3s$, $3p$, $3d$ orbitals. Each configuration ($\ell\ell'$) leads to a multiplet levels $^{2S+1}L^{\pi}$. Figure 5.1 shows these multiplets. Because of extensive configuration mixing, these multiplets cannot be well described in terms of dominant configuration. The configuration mixing also perturbs the ordering of levels which may be expected from the Hund's rules. For example, one finds the ordering $^1D^e < {}^3P^e < {}^1S^e$ in Figure 5.1, in contrast to the prediction, $^3P^e < {}^1D^e < {}^1S^e$ by Hund's rule.

Herrick and Kellman [6, 15–18] applied O(4) group to classify doubly excited states of He atom. O(4) representations $[P, T]$ are constructed in two-electron atoms, labeling multiplets of O(3) terms with angular momentum $L = T, T + 1, ..., P$. P and T are two O(4) quantum numbers, which replace the two quantum numbers ℓ and ℓ' in electron configuration picture. Possible values of P and T are $P = T, T + 2, 2N - 2 - T$, and $T = 0, 1, ..., N - 1$. O(4) generators are $\vec{L} = \vec{\ell}_1 + \vec{\ell}_2$ and $\vec{B} = \vec{b}_1 - \vec{b}_2$, where $\vec{b}$ designates the energy-weighted Lenz vector. Configuration mixing is interpreted in terms of this O(4) symmetry group.

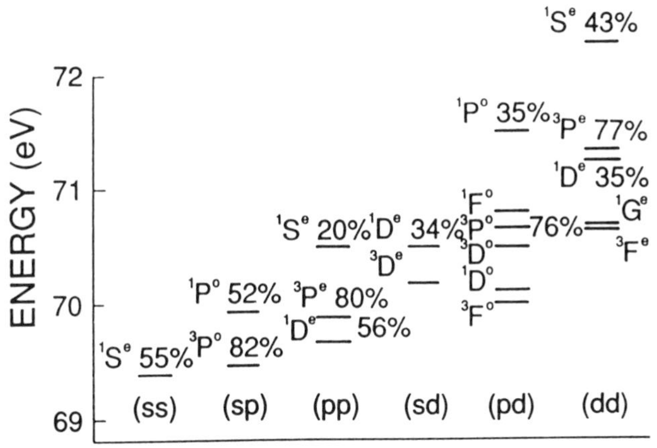

Figure 5.1 Configurational classification of helium-atom doubly excited states with two electrons in the same shell $n = 3$. Percentages indicate contribution of configuration to the wave function. Energies are in electron volts above the ground state. *Source:* from [15].

Figure 5.2 presents the O(4) classification of $N = 3$ intrashell doubly excited states of He. It can be recognized that this classification scheme provides a more ordered picture of the atomic states. Energies increase with L in each multiplet. The parity π of each level satisfies $\pi(-1)^L = +1$ when $T = 0$. For $T \neq 0$, there are two nearly degenerate levels for each value of L. The level with $\pi(-1)^L = +1$ lies slightly higher in energy than the level with $\pi(-1)^L = -1$ (T doubling). Also in Figure 5.2 the rigid rotor-like structure is clearly seen for [4, 0] multiplet. Thus, this classification scheme organizes the doubly excited states in a clearer form. Herrick and Kellman noticed *level clustering* in their classification of $N = 3$ intrashell doubly excited states. For example, the T-doublet pair $^1D^e$ and $^3D^e$ belonging to [2, 2] multiplet, together with the $^1S^e$ state from the [2, 0] multiplet are nearly degenerate. This can be explained in channel description of doubly excited states using hyperspherical coordinates, as will be reviewed below. When the electron correlation decreases with the nuclear charge Z, it is possible that the levels in each multiplet will show nonrotor character in their energies. In fact it was found that [4, 0] multiplet spectrum in Figure 5.2 becomes a less rotorlike structure with increasing Z. Herrick and Kellman constructed two types of supermultiplets, the d supermultiplets and I supermultiplets, that took these

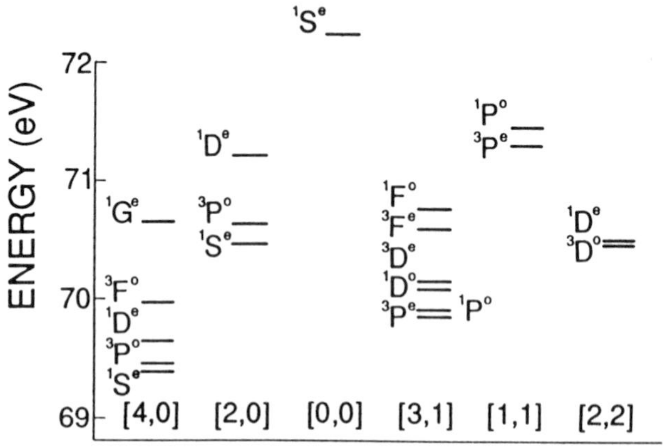

Figure 5.2 O(4) classification of helium atom doubly excited states with two electrons in the same shell $n = 3$. Each O(4) multiplet $[P, T]$ contains levels with $T \leq L \leq P$. *Source:* from [15].

nonrotational parts of the correlation energy into consideration. We will not elaborate on this further here.

B. Hyperspherical Coordinate Approach

Hyperspherical coordinates [7, 8, 19–21] proved very useful in classifying the atomic doubly excited states, yielding approximate photoabsorption selection rules and predicting relative magnitudes of the autoionization widths. The hyperspherical coordinates (R, α) are defined as

$$R = (r_1^2 + r_2^2)^{1/2} \text{ and } \alpha = \tan^{-1} \frac{r_2}{r_1} \tag{5.1}$$

where r_1 and r_2 are distances of the electrons from the nucleus. The coordinate R specifies the *size* of the atom, while two angles α and θ_{12} describe electron correlation. The correlation depicted by the angle α is referred to as radical correlation, while the interelectronic angle describes angular correlation. The Schrödinger equation for two-electron atoms are written in hyperspherical coordinates as

$$\left(-\frac{d^2}{dR^2}+\frac{\Lambda^2+15/4}{R^2}+\frac{2C}{R}-2E\right)\left(R^{5/2}\Phi\right)=0 \tag{5.2}$$

where

$$\Lambda^2=-\frac{1}{\sin^2\alpha\cos^2\alpha}\frac{d}{d\alpha}\left(\sin^2\alpha\cos^2\alpha\frac{d}{d\alpha}\right)+\frac{l_1^2}{\cos^2\alpha}+\frac{l_2^2}{\sin^2\alpha}$$

is the square of the grand angular momentum operator and

$$C=-\frac{Z}{\cos\alpha}-\frac{Z}{\sin\alpha}+\frac{1}{(1-\sin 2\alpha\cos\theta_{12})^{1/2}} \tag{5.3}$$

is the effective charge and Z is the charge of the nucleus. Macek [20] assumed that the R coordinate is approximately separable from the angle variables and constructed the approximate adiabatic eigenfunctions $\Phi_\mu(R;\alpha,\theta_{12})$ and the potentials $U_\mu(R)$ at each R. The Schrödinger equation is solved by expanding the total wavefunction as

$$\Psi=\sum_\mu\frac{\Phi_u(R;\alpha,\theta_{12})F_n^\mu(R)}{R^{5/2}\sin\alpha\cos\alpha} \tag{5.4}$$

when μ identifies the channel and n denotes the nth state within that channel (n corresponds to the principal quantum number of the outer electron). By dropping all the coupling terms and keeping only diagonal terms, $F_n^\mu(R)$ is obtained by solving

$$\left(\frac{d^2}{dR^2}+\frac{1}{4R^2}+U_\mu(R)+W_{\mu\mu}(R)+2E_n^\mu\right)F_n^\mu(R)=0 \tag{5.5}$$

where

$$W_{\mu\mu}(R)=2<\Phi_\mu\left|\frac{d}{dR}\right|\Phi_\mu>\frac{d}{dR}+<\Phi_\mu\left|\frac{d^2}{dR^2}\right|\Phi_\mu> \tag{5.6}$$

Macek [20] found that each sequence of eigenvalues E_n^{μ} with the same (μ,s,L,parity), obtained by employing the adiabatic approximation as described previously, coincides very well with an observed series of doubly excited He^{**} levels. While the group theoretical approach and rovibrator model describe the electron correlation in each doubly excited state of atoms, the hyperspherical coordinate approach has the advantage of describing electron correlation patterns of a *channel* rather than each level. Hence, doubly excited states belonging to the same channel exhibit similar electron correlation pattern and also have nearly identical potential curves. In hyperspherical approach a given state of a two-electron atom is designated [8] as ${}_n(K,T)_N^{A}\,{}^{2S+1}L^{\pi}$, where L and S are orbital and spin quantum numbers, respectively, π is the parity, $N(n)$ denotes the principal quantum number of the inner (outer) electron. The correlation quantum numbers K and T are the same as in group theoretical methods and A is the radial correlation quantum number which describes the symmetry of the wave function with respect to $\alpha = 45°$. The possible values of K, T, and A are

$$T = 0, 2, \ldots, \min(L, N-1)$$
$$K = N-1-T, N-3-T, \ldots, -(N-1-T)$$
$$A = +1, -1, 0$$

A given channel μ is described by the notation $\mu = (K,T)_N^{A}\;{}^{2S+1}L^{\pi}$, where the principal quantum number N corresponds to the hydrogenic principal quantum number in the dissociation limit ($R = \infty$).

This classification scheme of atomic states has several advantages. First, it describes the photoabsorption selection rule in an elegant manner. The strong absorption in Madden and Codling's photoabsorption spectra is denoted as $(0,1)_2^+$, while in Woodruff and Samson's spectra [22] the prominent series below each of the N = 2, 4, and 5 series, are classified as $(1, 1)_3^+$, $(2, 1)_4^+$, and $(0, 1)_4^+$ channels, respectively. Since the ground state of He belongs to $(0, 0)_1^+$ channel, photoabsorption selection rules in these experiments is $\Delta A = 0$, $\Delta T = 1$, and $K = N - 2$. The scheme also predicts that the states belonging to the same channel would exhibit similar electron correlation patterns. Figure 15 of Lin [8] shows the electron correlation patterns of several states belonging to $(2, 0)_3^+$ channel of He, demonstrating the effectiveness of the hyperspherical coordinate approach. Finally, this approach explains the origin of the near degeneracy of the levels from different O(4) multiplets [15]. This near degeneracy results from the fact that channels with

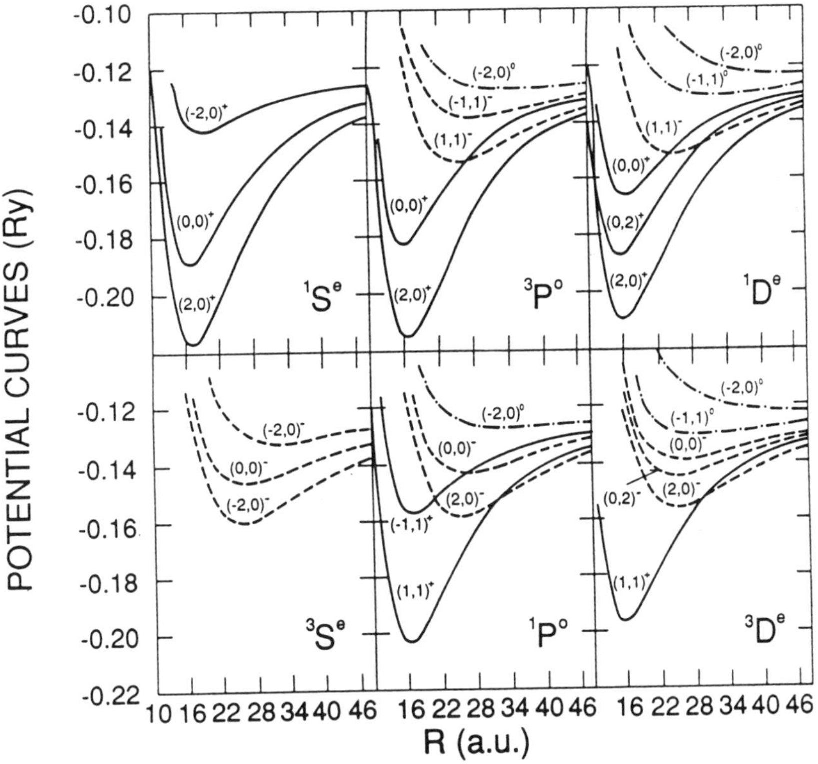

Figure 5.3 Potential curves for all the $^{1,3}S^e$, $^{1,3}P^0$, and $^{(1,3)}D^e$ channels for He which converge to He^+ ($n = 3$). Curves are labeled in terms of K, T, and A correlation quantum numbers. *Source:* from [8].

identical correlation quantum numbers have nearly identical potential waves and thus, nearly identical eigenenergies (Figure 5.3).

C. Rovibrator Model and Its Microscopic Foundation

The unexpected feature of collective rotation and vibration of electrons in doubly excited states were noticed by Kellman and Herrick [16] when the manifold of two-electron states were reorganized as supermultiplets. The energy levels of intrashell doubly excited states showed patterns that were very similar to the rotation-vibration spectra of linear triatomic molecules, although the manifold of states were truncated because of the finite number of states involved. A one-to-one correspondence between the supermultiplet labels and the quantum numbers of bending and rotating degrees of freedom of triatomic molecules was found and it was recognized that atomic doubly excited states could be classified in a physically appealing scheme using these quantum numbers.

Subsequently, Berry et al. analyzed the correlation of electrons in doubly excited states using simple models of two particles on a sphere [23] and two particles on concentric spheres [24]. They found that two particles on a sphere, mimicking the intrashell doubly excited states, exhibit angular correlation. The energy levels of $N=3$ He doubly excited states, according to the rovibrator scheme, are shown in Figure 5.4. The quantum number ν_2 is the number of quanta of bending, ℓ is the projection of the vibrational angular momentum due to the degenerate bending mode along the molecular axis, and J is the total angular momentum. The hierarchical relations between them are:

$$
\begin{aligned}
&\nu_2 = 0, 1, 2, \ldots \\
&\ell = \nu_2, \nu_2 - 2, \nu_2 - 4, \ldots, 1, \text{or } 0 \\
&J = l, l+1, \ldots
\end{aligned}
$$

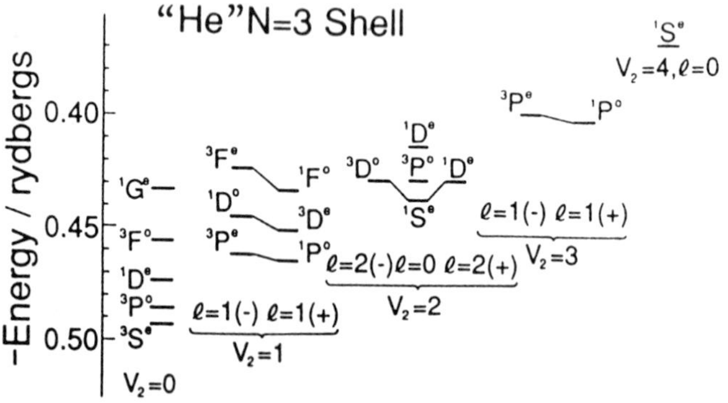

Figure 5.4 Calculated energy levels for two particles on a sphere of radius 13.7 a_0 interacting via a Coulomb repulsion. The 19 states shown correspond to the He $n = 3$ doubly excited shell and are arranged according to the rotation vibration pattern. *Source:* from [23].

The five atomic states $^1S^e$, $^3P^0$, $^1D^e$, $^3F^0$, and $^1G^e$ on the left side of Figure 5.4 form a vibrationless rotor series. The three $^1S^e$ states with 0, 2, and 4 bending quanta exhibit a progression of bending vibrational levels. The rotational constants and bending frequencies were calculated. Characteristic ratios of rotational to vibrational energies were found to be significantly larger than those of rigid triatomic molecules, indicating that the doubly excited states can be described as a *floppy triatomic molecule*. A dominant feature is a marked angular correlation around $\theta_{12} \sim \pi$. For intershell doubly excited states modeled as two particles on concentric spheres, the collective behavior of the particles transform to independent particle picture as the relative radius of the two spheres increases. The transition from collective behavior to independent particle behavior is not particularly dependent on the interacting potentials, suggesting that the degree of collective or independent particle behavior might be largely determined by kinematics (relative magnitudes of kinetic energies and interaction potentials) rather than by the details of dynamics (potential functions) [24].

Ezra and Berry also employed [25] high quality CI functions to compute the conditional density $\rho(r_1, \theta_{12}|r_2 = \alpha)$ for $N = 2$ intrashell doubly excited states of He and isoelectronic ions Li^+, Be^{2+}, and Ne^{8+}. The function $\rho(r_1, \theta_{12}|r_2 = \alpha)$ denotes the probability of finding an electron at distance r_1 from the nucleus and with interelectronic angle of θ_{12}, when another electron is located at distance α from the nucleus. Figure 5.5 shows the

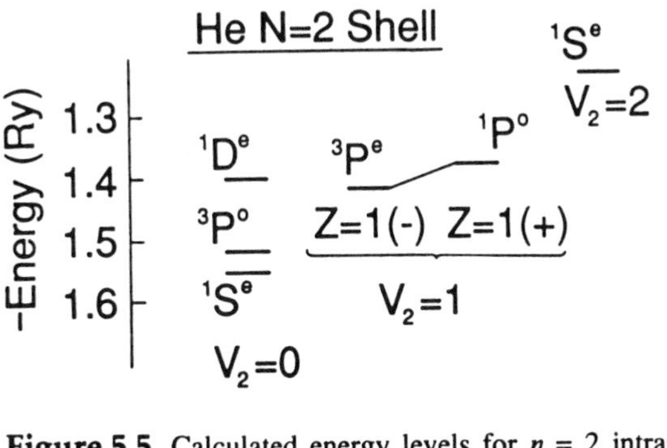

Figure 5.5 Calculated energy levels for n = 2 intrashell doubly excited states of He (Strumian basis). Assignment of molecular quantum numbers is shown. *Source:* from [23].

molecular classification of $N = 2$ intrashell doubly excited states of He. $^1S^e$, $^3P^0$, and $^1D^e$ states on the left side of Figure 5.5 form a vibrationless rotor series. The $^3P^e$ and $^1P^0$ states are the states with one quantum of bending vibration, which split due to T-doubling (analogous to ℓ-doubling in linear polyatomics). The highest $^1S^e$ state is a rotationless state with two quanta of bending vibration. These atomic doubly excited states correspond to the united atom limit of the molecular doubly excited states of H_2 molecule, which will be described in Sections 5.3 and 5.4.

Electron correlation patterns for these atomic doubly excited states were extensively studied by Ezia and Berry [25]. As expected, the rotor series $^1S^e$, $^3P^0$, and $^1D^e$ exhibit similar correlation pattern, that is, localization of electrons around $\theta_{12} \sim \pi$. Correlation becomes less evident as the angular momentum increases and/or nuclear charge increases. The correlation patterns of $^1P^0$ and $^3P^e$ states with one quantum of bending motion exhibit radial nodes at $\theta_{12} = 0$ and π, and the conditional density rises to a maximum near $\theta_{12} = \pi/2$. Conditional densities for the highest $^1S^e$ state with two bending quanta exhibit maxima near $\theta_{12} = 0$ and 60° in the case of low nuclear charge, while the maximum appears near $\theta_{12} = 0$ in the case of high nuclear charge. This suggests that an antirotor state develops with increasing nuclear charge. Again, these electron correlation patterns represent the united atom limit of those for molecular doubly excited states, as will be discussed in Sections 5.3 and 5.4.

As explained previously, the rovibrator model has been very successful in classifying the doubly excited states and providing a qualitative understanding of correlation effects. There remain, however, the following two basic questions:

1) How can this model be microscopically founded and justified, and
2) Why is the energy spacing wider than that of the ideal rotor spectra?

The appearance of a rotorlike progression of energy levels suggests the existence of some kind of stable intrinsic rotor states that violate the spherical symmetry; hence, there should be an explicit quantum mechanical formulation for this. Also, the molecular model can not predict a priori the molecular constants of the rovibrational motion. What is actually the rovibrating substance and what causes the wider energy spacing? This last effect, called *rotational contraction*, indicates that atom shrinks with increasing angular momentum (or angular velocity), which obviously contradicts the conventional classical picture. These pose serious questions concerning the naive collective rotation model. Answers to these questions were partly given by Iwai and Nakamura [9, 26]. Using the generator-coordinate representation of the O(4) supermultiplet, they obtained an explicit expression of the intrinsic rotor state [9]. A complete separation of the collective and internal intrinsic motions was shown to be realized, especially in the high-excitation (high principal quantum number) limit. The collective rotational motion of the system as a whole is, of course, to restore the broken symmetry. Furthermore, a new explicit expression was derived for the collective bending vibration in terms of quasi-spin operators. The rovibrator model was thus justified and embodied quantum mechanically. It should be emphasized that the method devised by this work would provide a nice guideline to investigate general strongly coupled many-particle systems not only of electrons but also of atoms and molecules. On the other hand, by employing the self-consistent cranking model the rotational contraction was shown to be caused by the quantum mechanical effect of shell structure [26]. The symmetry-violating intrinsic deformed state has a classical nature and can be represented by a wave packet expanded in terms of the angular momentum eigenstates. In the cranking model the classical rotation is applied to this wave packet with the constraint of the semiclassical quantization of angular momentum. The cutoff of rotational band (the quantum mechanical shell effect) was found to cause a drastic change of the moment of inertia (i.e., the rotational contraction). In other words, when n is very large and the nuclear charge is effectively very small, the intrashell spectrum of doubly excited states would exhibit the almost ideal rotorlike spectrum owing to the mixing of states with higher angular momentum ℓ in the higher principal shell.

5.3 DOUBLY EXCITED STATES OF HYDROGEN MOLECULE

Doubly (or inner-shell-) excited states of molecules are the first kind SES and they play an important role together with rovibrationally excited Rydberg states (the second kind of SES), as intermediate resonance states in many molecular dynamic processes such as autoionization [2–4, 27–29], dissociative recombination [4, 30–31], associative ionization [32–34], and photodissociation [27, 35–37]. Various kinds of couplings between electronic motion and nuclear motion in these molecular superexcited states result in different kinds of dynamic processes. Studying these highly excited states can thus lead to a unified understanding of a variety of molecular dynamic processes. Progress in laser technology has made these highly excited states more easily accessible and thus the information on their energies and widths is eagerly desired.

Although there have been some published studies on doubly excited states of molecules, the characteristics of these states and their effects on the dynamic processes are still not well understood. It is not well known, for instance, what kind of electron correlation patterns exist and how they vary as a function of nuclear degrees of freedom. What determines the magnitude of autoionization width Γ is a very basic interesting question to be investigated.

Determination of resonance energies and widths are of course most essential in characterizing the doubly excited states. There have been several calculations of the potential energies for some first kind of superexcited states of small molecules, such as H_2, O_2, HeH, and CH [30, 31]. In this section we present more extensive results for all the doubly excited states of H_2 molecule converging to $n = 2$ intrashell atomic limit and also for the lowest doubly excited states converging to the ($n = 2$, $n' = 3$)-intershell states in the atomic limit. Electron correlation patterns and their effects on resonance widths and autoionization mechanism will be discussed in Section 5.4.

A. Method of Calculation

The Feshbach projection operator method [38, 39] is employed here to calculate the resonance energies of the doubly excited states. Two projection operators P and Q are introduced. Q projects out the bound-state part of the wavefunction $Q\Psi$ from a resonance state Ψ. Its orthogonal component $P\Psi$ has the same asymptotic form as the resonance state Ψ. The Schrödinger equation becomes

$$(QHQ - E)(Q\Psi) = QHP(P\Psi) \tag{5.7}$$

and

$$(PHP - E)(P\Psi) = PHQ(Q\Psi) \tag{5.8}$$

Defining the optical potential $\overline{V}_{opt}$,

$$\overline{V}_{opt} = (PHQ)(E - QHQ)^{-1}(QHP) \tag{5.9}$$

we are led to the formally uncoupled equation,

$$(PHP + \overline{V}_{opt} - E)P\Psi = 0 \tag{5.10}$$

where $\overline{V}_{opt}$ is nonlocal and energy-dependent. The projection operator Q is constructed by using the eigenvectors $\{\Psi_n^{(0)}\}$ of QHQ as

$$Q = \sum_n \left| \Psi_n^{(0)} \right> \left< \Psi_n^{(0)} \right| \tag{5.11}$$

If we are interested in an isolated resonance whose position is close to an eigenvalue $E_n^{(0)}$ of QHQ, then Equation 5.10 can be rewritten as

$$\begin{aligned}(H_{NR} - E)P\Psi &\equiv \left(PHP + \sum_{m \neq n} \frac{PHQ|\Psi_m^{(0)}\rangle\langle\Psi_m^{(0)}|QHP}{E - E_m^{(0)}} - E \right) P\Psi \\ &= -\frac{PHQ|\Psi_n^{(0)}\rangle\langle\Psi_n|QHP}{E - E_n^{(0)}} P\Psi\end{aligned} \tag{5.12}$$

The solution of this equation is

$$P\Psi = \varphi_E + \frac{1}{E^+ - H_{NR}} \frac{PHQ|\Psi_n^{(0)}\rangle\langle\Psi_n^{(0)}|QHP|\varphi(E)\rangle}{E - E_n^{(0)} - \langle\Psi_n^{(0)}QHP\left| 1/(E^+ - H_{NR}) \right| PHQ\Psi_n^{(0)}\rangle} \tag{5.13}$$

where φ_E denotes the nonresonant part of the scattering and satisfies the equation

$$(H_{NR} - E^+)\varphi_E = 0 \tag{5.14}$$

The shift Δ_n of the resonance position and the resonance width Γ_n are obtained from Equation 5.12 as

$$\Delta_n = \mathrm{Re} <\Psi_n^{(0)}|QHP\, G_{NR}^{(+)}\, PHQ|\Psi_n^{(0)}> \tag{5.15}$$

and

$$\Gamma_n = -2Im <\Psi_n^{(0)}|QHP\, G_{NR}^{(+)}\, PHQ|\Psi_n(0)> \tag{5.16}$$

where

$$G_{NR}^{(+)} = (E - H_{NR} + i\varepsilon)^{-1} \tag{5.17}$$

and Re[Im] means the real [imaginary] part. When the level shift Δ_n and the nonlocal potential in H_{NR} can be neglected, the width is approximated as

$$\Gamma = 2\pi|<\varphi_E^{(-)}|PHQ|\Psi_n^{(0)}>|^2 \equiv 2\pi|V(R)|^2 \tag{5.18}$$

where $\varphi_E^{(0)}$ satisfies

$$(PHP - E)\varphi_E^{(0)} = 0 \tag{5.19}$$

We are interested here in the energy region where only the elastic scattering channel of the H_2 + e system is open. In this case, following O'Mally and Geltman [40], the projection operator Q is defined as

$$Q = [1 - p(1)][1 - p(2)] \tag{5.20}$$

with

$$p(i) = |\varphi_{000}(i)> <\varphi_{000}(i)| \tag{5.21}$$

where i denotes i^{th} electron and $\varphi_{000}(i)$ is the ground state wavefunction of the H_2^+ ion. The eigenfunctions of QHQ, $\{\Psi_n^{(0)}\}$ are then expanded by the eigenfunctions of the H_2^+ ion, $\varphi_{n\ell m}(i)$, as

$$\Psi_n^{(0)}(1, 2; \Lambda, S) = \sum_{n_1 l_1 m_1 n_2 l_2 m_2} C^{(n)}_{n_1 l_1 m_1 n_2 l_2 m_2} \{\varphi_{n_1 l_1 m_1}(1)\, \varphi_{n_2 l_2 m_2}(2) \pm (-1)^S \varphi_{n_2 l_2 m_2}(1) \varphi_{n_1 l_1 m_1}(2)\} \tag{5.22}$$

where Λ and S are the projection of the total orbital angular momentum onto the molecular axis and total electron spin, respectively. The summation over m_1 and m_2 is restricted to $m_1 + m_2 = \Lambda$. The QHQ matrix is diagonalized to obtain the expansion coefficients $\{C^{(n)}\}$ and eigenvalues $\{E_n^{(0)}\}$. The prolate spheroidal coordinate system $\{\xi, \eta, \varphi\}$ are adopted here. The basis function $\{\varphi_\alpha\}$'s [α donates $(n_\alpha, l_\alpha, m_\alpha)$] are separable in these coordinates as [41, 42]

$$\varphi_\alpha = \Lambda^\alpha(\xi)\mu^\alpha(\eta)\Phi^\alpha(\varphi) \quad \text{(see [41] for notation)} \tag{5.23}$$

The continuum wave function $\varphi_E^{(0)}$ in Equation 5.20 is approximated in the present work by

$$\varphi_E^{(0)} = \frac{1}{\sqrt{2}} [\varphi_{000}(1)\psi_k(2) + (-1)^S \psi_{000}(2)\psi_k(1)] \tag{5.24}$$

where $k^2 = 2(E_n^{(0)} - \varepsilon_{1\sigma_g})$ with $\varepsilon_{1\sigma_g}$ the ground state energy of H_2^+. We employ two-center Coulombic wave function for the continuum orbital ψ_k, which can be written as

$$\psi_k(\xi, \eta, \varphi) = N \sum_{lm} \frac{g_{lm}(\xi)}{\sqrt{\xi^2 - 1}} S_{ml}(\eta) \frac{e^{im\varphi}}{\sqrt{2\pi}} \tag{5.25}$$

where $S_{lm}(\eta)$ is the prolate spheroidal angle function of the first kind. By employing the two-center Coulombic wave function for ψ_k, we assume the perfect screening of the nuclear charge by the core electron. The Q-space is constructed by employing eigenfunctions of H_2^+ with the principal quantum numbers n = 2, 3, 4, and 5. The conditional density $\rho(\xi_1\eta_1;\xi_2\eta_2;\Delta\varphi)$ is depicted to describe electron correlation in molecular doubly excited states. $\rho(\xi_1\eta_1;\xi_2\eta_2\Delta\varphi)$ represents the relative probability of finding another electron at $(\xi_2\eta_2\varphi_2)$ with $\varphi_2 = \varphi_1 - \Delta\varphi$, when one electron is put at a position $(\xi_1\eta_1\varphi_1)$. Autoionization mechanism is described by visualizing the two-body transition density matrix $\rho_{n\to\varepsilon}(\xi_1\eta_1;\xi_2\eta_2;\Delta\varphi)$. Here n denotes the bound-state and ε represents electronic continuum. $\rho_{n\to\varepsilon}(\xi_1\eta_1;\xi_2\eta_2;\Delta\varphi)$ represents the contribution of the configuration $(\xi_1\eta_1;\xi_2\eta 2;\Delta\varphi)$ to the resonance width Γ. Integrating $\rho_{n\to\varepsilon}$ with respect to all variables results in Γ. The explicit expression for these two functions are obtained, after a straight algebra, as

$$\begin{aligned}\rho_n(\xi_1\eta_1;\xi_2\eta_2;\Delta\varphi) = \sum C^{(n)*}_{\{j_1j_2\}} C^{(n)}_{\{k_1k_2\}} \\ [\{\Lambda^{j_1}(\xi_1)M^{j_1}(\eta_1)\Lambda^{j_2}(\xi_2)M^{j_2}(\eta_2)\Lambda^{k_2}(\xi_1)M^{k_2}(\eta_1)\Lambda^{k_1}(\xi_2)M^{k_1}(\eta_2) \\ + \Lambda^{j_1}(\xi_2)M^{j_1}(\eta_2)\Lambda^{j_2}(\xi_1)M^{j_2}(\eta_1)\Lambda^{k_2}(\xi_2)M^{k_2}(\eta_2)\Lambda^{k_1}(\xi_1)M^{k_1}(\eta_1)\} \\ \times \cos[(m_{k_1} - m_{j_1})\Delta\varphi] \pm (-1)^S\{\Lambda^{j_1}(\xi_1)M^{j_1}(\eta_1)\Lambda^{j_2}(\xi_2)M^{j_2}(\eta_2) \\ \times \Lambda^{k_1}(\xi_1)M^{k_1}(\eta_1)\Lambda^{k_2}(\xi_2)M^{k_2}(\eta_2) + \Lambda^{j_1}(\xi_2)M^{j_1}(\eta_2)\Lambda^{j_2}(\xi_1)M^{j_2}(\eta_1) \\ \times \Lambda^{k_1}(\xi_2)M^{k_1}(\eta_2)\Lambda^{k_2}(\xi_1)M^{k_2}(\eta_1)\}\} \cos[(\Lambda - m_{k_1} - m_{j_1})\Delta\varphi]\}]\end{aligned} \tag{5.26}$$

$$\begin{aligned}\rho_{n\to\varepsilon}(\xi_1\eta_1;\xi_2\eta_2;\Delta\varphi) = \sum C^{(n)}_{\{j_1j_2\}} \\ [\{\Lambda^{j_1}(\xi_1)M^{j_1}(\eta_1)\Lambda^{j_2}(\xi_2)M^{j_2}(\eta_2)\psi_0(\xi_1,\eta_1)\psi_c(\xi_2,\eta_2)\exp[i(m_0 - m_{j_1})\Delta\varphi] \\ + \Lambda^{j_1}(\xi_2)M^{j_1}(\eta_2)\Lambda^{j_2}(\xi_1)M^{j_2}(\eta_1)\psi_0(\xi_2,\eta_2)\psi_c(\xi_1,\eta_1)\exp[-i(m_0 - m_{j_1})\Delta\varphi]\} \\ +(-1)^S\{\Lambda^{j_1}(\xi_1)M^{j_1}(\eta_1)\Lambda^{j_2}(\xi_2)M^{j_2}(\eta_2)\psi_0(\xi_2,\eta_2)\psi_c(\xi_1,\eta_1)\exp[-i(m_0 - m_{j_2})\Delta\varphi] \\ +\Lambda^{j_1}(\xi_2)M^{j_1}(\eta_2)\Lambda^{j_2}(\xi_1)M^{j_2}(\eta_1)\psi_0(\xi_1,\eta_1)\psi_c(\xi_2,\eta_2)\exp[i(m_0 - m_{j_2})\Delta\varphi]\}]\end{aligned} \tag{5.27}$$

where $\psi_0(\xi,\eta)e^{im_0\varphi}$ and $\psi_c(\xi,\eta)e^{i(\Lambda - m_0)\varphi}$ represent the ground state wave function of H_2^+ and the two-center Coulombic continuum wave function, respectively. j and k collectively denote $(n_j l_j m_j)$ and $(n_k l_k m_k)$, respectively.

B. Resonance Energies

There have been reported potential energies of doubly excited states of H_2, mostly for $R \geq 1.0a_0$ [43–53]. Bottcher and Docken [43] calculated the potential energy curve and width of the lowest doubly excited state of $^1\Sigma_g^+(2p\sigma_u)^2$ of H_2, employing the Feshbach projection operator technique. More elaborate methods, such as Kohn variational principle [45], the *R*-matrix method [46, 47] and the linear algebraic approach [48] have been employed also. We present here the potential energy curve of the doubly excited states converging to $n = 2$ intrashell doubly excited states of He, and also those states converging to ($n = 2$, $n' = 3$) doubly excited states from the united atom limit to intermediate *R*.

Figure 5.6 presents the potential curve of the three lowest $^1\Sigma_g^+$ states [10] which converge to the $|2s^2\ ^1S^e$, $|2p^2\ ^1D^e$, and $|2p^2\ ^1S^e$ states of He at the united atom limit, respectively. The two lowest $^1\Sigma_g^+$ states experience an avoided crossing near $R \cong 0.6a_0$, while the second and third lowest $^1\Sigma_g^+$ states have another avoided crossing at $R \cong 1.2a_0$. These avoided crossings significantly effect electron correlation in these doubly excited states, as will be discussed in next section. A comparison can be made with more elaborate results by Shimamura et al. using the *R*-matrix method [47]. For $R \geq 1.4a_0$ the two calculations agree very well, but at $R \leq 1.2a_0$, the *R*-matrix method results are much larger than the results by the present calculation. It turned out that the former actually corresponds to the diabatically connected states. This is simply because the *R*-matrix calculation did not include those functions which converge to $2s$ orbital at the united atom limit. Thus we can conclude that the present calculation is accurate enough.

The numerical values of the potential energies of the three lowest $^1\Sigma_g^+$ states, correlating to $n = 2$ intrashell doubly excited states of He, are presented in Table 5.1; comparisons are also made with other results in this Table. Figures 5.7 and 5.8 show the potential energies of all the doubly excited states correlating to $n = 2$ intrashell doubly excited states of He; the corresponding numerical values of these states are presented in Tables 5.2 and 5.3. Although there have been reported several studies on intershell doubly excited states of H_2, the information on these states is mostly fragmentary. Tables 5.4 and 5.5 present potential curves of the lowest singlet and triplet states of H_2, correlating to ($n = 2$, $n' = 3$) intershell doubly excited states of He. The same potential curves are shown in Figures 5.9 and 5.10. The $^1\Delta_g(2)$ state is shown to lie above the $2p\sigma_u$ state of H_2^+ at $R \geq 1.4a_0$; hence this state can autoionize both to the $1s\sigma_g$ and to the $2p\sigma_u$ states of H_2^+. The potential energies of these doubly excited states are invaluable in analyzing the photoionization and/or photodissociation spectra involving

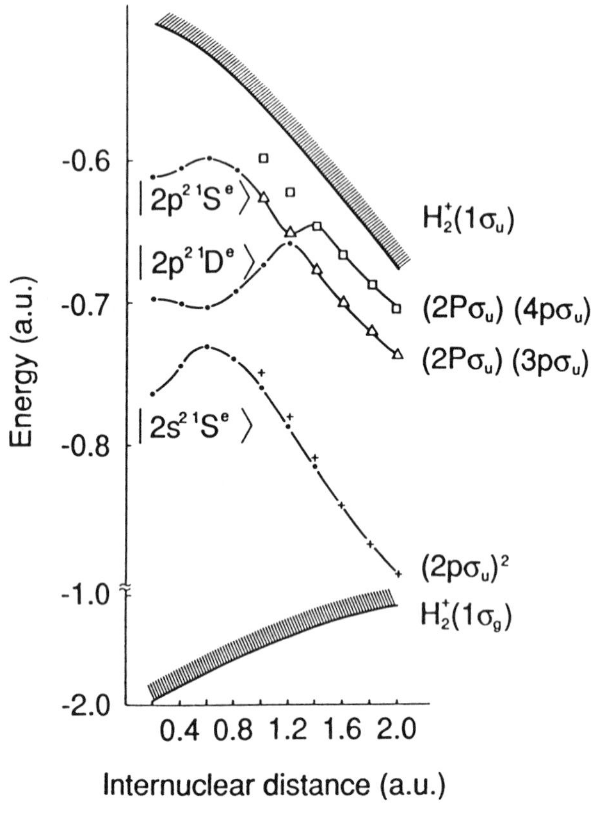

Figure 5.6 Potential energy curves (nuclear repulsion $1/R$ is excluded) for three lowest $^1\Sigma_g^+$ states: (— · —): present results, (+): R-matrix calculation for $^1\Sigma_g(1)$, ($\triangle$): R-matrix calculation for $^1\Sigma_g(2)$, ($\square$): R-matrix calculation for $^1\Sigma_g(3)$. Potential curves of the ground state ($1\sigma_g$) and the first excited state ($1\sigma_u$) of H_2^+ are also shown. *Source:* from [10].

Table 5.1 Resonance Positions (a.u.) of the Three Lowest $^1\Sigma_g^+$ States that Correlate to $n = 2$ Intrashell Doubly Excited States of He (see Figure 5.6)

R (a.u.)	$^1\Sigma_g^+(1)$						$^1\Sigma_g^+(2)$			$^1\Sigma_g^+(3)$	
0.2	−0.7637						−0.6972			−0.6115	
0.4	−0.7448						−0.7004			−0.6059	
0.6	−0.7318						−0.7026			−0.5987	
0.8	−0.7397						−0.6918			−0.6069	
1.0	−0.7606	−0.7628 [a]	−0.7599 [b]	−0.7259 [c]	−0.7566 [d]	−0.7515 [e]	−0.6753	−0.6431 [d]	−0.6289 [e]	−0.6290	−0.5998 [e,f]
1.2	−0.7869	−0.7901	−0.7860	−0.7625	−0.7851	−0.7821	−0.6594	−0.6521	−0.6537	−0.6527	−0.6239
1.4	−0.8150	−0.8198	−0.8142		−0.8159	−0.8137	−0.6780	−0.6783	−0.6784	−0.6460	−0.6479
1.6	−0.8426	−0.8484	−0.8419	−0.8299	−0.8458	−0.8444	−0.7003	−0.7011	−0.7012	−0.6667	−0.6699
1.8	−0.8682	−0.8746			−0.8731	−0.8730	−0.7198	−0.7207	−0.7210	−0.6857	−0.6889
2.0	−0.8908	−0.9000	−0.8900	−0.8838	−0.8969	−0.8935	−0.7360	−0.7369	−0.7372	−0.7015	−0.7045

[a] *Source*: [51]
[b] *Source*: [43]
[c] *Source*: [44]
[d] *Source*: [50]
[e] *Source*: [47]
[f] More explanation in the text.

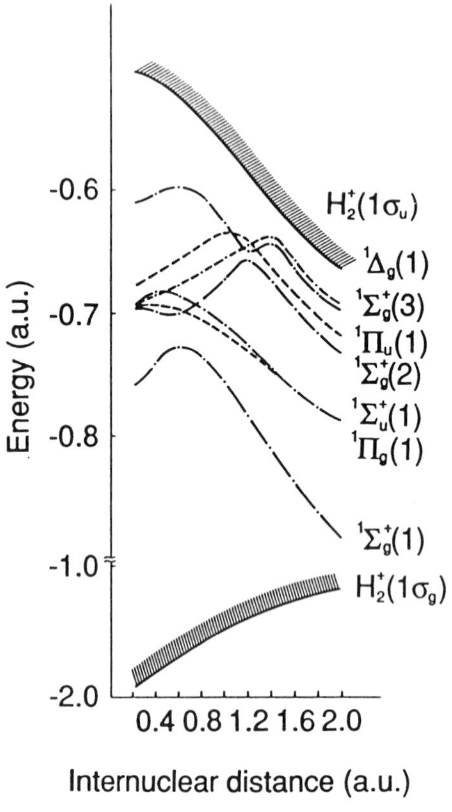

Figure 5.7 Potential energy curves (nuclear repulsion $1/R$ is excluded) of singlet states correlating to $n = 2$ intrashell doubly excited states of He. (— · —): Σ-states, (- - -): Π-states, (- - · - -): Δ-states.

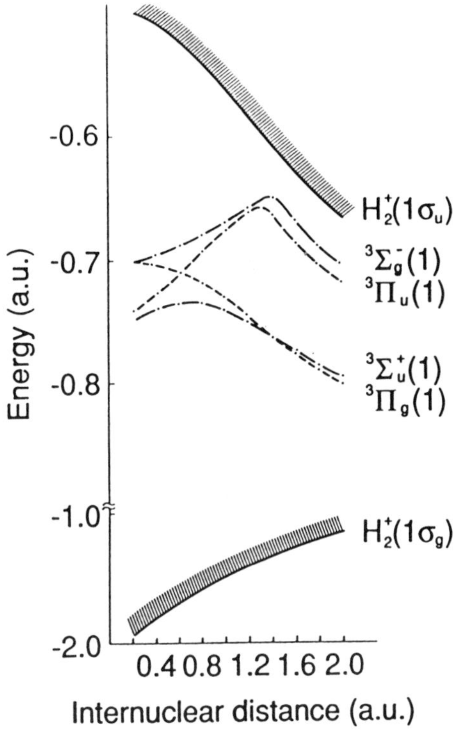

Figure 5.8 Potential energy curves (nuclear repulsion $1/R$ is excluded) of triplet states correlating to $n = 2$ intrashell doubly excited states of He. (— · —): Σ-states. (– – · – –): Π-states.

Table 5.2 Resonance Positions (a.u.) of Singlet Doubly Excited States Other than $^1\Sigma_g^+$ States that Correlate to n = 2 Intrashell Doubly Excited States of He

R (a.u.)	$^1\Sigma_u^+(1)$			$^1\Pi_g^+(1)$			$^1\Pi_u^+(1)$		$^1\Delta_g(1)$	
0.2	−0.6853			−0.6964			−0.6824		−0.6942	
0.4	−0.6842			−0.6977			−0.6721		−0.6884	
0.6	−0.6884			−0.7017			−0.6602		−0.6802	
0.8	−0.6985			−0.7096			−0.6480		−0.6704	
1.0	−0.7134	−0.7148 [a]	−0.7140 [b]	−0.7215	−0.7234 [a]	−0.7192 [b]	−0.6376	−0.6217 [b]	−0.6595	−0.6681 [a]
1.2	−0.7311	−0.7325	−0.7320	−0.7365	−0.7394	−0.7345	−0.6465	−0.6443	−0.6480	−0.6499
1.4	−0.7493	−0.7507	−0.7503	−0.7530	−0.7557	−0.7512	−0.6685	−0.6674	−0.6422	−0.6373
1.6	−0.7662	−0.7673	−0.7672	−0.7691	−0.7715	−0.7674	−0.6897	−0.6889	−0.6637	
1.8	−0.7807	−0.7818	−0.7816	−0.7834	−0.7855	−0.7817	−0.7082	−0.7075	−0.6823	
2.0	−0.7923	−0.7926	−0.7931	−0.7952	−0.7972	−0.7934	−0.7234	−0.7228	−0.6976	

[a] *Source*: [51]
[b] *Source*: [50]

Table 5.3 Resonance Positions (a.u.) of Triplet States that Correlate to $n = 2$ Intrashell Doubly Excited States of He

R (a.u.)	$^3\Sigma_g^+(1)$	$^3\Sigma_u^+(2)$		$^3\Pi_g(1)$		$^3\Pi_u(1)$		
0.2	−0.7045	−0.7523		−0.7071		−0.7496		
0.4	−0.6997	−0.7445		−0.7101		−0.7339		
0.6	−0.6923	−0.7402		−0.7159		−0.7160		
0.8	−0.6830	−0.7409		−0.7252		−0.6977		
1.0	−0.6724	−0.7468	−0.7461 [a]	−0.7379	−0.7359 [b]	−0.6798	−0.6209 [b]	−0.6569 [a]
1.2	−0.6609	−0.7568	−0.7558	−0.7530	−0.7513	−0.6628	−0.6472	−0.6545
1.4	−0.6526	−0.7690	−0.7679	−0.7690	−0.7678	−0.6702	−0.6694	−0.6716
1.6	−0.6745	−0.7817	−0.7804	−0.7844	−0.7834	−0.6913	−0.6907	−0.6919
1.8	−0.6933	−0.7932		−0.7098	−0.7970	−0.7908	−0.7092	
2.0	−0.7088	−0.8028	−0.8013	−0.8091	−0.8080	−0.7251	−0.7245	−0.7253

[a] *Source*: [43]
[b] *Source*: [50]

Table 5.4 Resonance Positions (a.u.) of Some of the Lowest Singlet States that Correlate to (n = 2, n' = 3) Intershell Doubly Excited States of He

R (a.u.)	$^1\Sigma_u^+(2)$[a]	$^1\Pi_g(2)$		$^1\Pi_u(2)$		$^1\Delta_g(2)$
0.2		−0.5805		−0.5908		−0.5646
0.4		−0.5847		−0.5814		−0.5595
0.6		−0.5945		−0.5816		−0.5549
0.8		−0.6097		−0.5980		−0.5731
1.0	−0.6181	−0.6290	−0.6283 [a]	−0.6174	−0.5974 [a]	−0.5950
				−0.6230 [b]		
1.2	−0.6390	−0.6504	−0.6498	−0.6240	−0.6199	−0.6187
1.4	−0.6609	−0.6720	−0.6714	−0.6414	−0.6431	−0.6361
1.6	−0.6816	−0.6920	−0.6914	−0.6622	−0.6646	−0.6046
1.8	−0.6998	−0.7094	−0.7087	−0.6806	−0.6832	−0.6243
2.0	−0.7150	−0.7237	−0.7229	−0.6958	−0.6985	−0.6390

[a] *Source*: from [50]
[b] R = 1.1 a.u.

Table 5.5 Resonance Positions (a.u.) of Some of the Lowest Triplet States that Correlate to (n = 2, n' = 3) Intershell Doubly Excited States of He

R (a.u.)	$^3\Sigma_u(2)$ [a]	$^3\Pi_g(2)$		$^3\Pi_u(2)$	
0.2		−0.5819		−0.5787	
0.4		−0.5848		−0.5750	
0.6		−0.5950		−0.5852	
0.8		−0.6110		−0.6023	
1.0	−0.6233	−0.6310	−0.6304 [a]	−0.6234	−0.6209 [a]
1.2	−0.6425	−0.6530	−0.6525	−0.6462	−0.6195
				−0.6524 [b]	
1.4	−0.6621	−0.6749	−0.6745	−0.6463	−0.6442
1.6	−0.6821	−0.6951	−0.6947	−0.6636	−0.6656
1.8	−0.7006	−0.7125	−0.7122	−0.6821	−0.6841
2.0	−0.7161	−0.7268	−0.7264	−0.6973	−0.6994

[a] *Source*: [50]
[b] R = 1.3 a.u.

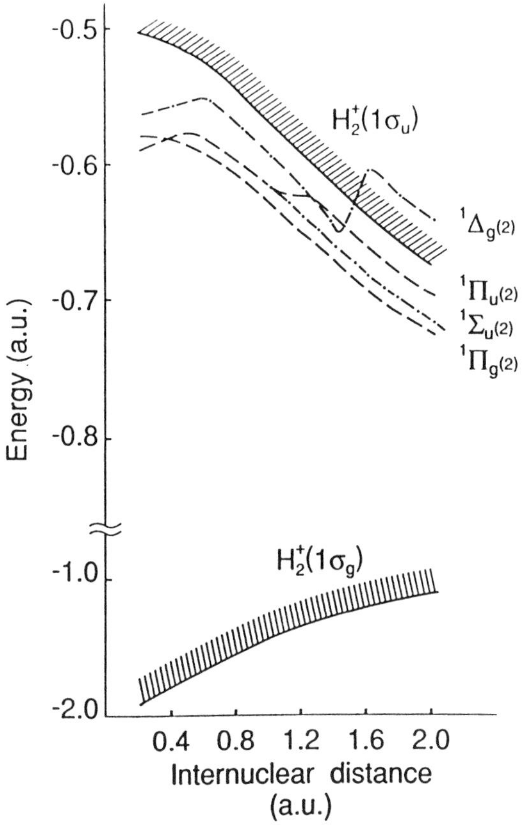

Figure 5.9 Potential energy curves (nuclear repulsion $1/R$ is excluded) of the lowest singlet states correlating to $n = 2$, $n' = 3$ intershell doubly excited states of He. (— · —): Σ^+-states, (– – · – –): Δ-states, (– – – –): Π-states.

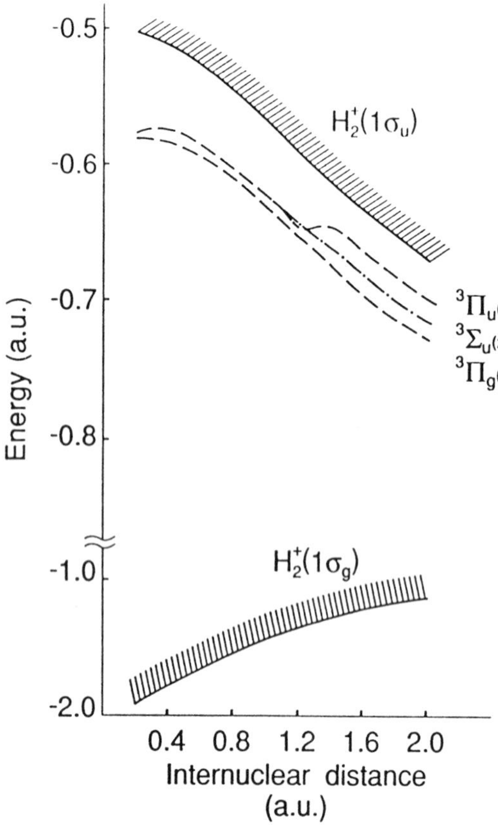

Figure 5.10 Potential energy curves (nuclear repulsion $1/R$ is excluded) of the lowest triplet states correlating to $n = 2$, $n' = 3$ intershell doubly excited states of He. (— · —): Σ-states, (– – – –): Π-states.

these states, since the small shifts in these potential curves may result in substantial changes in the Franck-Condon factor of the transitions from the lower states (say $B^1\Sigma_u^+$ or $C^1\Pi_u$ states) in REMPI experiments. Hopefully, our results will encourage more elaborate experiments on photoionization and photodissociation involving these very interesting states.

C. Resonance Widths

The resonance widths of the molecular doubly excited states, or more specifically, the electronic coupling matrices $V(R)$ (see Equation 5.18) measure the tendency of the states to autoionize by the coupling to the electronic continuum. The determination of this parameter is essential, as was explained before, to characterize the first kind of SES, particularly the competition among the various decay channels (i.e., autoionization, dissociation, and radiative transition). Information on this important parameter for the doubly excited states of H_2 is summarized here, together with data obtained by the present calculation.

Figure 5.11 shows the resonance widths of the three lowest $^1\Sigma_g^+$ states as a function of internuclear distance [10] and a comparison is made with more elaborate results by the R-matrix method [47]. Although single two-center Coulombic function (Equation 5.25) has been used for the continuum wave function, the resulting resonance widths agree quite well with the R-matrix results. This indicates that the accuracy of the resonance widths does not depend much on the quality of the continuum function but on that of the bound-state wave function. Table 5.6 presents the numerical values of the resonance widths of these states. The results of Bottcher and Docken [43] are not listed here since they are known to be less accurate [45]. The lowest $^1\Sigma_g^+$ state is expected to play a significant role in photoionization process of H_2 via the $B^1\Sigma_u^+$ state [54, 55]. The non-Franck-Condon behavior of the ionization is attributed to the autoionization of this doubly excited states. (This will be reviewed in Section 5.5.)

Table 5.7 presents the resonance widths of the lowest Σ_u^+ states as a function of R and comparisons are made with other results; resonance widths for $R < 1.0a_0$ are reported here for the first time. Table 5.8 also presents resonance widths of the Π states. For $^1\Pi_g$ states, our values are larger than those calculated by Sato and Hara [52] at all R. However, Collins and Schneider [48] and Tennyson et al. [49] have reported the resonance width

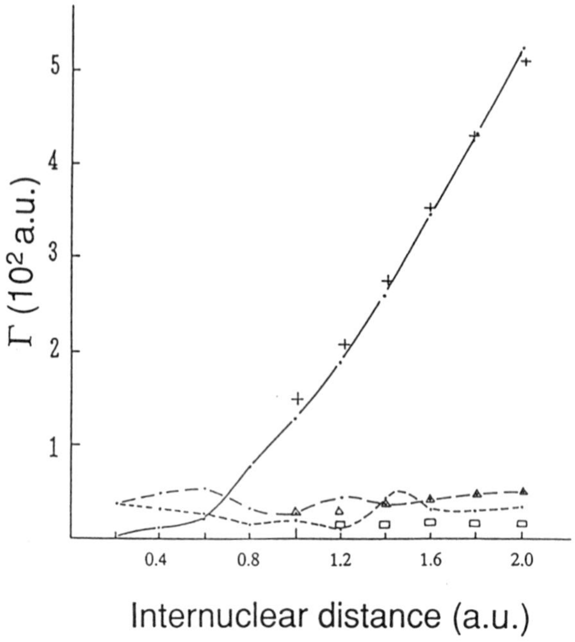

Figure 5.11 Resonance widths as a function of R. (— · —): present results; R-matrix calculations: (+) for $^1\Sigma_g(1)$, (△) for $^1\Sigma_g(2)$, (□) for $^1\Sigma_g(3)$. *Source*: from [10].

Table 5.6 Resonance Widths (a.u.) of the Three Lowest Singlet $^1\Sigma_g^+$ States as a Function of Internuclear Distance

R (a.u.)	$^1\Sigma_g^+(1)$				$^1\Sigma_g^+(2)$		$^1\Sigma_g^+(3)$	
0.2	6.80×10^{-6}				3.69×10^{-3}		3.59×10^{-3}	
0.4	1.37×10^{-4}				4.93×10^{-3}		3.25×10^{-3}	
0.6	2.08×10^{-3}				5.30×10^{-3}		2.67×10^{-3}	
0.8	7.65×10^{-3}				3.17×10^{-3}		1.49×10^{-3}	
1.0	1.29×10^{-2}	1.50×10^{-2} [a]		1.19×10^{-2} [c]	2.80×10^{-3}	2.57×10^{-3} [a]	1.99×10^{-3}	
1.2	1.88×10^{-2}	2.07×10^{-2}	1.9×10^{-2} [b]	1.91×10^{-2}	4.51×10^{-3}	3.01×10^{-3}	1.13×10^{-3}	1.30×10^{-3} [a]
1.4	2.59×10^{-2}	2.76×10^{-2}		2.78×10^{-2}	3.88×10^{-3}	3.58×10^{-3}	4.51×10^{-3}	1.45×10^{-3}
1.6	3.41×10^{-2}	3.52×10^{-2}	3.6×10^{-2}	3.50×10^{-2}	4.38×10^{-3}	4.13×10^{-3}	3.19×10^{-3}	1.62×10^{-3}
1.8	4.30×10^{-2}	4.31×10^{-2}		4.24×10^{-2}	4.85×10^{-3}	4.58×10^{-3}	3.36×10^{-3}	1.77×10^{-3}
2.0	5.20×10^{-2}	5.09×10^{-2}	5.9×10^{-2}	4.98×10^{-2}	5.44×10^{-3}	4.96×10^{-3}	3.52×10^{-3}	1.89×10^{-3}

[a] *Source*: [47]
[b] *Source*: [45]
[c] *Source*: [51]

Table 5.7 Resonance Widths (a.u.) of the Lowest $^1\Sigma_u^+$ and $^3\Sigma_u^+$ States that Correlate to $n = 2$ Intrashell Doubly Excited States of He

R (a.u.)	$^1\Sigma_u^+(1)$			$^3\Sigma_u^+(1)$
0.2	1.54×10^{-3}			8.80×10^{-4}
0.4	2.26×10^{-3}			9.38×10^{-4}
0.6	3.44×10^{-3}			9.94×10^{-4}
0.8	5.07×10^{-3}			1.02×10^{-3}
1.0	7.04×10^{-3}	8.20×10^{-3} [a]	8.30×10^{-3} [b]	1.00×10^{-3}
1.2	9.07×10^{-3}	1.13×10^{-2}	1.29×10^{-2}	9.35×10^{-4}
1.4	1.13×10^{-2}	1.43×10^{-2}	1.50×10^{-2}	8.35×10^{-4}
1.6	1.33×10^{-2}	1.72×10^{-2}		7.14×10^{-4}
1.8		1.99×10^{-2}	2.20×10^{-2}	5.85×10^{-4}
2.0	1.68×10^{-2}	2.10×10^{-2}	2.70×10^{-2}	4.55×10^{-4}

[a] *Source*: [52]
[b] *Source*: [45]

Table 5.8 Resonance Widths (a.u.) of the Lowest Π States that Correlate to n = 2 Intrashell Doubly Excited States of He

			$^1\Pi_u^+(2)$				$^3\Pi_u(1)$		
R (a.u.)	$^1\Pi_g^+(1)$		Pπ		$f\pi$	$^3\Pi_g(1)$	pπ		$f\pi$
0.2	3.56×10^{-3}		1.43×10^{-3}		5.95×10^{-9}	1.16×10^{-7}	8.61×10^{-4}		–
0.4	4.67×10^{-3}		1.79×10^{-3}		1.37×10^{-7}	2.15×10^{-6}	8.73×10^{-4}		–
0.6	6.29×10^{-3}		2.22×10^{-3}		1.11×10^{-6}	1.26×10^{-5}	8.72×10^{-4}		8.20×10^{-9}
0.8	8.30×10^{-3}		2.67×10^{-3}		6.45×10^{-6}	4.55×10^{-5}	8.53×10^{-4}		8.55×10^{-8}
1.0	1.05×10^{-2}	0.82×10^{-3} [a]	2.72×10^{-3}	2.1×10^{-3} [c]	4.20×10^{-5}	1.22×10^{-4}	8.19×10^{-4}	2.0×10^{-4} [c]	6.62×10^{-7}
1.2	1.26×10^{-2}	0.60×10^{-3}	4.23×10^{-4}	3.8×10^{-4}	1.29×10^{-4}	2.70×10^{-4}	7.43×10^{-4}	1.9×10^{-4}	9.24×10^{-6}
1.4	1.46×10^{-2}	0.20×10^{-3}	1.29×10^{-4}	1.5×10^{-4}	1.77×10^{-4}	5.14×10^{-4}	5.76×10^{-6}	$\leq 1.0 \times 10^{-6}$	1.86×10^{-4}
1.6	1.64×10^{-2}	0.50×10^{-3}	8.81×10^{-5}	5.2×10^{-5}	2.48×10^{-4}	8.83×10^{-4}	1.66×10^{-7}	4.0×10^{-6}	2.70×10^{-4}
1.8	1.83×10^{-2}	0.10×10^{-4}	8.31×10^{-5}	9.1×10^{-5}	3.50×10^{-4}	1.40×10^{-3}	1.48×10^{-8}	8.6×10^{-6}	3.89×10^{-4}
2.0	2.02×10^{-2} [b]	$0.8\text{~}1.0 \times 10^{-3}$	9.09×10^{-5}	1.1×10^{-4}	4.92×10^{-4}	2.11×10^{-3}	4.90×10^{-8}	1.3×10^{-5}	5.59×10^{-4}

[a] *Source*: [51]
[b] Corresponding values in [48] and [49] are 2.3×10^{-2} and 1.9×10^{-2}, respectively
[c] *Source*: [45]

at $R = 2.0a_0$ as 2.30(−2) and 1.90(−2), respectively, in better agreement with our results; more calculations should certainly be made for this state. Takagi and Nakamura [45] reported resonance widths of the $^1\Pi_u$ state by using the Kohn variational method. Their results are based on the analysis of the $p\pi$-wave phase shifts in elastic scattering of electrons from H_2^+. Table 5.8 shows that the contributions from the $p\pi$ partial wave to the resonance widths of $^1\Pi_u$ state indeed agree well with the results reported by Takagi and Nakamura. However the $f\pi$ partial widths are seen to dominate at $R \geq 1.6a_0$, and thus, the total resonance widths will be larger than those given by Takagi and Nakamura. The same trend is also seen in $^3\Pi_u$ state. The resonance widths for the $^3\Pi_g$ state, to our knowledge, have not been reported elsewhere. These doubly excited states, especially the lowest $^1\Sigma_g^+$ state and the $^1\Pi_g$ state, play a significant role in the photoionization and photodissociation of H_2 at energies near the $1s\sigma_g$ ionization limit; this will be discussed in Section 5.5.

5.4 ELECTRON CORRELATION IN MOLECULAR DOUBLY EXCITED STATES

Autoionization of the first kind of SES is basically induced by electron-electron Coulombic repulsion or electron correlation, and can occur even if the nuclei are clamped. As mentioned before, this is called *electronic autoionization.* The mechanism is different from the autoionization of the second kind of SES, or rovibrationally excited Rydberg states, in which autoionization is caused by the coupling between electronic and nuclear motions (vibrational autoionization) [2, 4]. Thus, understanding of electron correlation in molecular doubly excited states is essential to elucidating autoionization mechanisms.

Many qualitative conclusions made for atomic doubly excited states may be applicable to molecular doubly excited states as well at least when the internuclear separation is small. However, electron correlation patterns must be naturally affected by nuclear conformation. In diatomic molecules, for instance, the two-center field or axial symmetry of the system is expected to influence the electron correlation significantly (discussed in Section 5.4.A). Also, the electron correlation in a particular electronic state might be affected by other electronic state(s) in a particular range of R. For instance, extensive configuration mixing between electronic states of the same symmetry near the avoided crossings may change the electron correlation pattern. Hence, as the nuclei separate from the united atom limit, new electron correlation patterns are expected to develop.

For atomic doubly excited states, two autoionization mechanisms have been proposed [56]. One mechanism, called *direct electron-electron binary collision*, requires that the two electrons be close enough so that the electrostatic repulsion between them propels one electron away from the nuclei. The other mechanism, called *generalized* S_N - 2 or *dynamical screening*, works when one electron gets close to the nuclei so that the screening of the nuclear charge allows the other electron to move away from the system. The relative contribution of these two mechanisms to the autoionization process is closely related to the electron correlation pattern in atomic doubly excited states. These two autoionization mechanisms were found also in molecular doubly excited states [10]. Their relative contribution can be clarified by analyzing the electron correlation patterns of molecular doubly excited states [10].

A. Electron Correlation Patterns in Doubly Excited States of H_2

There are the following natural intriguing questions concerning the electron correlation in molecular doubly excited states:

1. How will the axial symmetry affect the electron correlation as the nuclei separate from united atom limit?
2. What will be the influence of the avoided crossings between states with the same symmetry?
3. What is the connection between the electron correlation and autoionization (i.e., how does the autoionization occur?)

Generally speaking, when the internuclear distance is small, the electron correlation in molecular doubly excited states will resemble those of the corresponding states in the united atom limit. When the nuclei are far apart, the electrons are confined to each of the nuclei, and thus the usual independent particle picture will develop. At intermediate distances, however, the symmetry of the electronic states will influence electron correlation in various ways. Figure 5.12 shows the conditional density ρ_n (see Equation 5.26) of the lowest ${}^1\Sigma_g^+$ state at $R = 0.2a_0$. In Figure 5.12(a) the first electron is located at a point along the line bisecting the two nuclei, with the distance of $2.9a_0$ from the center of the molecule, with the two electrons on the same

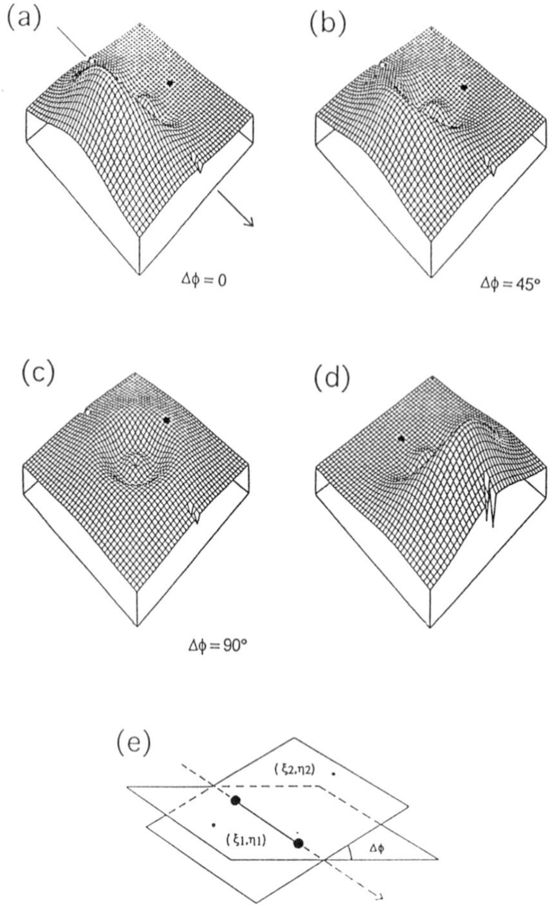

Figure 5.12 Electron density maps for the lowest $^1\Sigma_g^+$ state. Dots in the figures denote positions of the first electron. Arrow indicates the direction of the molecular axis. The size of the square is $10a_0 \times 10a_0$. This is the same for all other figures. The electron densities here are drawn to the same scale. (a) $\Delta\varphi = 0$, (b) $\Delta\varphi = \pi/4$, (c)$\Delta\varphi = \pi/2$, (d) $\Delta\varphi$ need not be specified because the first electron is located along the molecular axis. $R = 0.2a_0$ for (a–d). Figure (e) shows the coordinate system used. *Source*: from [10].

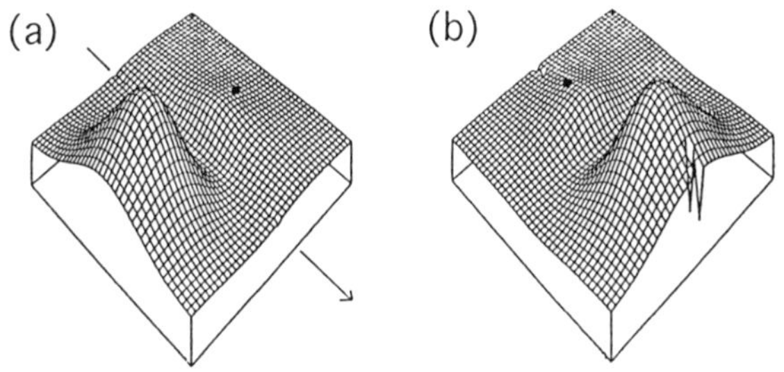

Figure 5.13 Electron density maps for the second lowest ${}^1\Sigma_g^+$ state. $R = 0.2a_0$. The arrow indicates the direction of molecular axis. $\Delta\varphi = 0$ for (a); $\Delta\varphi = 0$ for (b). *Source*: from [10].

plane ($\Delta\varphi = 0$). The dot indicates the position of the first electron. Figure 5.12(a) clearly shows that the electrons tend to be located on the opposite sides with respect to the nuclei. It was found that as $\Delta\varphi$ increases from 0 to $\pi/2$ (from Figure 5.12(a–c)), electron correlation decreases; this indicates that the electrons have the largest probability of being on the same plane as the two nuclei. When the first electron is put at a point along the molecular axis (as in Figure 5.12(d)), the second electron also has highest probability on the opposite side. Thus, it is concluded that the electrons have a strong tendency to be located on the other side of the nuclei, irrespective of the position of the first electron. This is similar to the electron correlation pattern in the atomic $|2s^2\ {}^1S^e\rangle$ state, in which the lowest ${}^1\Sigma_g^+$ state actually converges at united atom limit; it also indicates that the linear rotor character of the atomic doubly excited states $|2s^2\ {}^1S^e\rangle$ still remains at $R = 0.2a_0$.

Figure 5.13 show the two-body correlation function for the second lowest ${}^1\Sigma_g^+$ state, which converges to $|2p^2\ {}^1D^e\rangle$ at the united atom limit. The angular correlation is also obvious in this state, although it is less clear than the lowest ${}^1\Sigma_g^+$ state. This is consistent with the fact that in atomic $|2p^2\ {}^1D^e\rangle$ state, which is the highest rotor state with $v_2 = 0$ (see Figure 5.4), the angular correlation is less evident than the lowest rotor state $|2s^2\ {}^1S^e\rangle$. The conditional density maps of the third lowest ${}^1\Sigma_g^+$ state are presented in Figure 5.14. The electrons have large probability of being on the same side (i.e., $\theta_{12} \approx 0$), similar to the angular correlation pattern in $|2p^2\ {}^1S^e\rangle$ state to which this state converges. The $|2p^2\ {}^1S^e\rangle$ atomic state is a rotationless state with two bending quanta excited (see Figure 5.4).

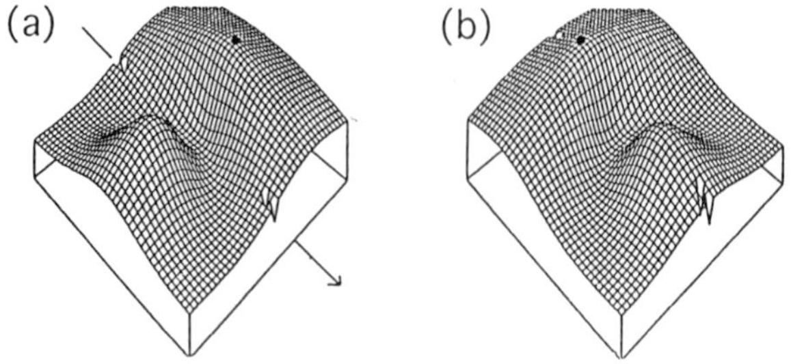

Figure 5.14 Electron density maps for the third lowest $^1\Sigma_g^+$ state. $R = 0.2a_0$. The arrow indicates the direction of molecular axis. $\Delta\varphi = 0$ for (a); $\Delta\varphi = 0$ for (b). *Source*: from [10].

As the nuclei move apart, the electrons begin to *feel* the axial symmetry of the system and this will be reflected in electron correlation patterns. Also some electronic states experience avoided crossings with other states of the same symmetry; this affects the correlation patterns in these states. When the nuclei are far apart, the electrons are confined to each nucleus and the system eventually transforms into the independent particle picture. These features are seen in Figures 5.15 and 5.16, which show the conditional density maps for the lowest $^1\Sigma_g^+$ state as a function of R. In Figure 5.15 the first electron is put along the line bisecting the nuclei. The angular correlation pattern survives until $R = 0.6a_0$, but as this state experiences an avoided crossing with the second lowest $^1\Sigma_g^+$ state at $R \approx 0.6a_0$, the correlated behavior of electrons suddenly disappears and independent particle picture develops after the avoided crossing. The effect of avoided crossing also can be seen when the first electron is put along the molecular axis (Figure 5.16). Figures 5.17 and 5.18 show the electron correlation in the second lowest $^1\Sigma_g^+$ state ($2p\sigma_u\, 3p\sigma_u$) as a function of R. When the first electron is put along the molecular axis, angular correlation pattern characteristic of the rotor states are strongly destroyed near the avoided crossing at $R \approx 0.6a_0$ (Figure 5.18(b)). On the other hand, the angular correlation survives even at $R \simeq 1.4a_0$ (Figure 5.17(e)) in a direction perpendicular to the molecular axis; we call this *m-polarization.* As the nuclei move apart, the electrons tend to be confined to each nucleus, and hence, do not feel the electrostatic repulsion from the other electron especially when the electrons are located along the molecular axis. When the electrons are located in a direction perpendicular to the

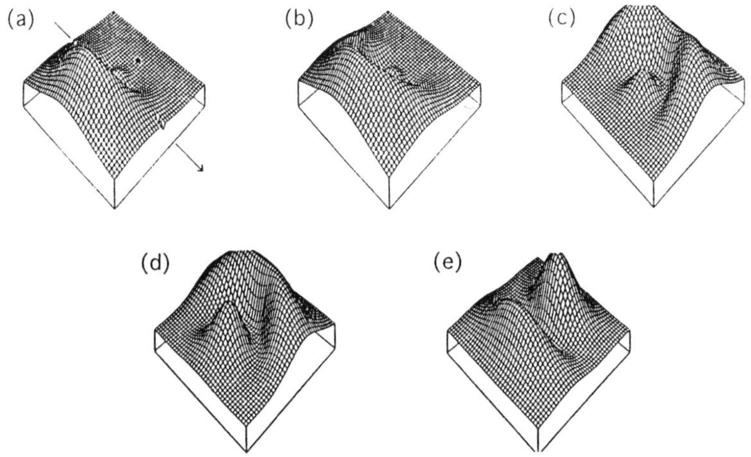

Figure 5.15 Change of electron correlation of the lowest $^1\Sigma_g^+$ state as a function of R. The first electron is located at $R = 2.9a_0$ from the center of the molecule along the line bisecting the nuclei. The arrow indicates the direction of molecular axis. $\Delta\varphi = 0$. (a) $R = 0.2a_0$, (b) $R = 0.6a_0$, (c) $R = 1.0a_0$, (d) $R = 1.2a_0$, (e) $R = 2.0a_0$. *Source*: from [10].

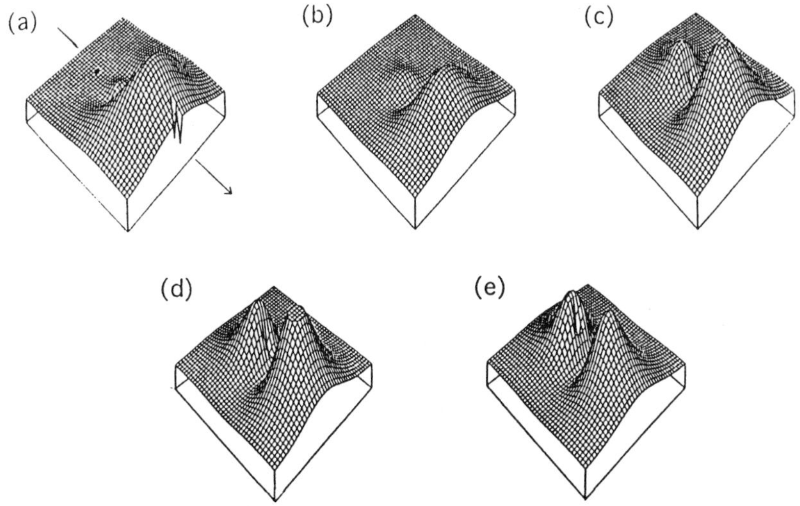

Figure 5.16 Change of electron correlation of the lowest $^1\Sigma_g^+$ state as a function of R. The first electron is located at $R = 2.9a_0$ from the center of the molecule along the molecular axis. The arrow indicates the direction of molecular axis. (a) $R = 0.2a_0$, (b) $R = 0.6a_0$, (c) $R = 1.0a_0$, (d) $R = 1.2a_0$, (e) $R = 2.0a_0$. *Source*: from [10]

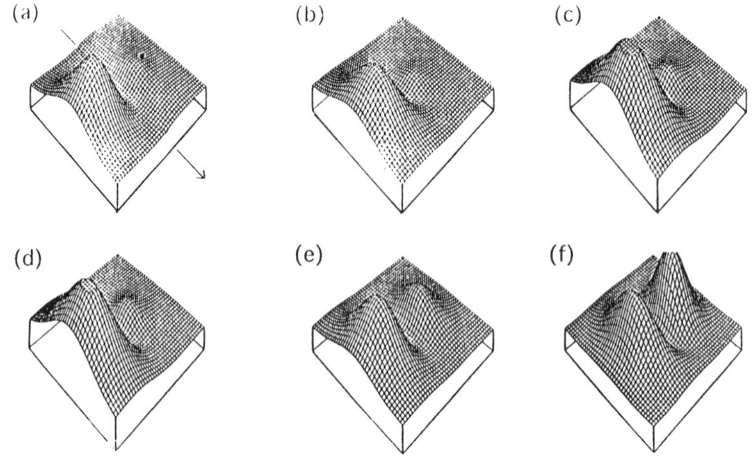

Figure 5.17 Same kind of plots as Figure 5.15 for the second lowest $^1\Sigma_g^+$ state. The arrow indicates the direction of molecular axis. $\Delta\varphi = 0$. (a) $R = 0.2a_0$, (b) $R = 0.6a_0$, (c) $R = 1.0a_0$, (d) $R = 1.2a_0$, (e) $R = 1.4a_0$, (f) $R = 2.0a_0$. *Source*: from [10]

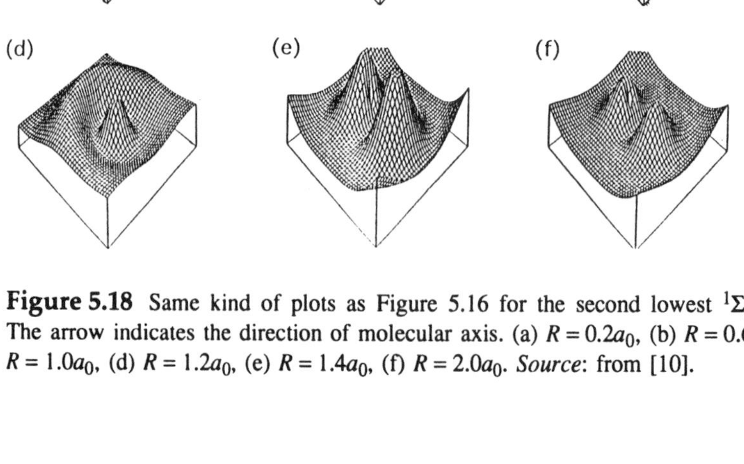

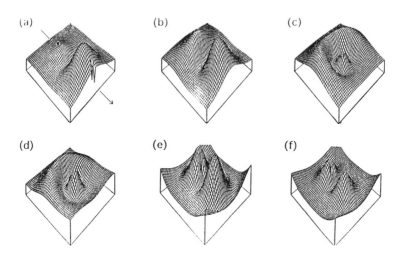

Figure 5.18 Same kind of plots as Figure 5.16 for the second lowest $^1\Sigma_g^+$ state. The arrow indicates the direction of molecular axis. (a) $R = 0.2a_0$, (b) $R = 0.6a_0$, (c) $R = 1.0a_0$, (d) $R = 1.2a_0$, (e) $R = 1.4a_0$, (f) $R = 2.0a_0$. *Source*: from [10].

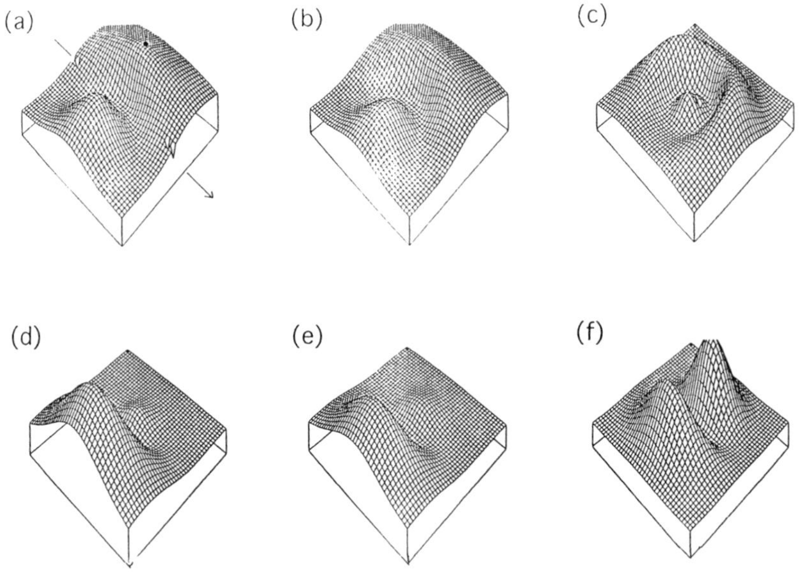

Figure 5.19 Same kind of plots as Figure 5.15 for the third lowest $^1\Sigma_g^+$ state. The arrow indicates the direction of molecular axis. $\Delta\varphi = 0$. (a) $R = 0.2a_0$, (b) $R = 0.6a_0$, (c) $R = 1.0a_0$, (d) $R = 1.2a_0$, (e) $R = 1.4a_0$, (f) $R = 2.0a_0$. *Source*: from [10].

molecular axis, however, there is much less screening by the nuclear charges. Therefore, electron correlation, if it survives at intermediate internuclear distances, will tend to appear in a direction perpendicular to the molecular axis. The second lowest $^1\Sigma_g^+$ state encounters another avoided crossing with the third lowest $^1\Sigma_g^+$ state at $R \simeq 1.2a_0$ (Figure 5.6). Figures 5.19(d) and (e) show that this avoided crossing seems to produce a strong *m*-polarization in the third lowest $^1\Sigma_g^+$ state. This is another evidence of the significant influence of avoided crossing on the electron correlation patterns. Interestingly, this indicates that the strong *m*-polarization at $R \leq 0.6a_0$ in the lowest $^1\Sigma_g^+$ state survives conspicuously, if we follow the state diabatically from Figures 5.15(a) and (b), Figures 5.17(c) and (d) to Figures 5.19(d) and (e). No such drastic changes happen along the internuclear axis.

Although electron correlation in molecular doubly excited states tends to survive in a perpendicular direction to the internuclear axis at intermediate

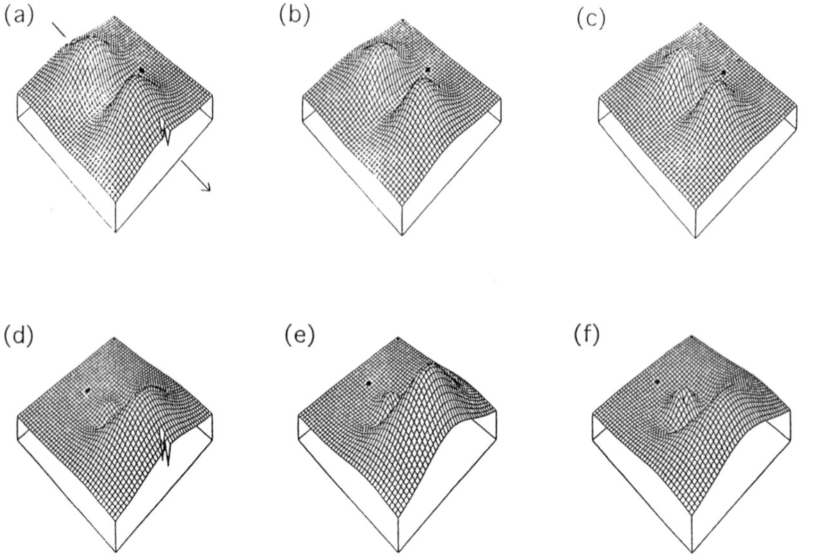

Figure 5.20 Density plots for the lowest $^3\Sigma_u$ state. In (a)–(c) the first electron is put at $2.9a_0$ from the center of the molecule along the line bisecting the nuclei. In (d)–(f) the first electron is located at $2.9a_0$ from the center of the molecule along the molecular axis. The arrow indicates the direction of the molecular axis. $\Delta\varphi = 0$ for (a)–(c). (a) $R = 0.2a_0$, (b) $R = 0.6a_0$, (c) $R = 1.2a_0$, (d) $R = 0.2a_0$, (e) $R = 0.6a_0$, (f) $R = 1.2a_0$.

R, it may appear parallel to the internuclear axis when there is a node in a plane perpendicular to the molecular axis. Figure 5.20 shows the density maps for the $^3\Sigma_u$ state which converges to $|2s2p\,^3P^0\rangle$ of He. This state has a nodal plane perpendicular to the molecular axis and electron correlation clearly shows up parallel to the molecular axis for a simple reason that the second electron cannot be near the line bisecting the nuclei.

B. Autoionization Mechanisms

Rehmus and Berry have suggested two kinds of mechanisms of atomic autoionization [56]. One mechanism, called *direct electron-electron binary collision* or *two-body interactions*, contributes when the two electrons are close to each other, whether or not they are close to the nuclei. The other mechanism requires that at least one electron has appreciable probability to be close to the nuclei. The screening of the nuclear charge by one electron induces the escape of the other electron from the ion core. This latter mechanism, called *dynamical screening* or *generalized* S_N*-2 mechanism*, is directly related to the electron correlation in doubly excited state: if one electron can penetrate close to the nucleus while the other electron moves away from it, dynamical screening is expected to dominate in autoionization processes. On the other hand, when the electrons have high probability of being on the same side of the nuclei, a large contribution from the direct electron-electron collision mechanism is expected. Actually, Renmus and Berry found that dynamical screening was important in the autoionization of the lowest rotor state $|2s^2\ {}^1S^e\rangle$, where electron correlation is very strong. For a radially bending state $|2p^2\ {}^1S^e\rangle$, where electrons have highest probability of being on the same side of the nuclei, direct electron-electron collision was found to dominate.

Figure 5.21 shows the two-body transition density matrix $\rho_{n\to\varepsilon}(\xi_1\eta_1;\xi_2\eta_2;\Delta\varphi)$ (see Equation 5.27) between the bound state $\psi_n^{(0)}$ and the continuum φ_E for the lowest ${}^1\Sigma_g^+$ state. At $R = 0.2a_0$, the feature of the dynamical screening is clearly seen. The dominant contribution to the resonance width results from the screening of the nuclear charges by one of the electrons because the strong angular correlation in the lowest rotor state $|2s^2\ {}^1S^e\rangle$ survives and the electrons have an appreciable probability of being close to the nucleus at this small internuclear distance. It is very interesting to notice the difference between Figure 5.21(b) and Figure 5.21(e). Figure 5.21(b) shows that dynamical screening is important when one electron is put perpendicular to the molecular axis, while Figure 5.21(e) shows that the direct electron-electron collision dominates when one electron is located along the molecular axis. This feature may be explained by *m*-polarization. Electron correlation tends to appear in perpendicular direction to the molecular axis at intermediate R and thus the dynamical screening mechanism dominates in Figure 5.17(b). As the nuclei separate further, however, the independent particle picture develops and the direct electron-electron collision dominates. It can also be seen that the absolute values of $\rho_{n\to\varepsilon}(\xi_1\eta_1;\xi_2\eta_2;\Delta\varphi)$ increase as R increases, suggesting that the magnitudes of the resonance widths are larger when contribution of the direct electron-electron collision mechanism increases. So very roughly speaking, strong electron correlation leads to small autoionization rate.

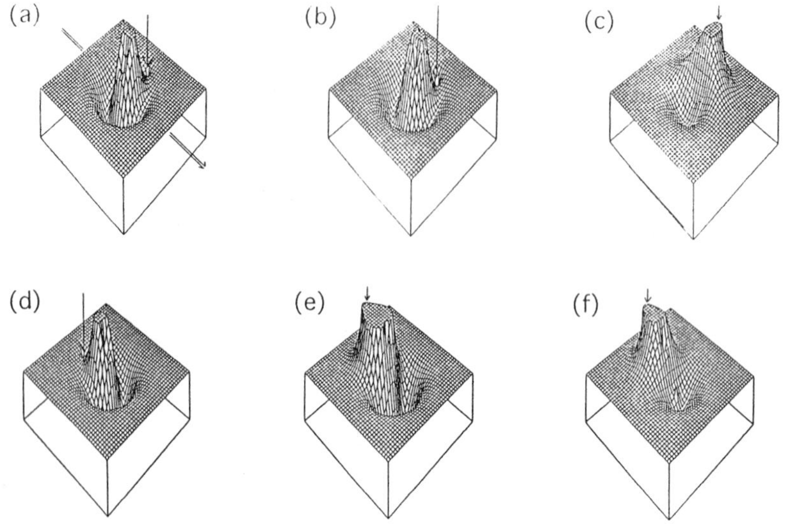

Figure 5.21 Plots of $\bar{\rho}$ for the lowest $^1\Sigma_g^+$ state. The arrows in the figure indicate the positions of the first electron. The double arrow indicates the direction of the molecular axis. In (a)–(c) the first electron is located at $2.9a_0$ from the center of the molecule along the line bisecting the nuclei. In (d)–(f) the first electron is located at $2.9a_0$ from the center of the molecule along the molecular axis. The figures are drawn by different scaling factors α. (a) $R = 0.2a_0$, $\alpha = 5.0$, (b) $R = 0.6a_0$, $\alpha = 0.5$, (c) $R = 1.4a_0$, $\alpha = 0.05$, (d) $R = 0.2a_0$, $\alpha = 5.0$, (e) $R = 0.6a_0$, $\alpha = 0.25$, (f) $R = 1.4a_0$, $\alpha = 0.0025$. *Source*: from [10]

Figure 5.22 presents the same kind of plots for the third lowest $^1\Sigma_g^+$ state which correlates to the radially bending $|2p^2\ {}^1S^e\rangle$ atomic state. Since the two electrons have the largest probability to be on the same side, direct electron-electron collision accounts for most of the resonance width, as expected. For the second lowest $^1\Sigma_g^+$ state, direct electron-electron collision accounts for most of the resonance width, as expected. For the second lowest $^1\Sigma_g^+$ state, direct electron-electron collision dominates at all internuclear distances (Figure 5.23), in contrast with the observation that this state shows a clear indication of angular correlation at small internuclear distances. This may be explained by refreshing the requirement of the dynamical screening mechanism: at least one electron must penetrate close to the nucleus. Since

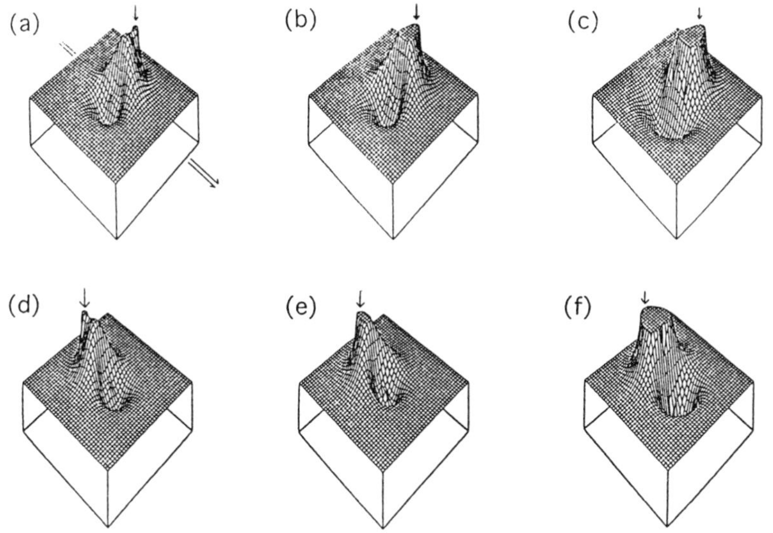

Figure 5.22 Same as Figure 5.21 for the third lowest $^1\Sigma_g^+$ state. The double arrow indicates the direction of the molecular axis. (a) $R = 0.2a_0$, $\alpha = 3.0$, (b) $R = 0.6a_0$, $\alpha = 0.2$, (c) $R = 1.4a_0$, $\alpha = 0.025$, (d) $R = 0.2a_0$, $\alpha = 3.0$, (e) $R = 0.6a_0$, $\alpha = 0.2$, (f) $R = 1.4a_0$, $\alpha = 0.025$. *Source*: from [10].

this state converges to $|2p^2\ ^1D^e\rangle$ atomic state, the electrons have less penetration to the nucleus than the lowest $^1\Sigma_g^+$ state.

5.5 DYNAMICS OF SUPEREXCITED STATES OF MOLECULES

The dynamics of electronically highly excited states of atoms and molecules is very fertile and quite different from that involving ground and lower excited states [4]. Since there are large number of these highly excited states interacting with each other by various kinds of coupling mechanisms, their dynamics is quite complicated. Many decay channels are competing and a unified view or treatment of their dynamic processes, which have been considered separately so far, is necessary. For example, the dissociative first kind of SES not only participates in dissociation but also plays an important role in autoionization processes. Actually, competition between autoioniza-

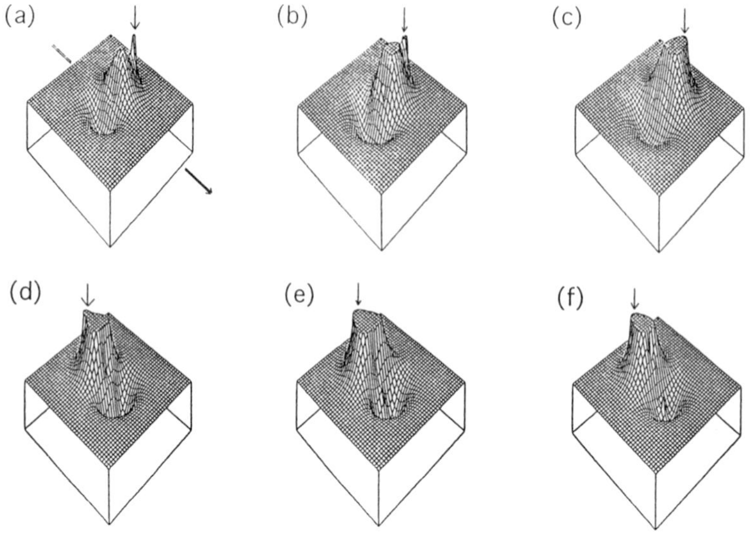

Figure 5.23 Same as Figure 5.21 for the second lowest $^1\Sigma_g^+$ state. The double arrow indicates the direction of the molecular axis. (a) $R = 0.2a_0$, $\alpha = 3.0$, (b) $R = 0.6a_0$, $\alpha = 0.2$, (c) $R = 1.4a_0$, $\alpha = 0.1$, (d) $R = 0.2a_0$, $\alpha = 3.0$, (e) $R = 0.6a_0$, $\alpha = 0.2$, (f) $R = 1.4a_0$, $\alpha = 0.01$. *Source*: from [10].

tion and dissociation is one of the most important features of SES. In the case of highly excited Rydberg states, there is a strong coupling between electronic and nuclear degrees of freedom and the Born-Oppenheimer approximation breaks down. The theoretical treatment of these states therefore must be different from the methods for lower states. Instead of incorporating the (radial or rotational) nonadiabatic coupling as perturbation, the deviation of the interaction between core and Rydberg electrons from pure Coulombic interaction is taken as perturbation. This is reasonable because the Rydberg electron is far away from the nuclei and experiences the Coulombic interaction with the nuclei most of the time. The effects of the occasional penetration into the core are taken as perturbation. In other words, the physics of the Rydberg electron is different in the inner and outer regions of coordinate space, and thus a kind of frame transformation between the two regions should be considered. This idea is actually the essence of the multichannel quantum defect theory (MQDT) as explained later.

Because of the high density of the highly excited states in high energy region, the states have large contribution to the optical oscillator strength and are quite sensitive to external perturbation such as collision with other particles or an external field. Progress in multiphoton ionization (MPI) techniques and synchrotron radiation technology has provided very efficient means to study these highly excited electronic states, and correspondingly, interest in these states is growing rapidly as well. Compared with atomic highly excited states, molecular states have nuclear degrees of freedom in addition to the electronic degrees of freedom. Hence highly excited states are expected to be involved in many dynamic processes such as,

$e + AB^+$	$\rightarrow AB^*$	$\rightarrow AB^+ + e$	Elastic scattering
$e + AB^+ (v)$	$\rightarrow AB^*$	$\rightarrow AB^+ (v') + e$	Vibrational excitation
$e + AB^+$	$\rightarrow AB^*$	$\rightarrow A + B^*$	Dissociative recombination
$e + AB^+$	$\rightarrow AB^*$	$\rightarrow A^+ + B^-$	Ion-pair formation
$\mathrm{A}^* + \mathrm{B}$	$\rightarrow AB^*$	$\rightarrow AB^+ + e$	Associative ionization
$A^* + B$	$\rightarrow AB^*$	$\rightarrow A + B^+ + e$	Penning ionization
$A^* + BC$	$\rightarrow ABC^*$	$\rightarrow AB^+ + C + e$	Chemical ionization
$AB + h\nu$	$\rightarrow \mathrm{AB}^*$	$\rightarrow \mathrm{A} + \mathrm{B}$	Photodissociation
$\mathrm{AB} + h\nu$	$\rightarrow \mathrm{AB}^*$	$\rightarrow \mathrm{AB}^+ + \mathrm{e}$	Photoionization
$\mathrm{AB} + h\nu$	$\rightarrow \mathrm{AB}^*$	$\rightarrow \mathrm{BA}$	Photoisomerization

and so on. The unstable highly excited states AB^* play a role as intermediate (resonance) states; and revealing their properties can provide a unified framework of the various processes. For instance, Figure 5.24 gives a unified view of the various dynamic processes in a system involving one manifold of Rydberg states and one dissociative SES [57]. There are essentially the two basic interactions: the electronic coupling V(R) and the rovibrational coupling represented by quantum defect. The essential physics governing the various dynamic processes takes place inside the rectangular area depicted by dashed line and the differences among these processes lie only in the boundary conditions.

Platzman [58] introduced the concept of *superexcited states* to point out the importance of these highly excited states in radiation chemistry. The superexcited states is defined as a state whose internal energy is higher than the first ionization potential (at least in certain nuclear configurations). Hence it autoionizes but can also end up with dissociation into neutral fragments.

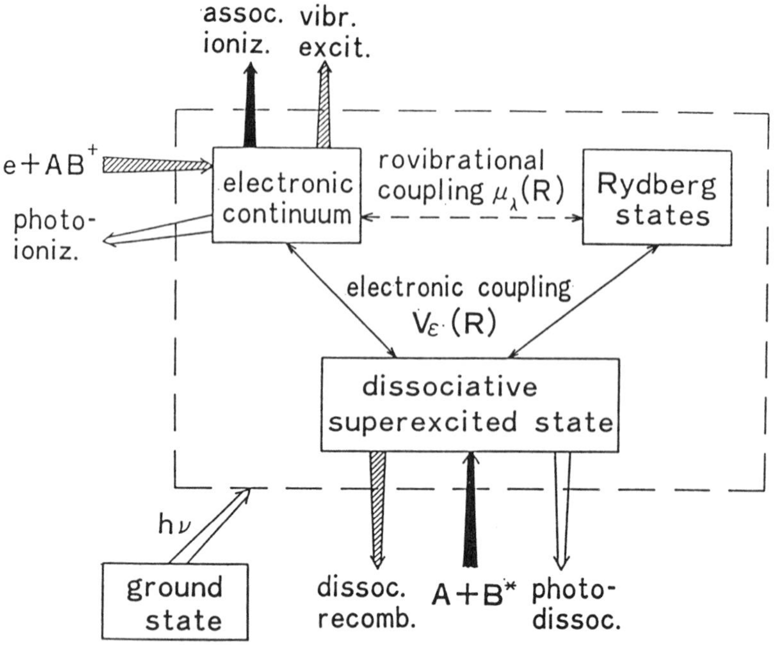

Figure 5.24 A unified view of various dynamic processes involving superexcited states. *Source*: from [57].

As was explained in the Introduction, the superexcited states are classified, according to the autoionization mechanisms, into the following two kinds:

1. Multiple- (or inner-shell-) excited states
2. Rovibrationally excited Rydberg states

In the first kind of SES, the electronic energy is higher than the ionization potential, at least in a certain range of nuclear configuration (Figure 5.25(a)). The autoionization occurs by interelectronic repulsion even if the nuclei are clamped as explained previously. The coupling to the electronic continuum is expressed by the electronic coupling $V(R)$ (Equation 5.16). The autoionization lifetime $\tau(R)$ at fixed R is $\tau(R) = \hbar/\Gamma(R)$ (Equation 5.18). This type of autoionization, called *electronic autoionization*, is closely related to the electron correlation patterns of SES (described in Section 5.4).

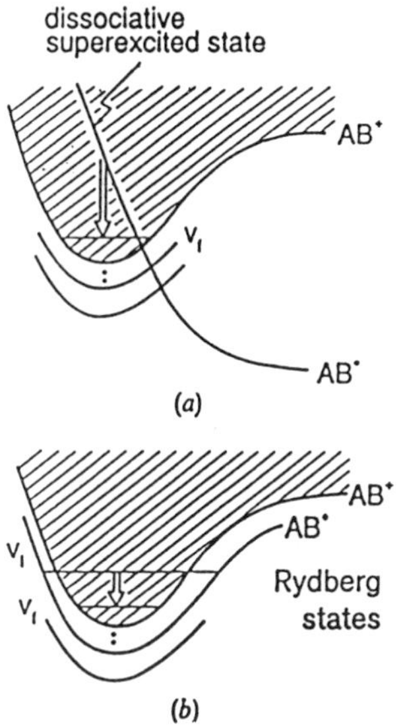

Figure 5.25 A schematic potential diagram for autoionization. (a) autoionization of the dissociative superexcited state of the first kind (electronic autoionization), (b) autoionization of the superexcited state of the second kind (vibrational autoionization). *Source*: from [4].

In the second kind of SES, autoionization results from the coupling between electronic motion and nuclear motion, that is, from the energy transfer from the nuclear degree of freedom to the electronic degree of freedom (Figure 5.25(b)). In this type of autoionization, called *vibrational autoionization*, the following propensity rule holds well: the transition with $\Delta v \ (\equiv v_i - v_f) = 1$ (or the possible minimum) is most probable, where $v_i(v_f)$ is the vibrational quantum number of the molecule (the ion). This is because the Rydberg states are well represented by the quantum defect $\mu(R)$ and the autoionization results from their R-dependence. The linear term dominates when the R-dependence of the quantum defect is not very strong. The quantum defect is the quantity to represent the overall electrostatic interactions other than Coulombic attraction. In general, both of these two kinds of SES are involved in the autoionization processes of many molecules, as will be demonstrated below.

A. Multichannel Quantum Defect Theory

We first outline the multichannel quantum defect theory (MQDT) to describe a theoretical framework for the dynamics of SES. Various dynamic processes involving SES will be discussed subsequently. Although there have been fragmentary studies on dynamic processes such as associative ionization [32–34], ion-pair formation [59], and transfer ionization [60], we will focus our attention here on photoionization, photodissociation, and dissociative recombination processes since these have been extensively studied.

The quantum defect theory (QDT) has emerged as a powerful tool to treat various dynamics in atomic and molecular physics. The theory was established by Seaton and coworkers to treat atomic Rydberg states and continua in a unified way [61]. Later, Fano extended it to molecules [62].

There are many versions of MQDT that have been adopted to specific problems; we briefly outline here the two-step formalism developed by Giusti [27, 63] since this version is most relevant to dynamic processes involving dissociative SES (i.e., dissociative recombination and photoionization/photodissociation processes). Two basic parameters, the quantum defect function $\mu(R)$ and electronic coupling matrix $V(R)$ between the dissociative SES and the electronic continuum interplay in these dynamic processes. The reactance matrix associated with the electronic coupling $V(R)$ is described by the integral equation

$$K_{jj'}(E,E') = \langle jE|V(R)|j'E'\rangle + P\int dE'' \frac{\langle jE|V(R)|j''E''\rangle}{E-E''} k_{j''j'}(E'',E') \tag{5.28}$$

where j, j', and j'' represent the vibrational (ν) or molecular dissociation (d) state. In the first-order perturbation, $K_{jj'}(E,E')$ is approximated as

$$K_{jj'}(E,E') = \begin{cases} \langle \nu/V(R)|F_d\rangle \equiv \xi_\nu/\pi & \text{for } j=\nu \text{ and } j'=d \\ 0 & \text{otherwise} \end{cases} \tag{5.29}$$

where F_d is the nuclear dissociation wave function. The eigenchannels $\{\alpha\}$ and eigenphase shift η_α are obtained by diagonalizing the reactance matrix,

$$\sum_{j'} K_{jj'} U_{j'\alpha} = -\frac{1}{\pi} \tan \eta_\alpha U_{j\alpha} \tag{5.30}$$

The phase shift η_α is given by

$$\eta_\alpha = \begin{cases} \pm \tan^{-1}\xi, & \alpha = 1 - I \\ 0 & ((I-2) \text{ fold degenerate}) \end{cases} \tag{5.31}$$

where I is the total number of channels (one dissociation channel and $I - 1$ vibrational channels) and

$$\xi^2 = \sum_v \xi_v^2 \tag{5.32}$$

Following the standard procedure of *QDT*, the scattering matrix S is obtained as

$$S = (1 + iR)(1 - iR)^{-1} \tag{5.33}$$

where the reactance matrix R is related to the *short-rage* reaction matrix $\mathcal{R}$ as

$$R = \mathcal{R}_{oo} - \mathcal{R}_{oc}[\mathcal{R}_{cc} + \tan(\pi\nu)]^{-1}\,\mathcal{R}_{co} \tag{5.34}$$

Here $o(c)$ denotes open (closed) channel. ν is a diagonal matrix with an element ν_j representing a parameter to measure energy from the jth closed channel (energy $= -1/2\nu_j^2$ in a.u.). Matrix $\mathcal{R}$ is obtained by

$$\mathcal{R} = \mathcal{S}\mathcal{C}^{-1} \tag{5.35}$$

where $\mathcal{S}$ and $\mathcal{C}$ are matrices defined as

$$\mathcal{C}_{v^+,\alpha} = \sum_v \langle v^+ | \cos[\pi\mu_\Lambda(R) + \eta_\alpha] | v \rangle U_{v\alpha} \tag{5.36}$$

and

$$\mathcal{S}_{v^+,\alpha} = \sum_v \langle v^+ | \sin[\pi\mu_\Lambda(R) + \eta_\alpha] | v \rangle U_{v\alpha}$$

for the ionization channel ($i = v^+$), and

$$\mathcal{C}_{d,\alpha} = U_{d\alpha} \cos \eta_\alpha \tag{5.37}$$

and

$$\mathcal{S}_{d,\alpha} = U_{d\alpha} \sin \eta_\alpha$$

for the dissociation channel; the rotational degree of freedom is neglected here for simplicity. Once the S-matrix is obtained, the various dynamic processes can be investigated in a unified way. Photoionization and photodissociation cross-sections are expressed as

$$\sigma_\beta = \frac{4\pi^2}{3} \alpha \hbar \omega |D_\beta|^2 \quad (\beta = \text{ionization or dissociation})$$

where α is the fine structure constant and $\hbar\omega$ is the photon energy. D_β represents the reduced dipole matrix element which is given by

$$D_\beta = D_\beta^{(open)} - \chi_{oc}[\chi_{cc} - \exp(-2\pi i \nu)]^{-1} D_\beta^{(closed)} \tag{5.38}$$

The first term represents the direct process and the second term describes the indirect process via the closed-channel Rydberg states. The matrix χ is related to the $\mathcal{R}$ matrix by the relation

$$\chi = (1 + i\mathcal{R})(1 - i\mathcal{R})^{-1} \tag{5.39}$$

This two-step formalism does not include interactions between the dissociation channels. These interactions will not affect the total cross-section for dissociation but will significantly influence the branching ratios of dissociation. Thus, these interactions should be considered in order to evaluate observable phenomena such as the branching ratios and angular distributions of the photodissociation products. Another assumption made in this approach is the first-order perturbation approximation for $K_{jj'}(E,E')$. Takagi [64] showed that this lowest-order approximation to $K_{jj'}$ is quite accurate in the case of the lowest $^1\Sigma_g$ doubly excited states of H_2. This problem, however, should be more carefully examined.

B. Photoionization and Photodissociation

When a molecule is excited into an energy region above ionization threshold, it is usually located above a dissociation limit also. Generally, there two kinds of SES coexist (as in Figure 5.24). Various mechanisms of ionization and dissociation should be taken into account at the same time. Autoionization of rovibrationally excited Rydberg states, for instance, can occur in the following two mechanisms:

1. Direct (vibrational) autoionization, where vibronic coupling plays an important role, and
2. Indirect (electronic) autoionization, where the dissociative superexcited states serve as an indirect channel between the rovibrationally excited Rydberg states and the electronic continuum.

There are some cases where only one mechanism dominantly contributes to the autoionization processes. Autoionization of H_2 molecule studied by Jungen et al. [65–69] is such an example. Only one-photon ionization process was considered and thus no dissociative superexcited state is involved since the lowest dissociative superexcited state near the ionization potential is of $^1\Sigma_g^+$ symmetry $((2p\sigma_u)^2)$ (Figure 5.26). A MQDT version developed by Jungen and Dill [66] was used to calculate the effect of vibrational-rotational autoionization on the total and partial oscillator strength distributions and photoelectron angular distribution in H_2 for excitation between 800–750Å; the agreement with high-resolution photoionization data of Dehmer and Chupka [70] was excellent. Specifically, it was found that vibrational and rotational preionization cannot be treated as separate processes in this spectral region. The importance of dissociative superexcited states have been noticed in autoionization process in many molecules. Morin et al. [71] interpreted a complicated partial cross-section for vibrational excitation of O_2 where electronic autoionization dominates. Raoult et. al. [72] presented a detailed study of electronic autoionization in the Hopfield series of N_2 where vibrational autoionization is not strong. In general, however, these two mechanisms cannot be distinguished in the autoionization process of the rovibrationally excited Rydberg states.

Progress in REMPI techniques [3] has enabled us to pinpoint the energy and the symmetry of the excited states; thus it is now possible to study the energy region where the interplay of the direct and indirect mechanisms of autoionization is important. NO is one of the best-known system for which

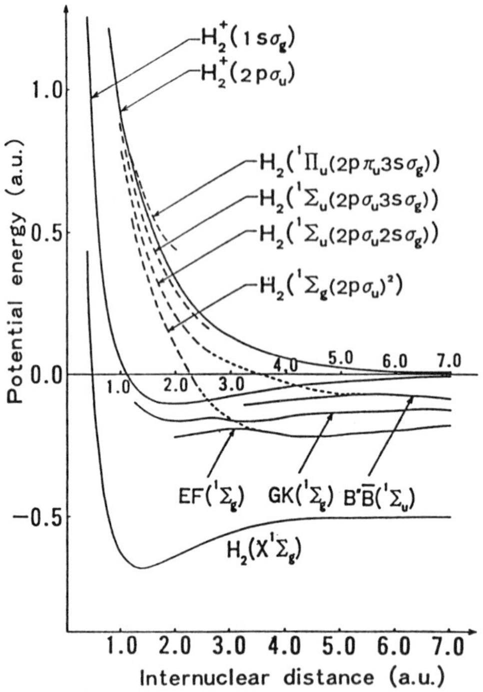

Figure 5.26 Potential energy curves of H_2 near the lowest ionic state. *Source*: from [4].

the role of the autoionization mechanism, involving the dissociative super excited states, is appreciated. Nakashima et al. [28] analyzed the REMPI experiment on NO done by Achiba and Kimura [29]. MQDT is used to interpret the photoionization and photoelectron kinetic energy spectra obtained by (2+1)-multiphoton process. Figure 5.27 shows the photoelectron kinetic energy spectra for $nd\sigma$ (n = 5–9) obtained by excitation from $v = 4$ vibrational level of the electronic ground state. Although $\Delta v = 1$ ($v^+ = 3$) transitions give the strongest peaks for $n \geq 8$, there is a substantial intensity of the transitions corresponding to $\Delta v > 1$ for $n < 8$. This suggests that indirect electronic autoionization via the repulsive superexcited state ($B'\,^2\Delta$) contributes substantially to the autoionization widths. Table 5.9 presents the comparison of the experimentally determined branching ratios of the final vibrational levels of NO^+ ion with those calculated by the Franck-Condon

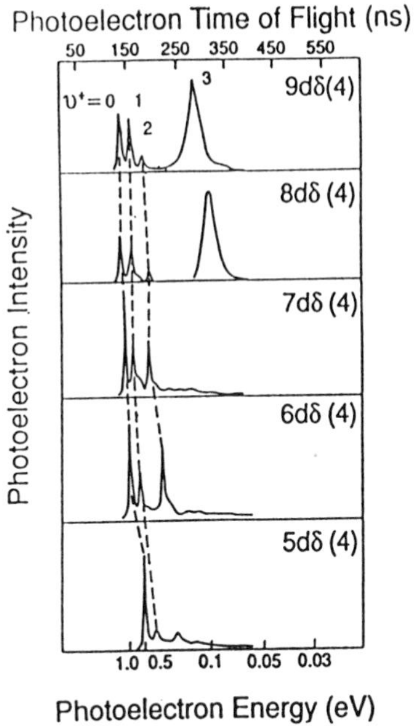

Figure 5.27 Photoionization kinetic energy spectra of NO for $nd\delta$ (ν = 4) (n = 5–9). *Source:* from [28].

Table 5.9 Comparison of the Experimental Values of Photoelectron Branching Ratios with the Franck-Condon Analysis (FCF) and the MQST Analysis

	$v_r = 0$			$v_r = 1$			$v_r = 2$		
State	Exp	FCF	MQDT	Exp	FCF	MQDT	Exp	FCF	MQDT
$5d\,\delta$	1.00	1.00	1.00	0.15	0.10	0.10	(0.15)	0.18	0.21
$6d\,\delta$	1.00	1.00	1.00	0.43	0.40	0.39	(0.61)	0.30	0.32
$7d\,\delta$	1.00	1.00	1.00	0.51	0.50	0.50	(0.58)	0.48	0.54
$8d\,\delta$	1.00	1.00	1.00	0.80	0.85	0.86	0.23	0.20	0.20
$9d\,\delta$	1.00	1.00	1.00	0.92	0.96	0.97	0.30	0.15	0.16

The numbers in parentheses are less accurate compared to the other experimental values.

analysis and the MQDT analysis, both of which were performed *without* considering the direct process. Good agreement between experimental and calculated branching ratios suggests that the vibrational autoionization can be disregarded for the $\Delta v \geq 2$ transitions. This dominance of the indirect autoionization mechanism is generally expected, because the electronic coupling $V(R)$ seems to be quite strong compared with the R-dependence of the quantum defects. When a molecule is excited to energies where both dissociation and ionization channels are open, a competition between dissociation and ionization naturally occurs; and simultaneous treatment of the two dynamic processes becomes necessary. Figure 5.28 shows relevant diabatic potential energy curves of $^2\Pi$ state in NO [27].

In energy region I, Rydberg-valence interactions between the $np\pi$ Rydberg state and $^2\Pi$ *valence* state cause perturbations of the vibronic level positions and band intensities in photoabsorption from the ground state NO($X^2\Sigma$). Region II denotes the energy region where predissociation occurs and Region III denotes the region where competition between predissociation and autoionization occurs. Interesting differences were observed in photoionization and photoabsorption spectra [73, 74]. Giusti-Suzor and Jungen [2, 27] employed the two-step MQDT formalism to analyze these spectra to obtain a good agreement between the experimental spectra and MQDT calculations. They reproduced the strong variation of the widths with vibrational quantum number of the $5p\pi$ state very well, by performing MQDT calculation including Rydberg-valence interaction only (i.e., setting the R-dependence of the quantum defect equal to zero, thus eliminating the *direct* mechanism of autoionization). They also reproduced the breakdown of the $\Delta v = 1$ selection rules, thereby concluding that:

1. The autoionization in NO is not primarily vibrational, but is partly induced by the *electronic* Rydberg-valence interaction, and
2. The predissociation-autoionization widths of the ($np\pi$, v) vibronic levels can also be interpreted in terms of the Rydberg-valence state interactions.

The importance of the dissociative first kind of SES was actually first pointed out by Nakamura; the simple qualitative analysis can easily predict a possible big effect of the state in photoelectron spectrum [75]. Since these kinds of states are not well known, care should be taken in the analysis of photoionization experiment. Giusti-Suzor and Jungen [27] also found that the

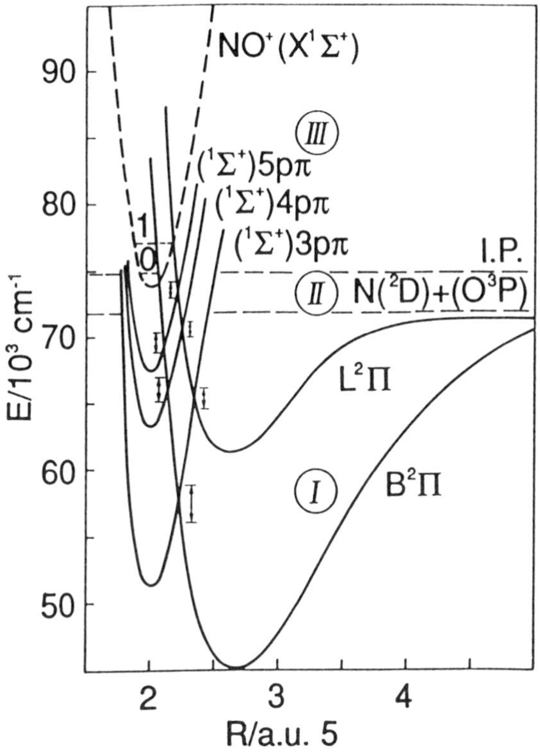

Figure 5.28 Diabatic potential energy curves of $^2\Pi$ of NO. The arrows indicate the avoided crossings in the adiabatic curves as a result of Rydberg-valence state interactions. Three energy ranges are marked and correspond to Zones I (spectral perturbations), II (predissociation), and III (competition between predissociation and preionization). *Source*: from [27].

step-like structure is expected to appear when the ratio of the magnitude of the R-dependence of the quantum defect to the Rydberg-valence state interaction is much less than unity.

The role of the first kind of SES in multiphotonionization processes has also been studied on H_2 molecule [54, 55]. In (3+1) REMPI processes [35, 37, 76–82], for example, via $B\ ^1\Sigma_u$ or $C\ ^1\Pi_u$ state, it has been observed that dissociation competes with ionization. Also the vibrational distribution of H_2^+ in ionization process showed strong deviation from the Franck-Condon

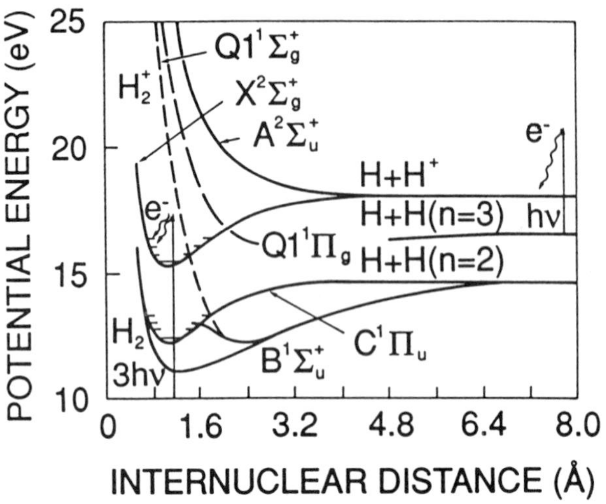

Figure 5.29 Potential energy curves of H_2 relevant to the multiphoton excitation scheme. The straight arrows represent the final photons absorbed. The wiggly arrows symbolize the emitted electrons and their corresponding kinetic energies. *Source*: from [35].

behavior. These observations were attributed to the significant role of the doubly excited states in these processes.

Pratt et al. [76] measured the photoelectron spectra of H_2 in the states $C\ ^1\Pi_u$, $v = 0$–4. Since $C\ ^1\Pi_u$ is the low-lying Rydberg state and its potential curve resembles that of the $1\sigma_g\ H_2^+$ state, direct ionization is expected to show Franck-Condon behavior (i.e., $\Delta v = 0$). Photoelectron spectra calculated by Dixit et al. [77], which assumed only direct dissociation mechanism, agreed well with experiment for $v = 0, 1$, but showed increasing disagreement for $v = 2, 3, 4$. By calculating cross-sections for direct dissociation of the $C\ ^1\Pi_u$ state and for excitation of the lowest $^1\Pi_g$ doubly excited states, Chupka [78] showed that autoionization of the $^1\Pi_g$ state (Figure 5.29) contributes more than the direct ionization for $v = 4$; Chupka attributed the non-Franck-Condon behavior of the ionization process to the $^1\Pi_g$ doubly excited state. Hickman [79] treated the competition between photodissociation and photoionization of the $C\ ^1\Pi_u$ state. The direct ionization pathway was neglected and it was shown that autoionization of the $^1\Pi_g$ doubly excited

Table 5.10 Ratio of Dissociation into H (n = 3) + H (n = 4) to Form H_2^+ in (3+1)-Photon Resonant Excitiation of H2 via the $C^1\Pi_u^+$ Pathway; Values are Given for the Percentage B-State Character Acquired in Mixing of the Π^+ Levels with the Nearest-Neighbor B-State Level

Transition level	Dissociation-ionization	B-State character	Excitation wavelength
Q1 v=0	~ 0	0	3029.31
Q1 v=1	< 0.02	0	2960.39
Q1 v=2	< 0.02	0	2898.29
Q1 v=3	0.03	0	2842.39
Q2 v=3	0.07	0	2845.85
Q1 v=4	~ 0.10	0	2791.72
R1 v=0	~ 0	0.08	3025.50
R2 v=0	~ 0	0.17	3027.09
R1 v=1	0.12	0.20	2956.95
R2 v=1	0.20	0.60	2959.74
R1 v=2	0.16	1.6	2895.20
R2 v=2	~ 0.95	12.0	2897.38
R1 v=3	0.55	13.0	2839.16
R1 v=4	0.04	1.1	2789.06

Source: from [35]

state is negligible compared to photodissociation for v = 0, 1, while it becomes important for $v \geq 2$. Although the relative magnitude of the autoionization and direct dissociation was not evaluated, these results suggest that non-Franck-Condon pattern of the photoelectron spectra of $C\,^1\Pi_u$ state originates from $^1\Pi_g$ doubly excited state. An examination of non-Franck-Condon distribution of ravibrational levels of the H_2^+ product resulting from (3+1) REMPI via $C\,^1\Pi_u$ state led O'Halloran et al. [80] to the similar conclusion. Xu et al. [35] examined photoelectron spectra obtained by Q and R transitions to v = 0–4, n = 1–3 states of the $C\,^1\Pi_u$ state, and found that Q and R transitions resulted in different dissociation/ionization ratios (Table 5.10), showing that the competition between dissociation and ionization depends significantly on the symmetry, and vibrational and rotational quantum numbers of the intermediate state. Q transitions led to significantly lower dissociation/ ionization ratios than R transitions. Xu et al. explained this difference in terms of the difference in symmetry of the C state: Q transition leads to the $C\,^1\Pi_u^-$ state while R transition results in the $C\,^1\Pi_u^+$ state. Xu et al. proposed that both the $^1\Pi_u^-$ and $^1\Pi_u^+$ states can couple radiatively to the $^1\Pi_g$ state (which can autoionize). However, the $^1\Pi_u^+$ state is rotationally coupled to the $B\,^1\Sigma_u^+$ state while $^1\Pi_u^-$ cannot and this would make the difference. Since the $B\,^1\Sigma_u^+$ state has a broad minimum with large R, the $^1\Pi_g$ doubly excited state radiatively coupled to the $B\,^1\Sigma_u^+$ state has a smaller tendency to autoionize (because of large R). Hence R transition to the $^1\Pi_u^+$ state will show larger dissociation/ ionization ratios.

The contribution of doubly excited states has also been recognized in multiphoton excitation of H_2 via the $B\,^1\Sigma_u^+$ state (Figure 5.29). Bonnie et al. [81] observed that one-photon excitation from $B\,^1\Sigma_u^+$, $v = 8, 9$ results mainly in dissociation, while direct ionization completely disappears. Verschuur et al. [82] found that two-photon dissociation from the $B\,^1\Sigma_u^+$ state into channels with $n \geq 3$ competes with one-photon ionization. Verschuur et al. [37] also observed significant dissociation into $n = 2$ channel following one-photon excitation of the $B\,^1\Sigma_u^+$ state. For excitations from lower vibrational levels of $^1\Sigma_u^+$, the $Q1\,^1\Sigma_g^+(2p\sigma_u)^2$ state is excited at regions isolated from the other doubly excited states. Thus, Verschuur et al. attributed dissociation to the $Q1\,^1\Sigma_g^+$ state. Excitation to the $^1\Sigma_g^+$ state was found to have a large oscillator strength compared to direct ionization. The H_2^+ vibrational distribution showed strong non-Franck-Condon behavior, as expected.

The contribution of the rovibrationally excited Rydberg states has not been elucidated in the context of the competition between dissociation and ionization; however, this kind of SES is expected to play a role in two occasions. First, beyond the stabilization point, the doubly excited states couple with high vibrational levels of the Rydberg states with high n. Thus latter states may cause vibrational autoionization. If the cross-sections for vibrational autoionization are large, the second kind of SES must be included in the treatment. When the molecule is excited to high vibrational levels of the Rydberg states to energies above the ionization threshold, vibrational autoionization of these levels may also contribute to ionization processes. In these cases, different kinds of ionization channels (namely direct ionization, electronic autoionization, and vibrational autoionization) will interfere with each other and compete with dissociation channel, yielding very rich and interesting dynamics.

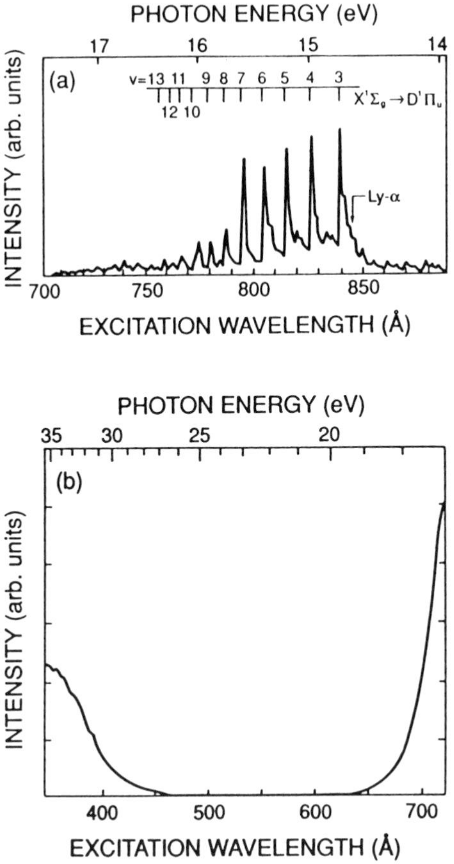

Figure 5.30 Lyman-α excitation spectra (a) and (b) in the wavelength ranges of 70–89nm and 35–72nm, respectively. *Source*: from [76].

Dissociation via SES of H_2 has been studied by (observing) fluorescence from excited fragments [36, 83]. Synchrotron radiation was used as excitation source and the Lyman-α excitation spectrum of the excited hydrogen atoms produced by the photodissociation of the doubly excited states of H_2 was observed. Lyman-α-fluorescence, resulting from the excitation wavelength 890 ~ 700Å, is attributed to the slow H(2*l*) atoms produced by the predissociation of vibrationally excited molecular *np*σ and *np*π Rydberg states (Figure 5.30(a)).

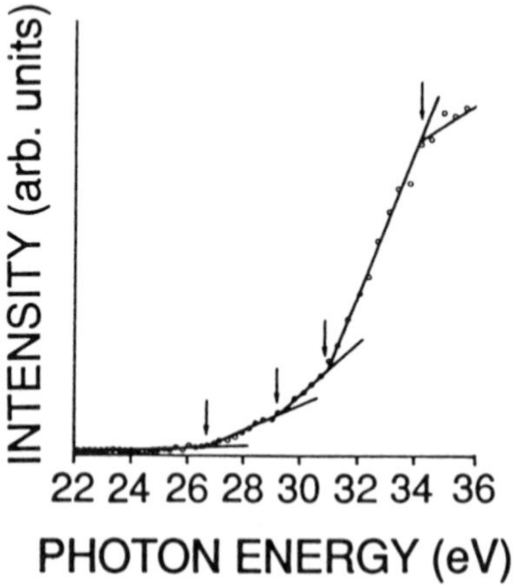

Figure 5.31 An attempt to find threshold energies in the Lyman-α excitation spectrum. *Source*: from [76].

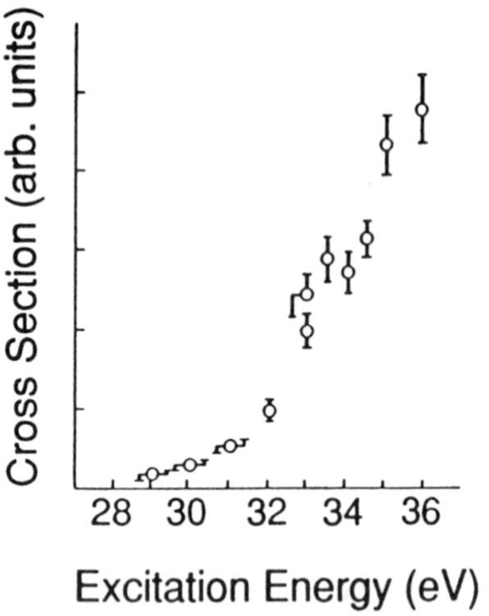

Figure 5.32 Excitation spectrum of the process: $H_2 + h\nu \rightarrow H_2^{**} \rightarrow H(2p) + H(2p)$. *Source*: from [36].

Lyman-α-fluorescence in the shorter wavelength region below 470Å (~26 eV) is due to the neutral fragmentation of the doubly excited states of H_2 (Figure 5.30b). This is the first observation of photodissociation via the doubly excited states. Three thresholds were found (26.6 eV, 29.2 eV, and 30.9 eV) in the Lyman-α excited spectrum and they were attributed to the $Q_1\ ^1\Sigma_u^+(1)$ $(2p\sigma_u, 2s\sigma_g)$, closely lying $Q_1\ ^1\Pi_u(1)(2p\sigma_u, 3d\pi_g)$, and/or $^1\Sigma_u^+(2)(2p\sigma_u, 3s\sigma_g)$ states, and $Q_2\ ^1\Pi_u(1)(2p\pi_u, 2s\sigma_g)$ state, respectively (Figure 5.31 and [45, 50]).

Lyman-α Lyman-α coincidence measurements by Arai et al. [36] on the dissociation process

$$H_2 + h\nu \rightarrow H^{**} \rightarrow H(2p) + H(2p)$$

also provide valuable information on the photodissociation of SES. Figure 5.32 presents the excitation spectrum, and shows a threshold just below 29 eV. This threshold is close to the second and third thresholds mentioned previously. Since the doubly excited $Q_2\ ^1\Pi_u(1)$ state dissociates into two $H(2p)$ atoms, this state is considered to be the precursor SES responsible for this photodissociation process. There have been some experimental studies on the photodissociation processes involving SES of polyatomic molecule, such as C_2H_2, SiH_4, H_2S, BF_3, BCl_3, CF_4, CF_3X (X = H, Cl, Br), GeH_4, and isomers of: C_3H_6 (cyclopropane and propylene), C_4H_8 (1-butene, iso-butene), C_2H_6O (ethyl alcohol and dimethyl ether) and C_3H_8O (*n*-propyl alcohol, *i*-propyl alcohol, and ethyl methyl ether). See the excellent review by Hatano [84] for further information on these studies.

Finally, it should be noted that because of the remarkable progress of laser technology, the various dynamic processes in a strong laser field have attracted much attention, as is reflected in this book. Various states can couple with each other by the field-induced nonadiabatic couplings. Although we do not go into the details here, this presents a nice target of semiclassical theory of nonadiabatic transitions [4, 85].

C. Dissociative Recombination

Dissociative recombination is another dynamic process where both kinds of SES play important roles. This is a process that produces neutral fragments from positive molecular ions by collision with electrons,

$$AB^+ + e \rightarrow A + B$$

Dissociative recombination is one of the key processes to produce neutral molecules in interstellar space and is also considered to be important in dense plasma. Again, two kinds of mechanisms are involved in dissociative recombination:

1. A direct process in which an incident electron is captured directly into the dissociative superexcited states, and
2. An indirect process in which an electron is first captured into a rovibrationally excited Rydberg states.

In the first type of mechanism, electronic coupling between the electronic continuum and the dissociative SES is the interaction responsible for the process. In the indirect process, rovibrational coupling represented by quantum defect function $\mu_\Lambda(R)$ plays a role. The Rydberg states are formed as intermediate states and decay by predissociation due to electronic configuration interaction (i.e., electronic coupling) with the dissociative SES. For both types of mechanisms, the dissociative SES should cross the ionic state near the equilibrium internuclear distance of the ion in order for the process to occur effectively. Thus, dissociative recombination process involves complicated interactions among the three kinds of states: electronic continuum, dissociative first kind of SES, and rovibrationally excited Rydberg states. The interplay of theory and experiment on dissociative recombination processes can provide very useful information on the properties of and mutual interactions among these states. Since nice reviews are available [30, 31], we do not go into the details here and just describe some intriguing essential features of the process.

The dissociative recombination cross-section behaves as a function of collision energy E, as is shown generally in Figure 5.33 [86]. This figure shows the theoretical results for the three ($B^2\Pi$, $B'^2\Delta$, and $L^2\Pi$) dissociative states for $v_+ = 0$. The overall energy dependence and magnitude of the cross-section are determined by the direct process which is essentially characterized by the ξ_v-factor defined by Equation 5.29. If we neglect the effects of closed Rydberg channels in the MQDT treatment (discussed in Section 5.5.A), then we obtain a simple formula for the cross-section

$$\sigma_{DR}(v_+,E) \cong g \frac{2\pi}{E} \frac{\xi^2_{v_+}}{(1+\xi^2)^2} \tag{5.40}$$

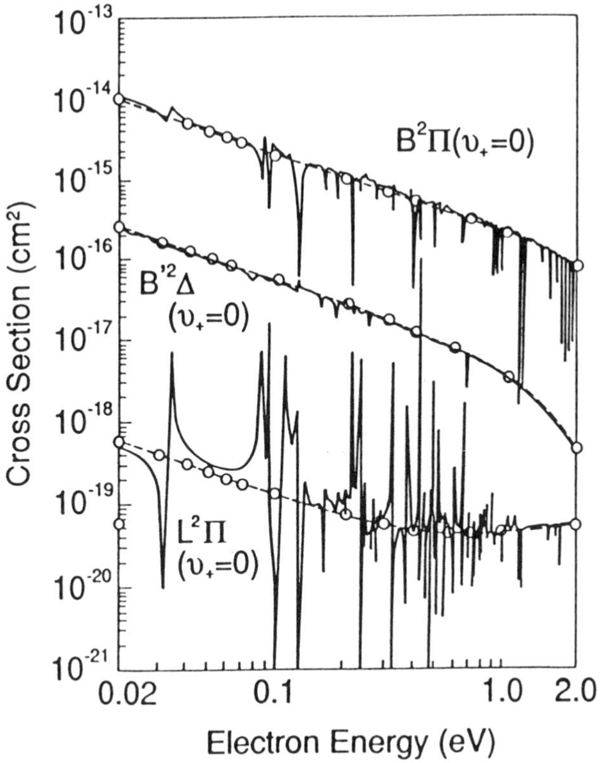

Figure 5.33 Dissociative recombination cross-sections in the case of $B^2\Pi(v_+ = 0)$, $B'^2\Delta(v_+ = 0)$ and $L^2\Pi(v_+ = 0)$. *Source*: from [86].

where g represents the statistical factor. The dashed lines with circles in Figure 5.33 are the results of this formula. The solid lines are the full MQDT calculations based on the method in Section 5.5.A. Since the Franck-Condon factor between the vibrational state v_+ and the dissociative state F_d essentially determines the energy dependence of ξ_{v_+}, the overall energy dependence of the cross-section is governed by this factor. The rich structures seen in Figure 5.33 are the resonances due to the indirect process via the second kind of SES. An infinite number of resonances can be reproduced by the MQDT and it is noticed easily that the appearance of the resonance structures varies from state to state. In the case of $L^2\Pi$, the resonances appear conspicuously as dips and peaks, while only weak dips appear in the case of $B'^2\Delta$. The $B^2\Pi$ state is intermediate. This is a reflection of the relative strength of the R-dependence of $\mu_\Lambda(R)$, as compared to $V(R)$. For instance, the very weak R-dependence of the $d\sigma$ quantum defect suppresses the coupling between the Rydberg states and the vibrational states of NO^+ and leads to the very weak resonances in the dissociative recombination of $B'^2\Delta$. In some cases, there appear windows in σ_{DR} versus E because of zeros of ξ_{v_+}. The effects of rotational states are not fully analyzed yet but are considered mainly to affect the resonance structures [64]. Experimental progress, especially that of the merging-beam technique, is remarkable and detailed comparison between theory and experiment will become feasible.

Vibrational transitions of hydrogen molecular ions by electron impact have also been investigated as a byproduct of the studies of dissociative recombination [87]. Interestingly, it was found that vibrational transition proceeds by the two-step mechanism via the dissociative first kind of SES. Here, the importance of dissociative SES is confirmed again.

5.6 SUMMARY

In this chapter, the concept of superexcited states of molecules was explained and their properties and important role in various dynamic processes were summarized. Emphasis was put on the doubly excited states (the first kind of SES). Electron correlation, an important feature to characterize these states, was analyzed by taking the low-lying doubly excited states of H_2, for example, and was compared with that in the corresponding atomic doubly excited states. Correlation patterns in the atomic doubly excited states survive at small internuclear distances ($R \leq 0.2a_0$) but they eventually (and naturally) transform into the independent particle picture as R increases. New interesting features peculiar to molecules were found in correlation patterns at intermediate R. For instance, avoided crossing was found to induce a drastic change in the correlation pattern. Interestingly, some patterns consistent with the axial symmetry survive if we follow the states diabatically. Autoionization mechanisms were also investigated by analyzing the discrete-to-continuum transition densities. The two mechanisms found in the atomic case (binary collision and dynamical screening) basically remain valid in molecules. Very qualitatively speaking, strong correlation leads to a small autoionization rate.

In the light of the natural correspondence in the united atom limit, a brief review was made for the various studies on the atomic doubly excited states such as the group theoretical analysis, the hyperspherical coordinate approach and the rovibrator model. It should be emphasized that the explicit quantum mechanical formulation can be obtained for the intrinsic collective motion of electrons and that the rotational contraction in the rovibrator model can be interpreted in terms of the quantum mechanical effect of shell structure.

We also presented the quantitative information on the resonance energies and autoionization widths for the doubly excited sates of H_2 that correspond to the $n = 2$ intrashell and ($n = 2$, $n' = 3$)-intershell doubly excited states in the united atom limit. These doubly excited states are becoming accessible by multiphoton processes in a strong laser field and the information on these states is now in a strong demand. In this connection, the dynamic processes involving molecular SES studied so far were briefly reviewed. Thanks to the progress of laser technology, studies of the dynamics of SES will flourish. In this sense, quantum chemical evaluations of the properties of SES will be strongly required. At least for H_2, quite a lot of information is already available and the various elaborate dynamics studies could now be carried out.

On the other hand, the following very fundamental question should also be investigated more extensively: what kind of collective motions do strongly interacting quantum-mechanical finite many-body systems exhibit and how can they be formulated in a unified way?

ACKNOWLEDGMENTS

One of the authors (S.L.) would like to thank the support by Japan-Korea cooperation program in molecular science and the partial support by Korea Science and Engineering Foundation. This work is partially supported by a Grant-in-Aid for Scientific Research on Priority Area "Theory of Chemical Reactions" from the Ministry of Education, Science and Culture of Japan. The numerical computations were carried out at the computer center of Institute for Molecular Science.

REFERENCES

1. K. P. Lawley (Ed.), *Adv. Chem. Phys.*, **60** (1985).
2. C. H. Greene and Ch. Jungen, *Adv. Atom. Mol. Phys.*, **21**, 51 (1985).
3. K. Kimura, *Int. Rev. Chem. Phys.*, **6**, 195 (1987).

4. H. Nakamura, *Int. Rev. Chem. Phys.*, **10**, 123 (1991).
5. K. Kuchitsu (Ed.), *Dynamics of Excited Molecules* Elsevier, New York, 1992.
6. D. R. Herrick, *Adv. Chem. Phys.*, **52**, 1 (1983).
7. U. Fano. Rep. Prog. Phys., **46**, 97 (1983).
8. C. D. Lin, *Adv. Atom. Mol. Phys.*, **22**, 77 (1986).
9. M. Iwai and H. Nakamura, *Phys. Rev.*, **A40**, 2247 (1989).
10. M. Iwai, S. Lee and H. Nakamura, *Phys. Rev.* **A47**, 2686 (1993).
11. R. P. Madden and K. Codling, *Phys. Rev. Lett.*, **10**, 516 (1963).
12. R. P. Madden and K. Codling, *Astrophys. J.*, **141**, 364 (1965).
13. J. W. Cooper, U. Fano and F. Prats, *Phys. Rev. Lett.*, **10**, 518 (1963).
14. U. Fano and J. W. Cooper, *Rev. Mod. Phys.*, **40**, 441 (1968).
15. D. R. Herrick and M. E. Kellman, *Phys. Rev.*, **A21**, 418 (1980).
16. M. E. Kellman and D. R. Herrick, *Phys. Rev.*, **A22**, 1536 (1980).
17. D. R. Herrick, M. E. Kellman and R. D. Poliak, *Phys. Rev.*, **A22**, 1517 (1980).
18. D. R. Herrick and O. Sinanoglu, *Phys. Rev.*, **A11**, 97 (1975).
19. C. D. Lin, *Phys. Rev.*, **A10**, 1986 (1974); C. D. Lin, *Phys. Rev.*, **A25**, 761 (1982); C. D. Lin, *Phys. Rev.*, **A26**, 2305 (1982); C. D. Lin, *Phys. Rev.*, **A27**, 22 (1983); C. D. Lin, *Phys. Rev.*, **A25**, 1535 (1982); C. D. Lin, *Phys. Rev.*, **A29**, 1019 (1984).
20. J. H. Macek, *J. Phys.*, **B1**, 831 (1968).
21. S. Watanabe and C. D. Lin, *Phys. Rev.*, **A34**, 823 (1986).
22. P. R. Woodruff and J. A. Samson, *Phys. Rev.*, **A25**, 848 (1982).
23. G. S. Ezra and R. S. Berry, *Phys. Rev.*, **A25**, 1513 (1982).
24. G. S. Ezra and R. S. Berry, *Phys. Rev.*, **A28**, 1989 (1983).
25. G. S. Ezra and R. S. Berry, *Phys. Rev.*, **A28**, 1974 (1983).
26. M. Iwai and H. Nakamura, *Phys. Rev.*, **A40**, 6695 (1989).
27. A. Giusti-Suzor and Ch. Jungen, *J. Chem. Phys.*, **80**, 986 (1984).
28. K. Nakashima, H. Nakamura, Y. Achiba and K. Kimura, *J. Chem. Phys.*, **91**, 1603 (1989).
29. Y. Achiba and K. Kimura, *J. Chem. Phys.*, **129**, 11 (1989).
30. J. B. A. Mitchell and S. L. Guberman (Eds.), *Disociative Recombination: Theory, Experiment and Applications*, World Scientific, Singapore, 1989; B. R. Rowe and J. B. A. Mitchell (Eds.), *Dissociative Recombination: Theory, Experiments and Applications*, Plenum, New York, 1992.
31. J. B. A. Mitchell, *Phys. Rep.*, **186**, 215 (1990).
32. H. Takagi and H. Nakamura, *J. Chem. Phys.*, **88**, 4552 (1988).
33. J. Weiner, F. Masnou-Seeuws and A. Giusti-Suzor, *Adv. Atom. Mol. Phys.*, **26**, 209 (1990).
34. X. Urbain, A. Cornet, F. Brouillard and A. Giusti-Suzor, *Phys. Rev. Lett.*, **66**, 1685 (1991).
35. E. T. Xu, T. Tsuboi, R. Kachru and H. Helm, *Phys. Rev.*, **A36**, 5645 (1987).
36. S. Arai, T. Kamosaki, M. Ukai, K. Shinsaka, Y. Hatano, Y. Ito, H. Koizumi, A. Yagishita, K. Ito and K. Tanaka, *J. Chem. Phys.*, **88**, 3016 (1988).

37. J. W. J. Verschuur and H. B. Van Linden Van den Heurell, *Chem. Phys.*, **129**, 1 (1989).
38. H. Feshbach, Ann. Phys., **78**, 1404 (1962).
39. A. Temkin and A. K. Bhatia, in *Autoionization: Recent Development and Applications*, (A. Temkin, Ed.) Plenum Press, New York, 1985.
40. T. F. O'Malley and S. Geltman, *Phys. Rev.*, **137A**, 1344 (1965).
41. R. D. Bates, K. Ledsham and A. L. Stewart, *Phil. Trans. R. Soc.*, **246**: 215 (1953).
42. D. R. Bates and R. H. G. Reid, *Adv. At. Mol. Phys.*, **4**, 13 (1968).
43. C. Bottcher and K. Docken, *J. Phys.*, **B7**, L5 (1974).
44. K. R. Dastidar and T. K. Dastidar, *J. Phys. Soc. Japan*, **46**, 1288 (1979).
45. H. Takagi and H. Nakamura, *Phys. Rev.*, **A27**, 691 (1983).
46. J. Tennyson and C. J. Noble, *J. Phys.*, **B18**, 155 (1985).
47. I. Shimamura, C. J. Noble and P. G. Burke, *Phys. Rev.*, **A41**, 3545 (1990).
48. L. A. Collins and B. I. Schneider, *Phys. Rev.*, **A27**, 101 (1983).
49. J. Tennyson, C. J. Noble and S. Salvini, *J. Phys.*, **B17**, 905 (1984).
50. S. L. Guberman, *J. Chem. Phys.*, **78**, 1404 (1983).
51. S. Hara and H. Sato, *J. Phys.*, **B17**, 4301 (1984).
52. H. Sato and S. Hara, *J. Phys.*, **B19**, 2611 (1986).
53. A. Macias, F. Martin, A. Riera and M. Yanez, *Phys. Rev.*, **A36**, 4203 (1987).
54. S. W. Allendorf, in *Coherence Phenomena in Atoms and Molecules in Laser Fields*, (A. D. Bandrauk, Ed.), NATO ASI Series B, vol. 287, Plenum, 1992. p. 99.
55. H. Helm, M. J. Dyer, H. Bissantz and D. L. Huestis, in *Coherence Phenomena in Atoms and Molecules in Laser Fields*, (A. D. Bandrauk, Ed.), NATO ASI Series B, vol.287, Plenum, 1992. p. 109.
56. P. Rehmus and R. S. Berry, *Phys. Rev.*, **A40**, 2247 (1989).
57. H. Nakamura, *J. Phys. Chem.*, **88**, 4812 (1984).
58. R. L. Platzman, *Vortex*, **23**, 372 (1962); *Rad. Res.*, **17**, 419 (1962).
59. B. Peart and K. T. Dolder, *J. Phys.*, **B8**, 1570 (1975).
60. W. Schon, S. Krudener, F. Melchert, K. Rinn, M. Wagner and E. Salzborn, *Phys. Rev. Lett.*, **59**, 1565 (1987).
61. M. J. Seaton, *Proc. Phys. Soc. London*, **88**, 801 (1966); M. J. Seaton, *Rep. Prog. Phys.*, **46**, 167 (1983).
62. U. Fano, *Phys. Rev.*, **A2**, 353 (1975); U. Fano, *J. Opt. Soc. Amer.*, **65**, 979 (1981); U. Fano, *Comments on Atom. Molec. Phys.*, **10**, 223 (1981); U. Fano, **13**, 157 (1983).
63. A. Giusti, *J. Phys.*, **B13**, 3867 (1980).
64. H. Takagi, in *Dissociative Recombination: Theory, Experiments and Applications*, (B. R. Rowe and J. B. A. Mitchell, Eds.), Plenum, New York, 1992.
65. D. Dill and C. Jungen, *J. Chem. Phys.*, **84**, 2116 (1980).
66. Ch. Jungen and D. Dill, *J. Chem. Phys.*, **73**, 3338 (1980).
67. M. Raoult, Ch. Jungen and D. Dill, *J. Chem. Phys.*, **77**, 599 (1980).
68. Ch. Jungen and M. Raoult, *Farad. Disc. Chem. Soc.*, **71**, 253 (1981).
69. M. Raoult and Ch. Jungen, *J. Chem. Phys.*, **74**, 3388 (1981).

70. P. M. Dehmer and W. A. Chupka, *J. Chem. Phys.*, **65**, 2243 (1976).
71. P. Morin, I. Nenner, M. Y. Adam, M. J. Hubin-Franskin, J. Delwiche, H. Lefebvre-Brion and A. Giusti-Suzor, *Chem. Phys. Lett.*, **92**, 609 (1982).
72. M. Raoult, H. LeRouzo, H. Raseev and Lefebvre-Brion, *J. Phys.*, **B16**: 4601 (1983).
73. E. Miescher, Y. T. Lee, and P. Gurtler, *J. Chem. Phys.*, **68**, 2753 (1978).
74. Y. Ono, S. H. Linn, H. F. Prest, C. Y. Ng, and E. Miescher, *J. Chem. Phys.*, **73**, 4855 (1980).
75. H. Nakamura, *Chem. Phys. Lett.*, **28**, 534 (1974); H. Nakamura, *Chem. Phys. Lett.*, **33**, 151 (1975); H. Nakamura, *Chem. Phys.*, **10**, 271 (1975).
76. S. T. Pratt, P. M. Dehmer and J. L. Dehmer, *Chem Phys. Lett.*, **105**, 28 (1984); S. T. Pratt, P. M. Dehmer and J. L. Dehmer, *J. Chem. Phys.*, **85**, 3379 (1986); S. T. Pratt, P. M. Dehmer and J. L. Dehmer, *J. Chem. Phys.*, **86**, 1727 (1987).
77. S. N. Dixit, D. L. Lynch and V. McKoy, *Phys. Rev.*, **A30**, 3332 (1984).
78. N. A. Chupka, *J. Chem. Phys.*, **87**, 1488 (1987).
79. A. P. Hickman, *Phys. Rev. Lett.*, **59**, 1553 (1987).
80. A. O'Halloran, S. T. Pratt, P. M. Dehmer and J. L. Dehmer, *J. Chem. Phys.*, **87**, 3288 (1987).
81. J. H. M. Bonnie, J. W. J. Verschuur, H. J. Hopman and H. B. Van Linden van den Heuvell, *Chem. Phys. Lett.*, **130**, 43 (1986).
82. J. W. J. Verschuur, L. D. Noordam, J. H. M. Bonnie and H. B. van Linden van der Heuvell, *Chem. Phys. Lett.*, **146**, 283 (1988).
83. S. Arai, T. Yoshimi, M. Morita, K. Hironaka, T. Yoshida, H. Koizumi, K. Shinsaka, Y. Hatano, A. Yagishita and K. Ito, *Z. Phys.*, **D4**, 65 (1986).
84. Y. Hatano, in *Dynamics of Excited Molecules*, (K. Kuchitsu, Ed.) Elsevier, New York, 1992.
85. H. Nakamura, in *State-Selected and State-to-State Ion-Molecule Reaction Dynamics, Part 2: Theory*, (M. Baer and C-Y. Ng, Eds.), John Wiley & Sons, New York, 1992. Adv. Chem. Phys., vol. LXXXII p. 243.
86. H. Nakamura, in *Dissociative Recombination: Theory, Experiments and Applications*, (B. R. Rowe and J. B. A. Mitchell, Eds.), Plenum, New York, 1992.
87. K. Nakashima, H. Takagi and H. Nakamura, *J. Chem. Phys.*, **86**, 726 (1987).

6

Coherence and Laser Control of Chemical Reactions

Paul Brumer

University of Toronto, Toronto, Ontario, Canada

Moshe Shapiro

The Weizmann Institute of Science, Rehovot, Israel

6.1 INTRODUCTION

Manipulating the yield of chemical reactions is the essence of chemistry and the capability to control reactions using lasers has been a goal for decades. We have previously demonstrated [1–19] how this goal could be achieved. An appreciation of this approach, termed *coherent control* of chemical reactions, opens up new avenues in chemistry by introducing chemical control concepts based upon previously unutilized quantum effects.

The purpose of this chapter is to provide an introduction to the concepts [20] underlying coherent control of chemical reactions and to review its current status. The paper is organized along the following lines. Section 6.1 provides an introduction to the basics of coherent control, followed by a detailed discussion of two control scenarios in Section 6.2. Some specific topics are dealt with in Section 6.3 where we discuss the issue of control in the presence of incoherence effects and the role of selection rules in designing control scenarios. Finally, Section 6.4 describes two interesting applications: control of symmetry breaking and the production of photocurrents in semiconductors. In all cases we provide meaningful examples to demonstrate the wide range of control possible. In most cases these results are obtained via fully quantum mechanical photodissociation calculations on reliable potential energy surfaces.

A. Aspects of Scattering Theory and Reaction Dynamics

We shall focus on unimolecular reactions where decomposition into more than one product is possible,

$$\mathrm{ABC} \rightarrow \mathrm{A} + \mathrm{BC} \tag{6.1a}$$

$$\rightarrow \mathrm{AB} + \mathrm{C} \tag{6.1b}$$

and remark as well on branching bimolecular reactions

$$\mathrm{A} + \mathrm{BC(n)} \rightarrow \mathrm{A} + \mathrm{BC(n')} \tag{6.2a}$$

$$\rightarrow \mathrm{AB(n'')} + \mathrm{C} \tag{6.2b}$$

where A, B, C are atoms or groups of atoms and n, n' denote the vibrational, rotational, etc., states of the reactant or product pair. Both inelastic (Equation 6.2a) and reactive scattering (Equation 6.2b) are indicated.

Treating the dynamics of a chemical reaction requires solving the Schrödinger Equation

$$H\Psi(t) = i\hbar\partial\Psi(t)/\partial t \tag{6.3}$$

for the wavefunction $|\Psi(t)\rangle$ associated with specific initial reactant conditions (i.e., $\Psi(0)$). The wavefunction at long times (i.e., when the products are well separated), then provides the probabilities of forming the products. The time-dependent Schrödinger equation is conveniently solved in two steps, first obtaining the stationary eigenfunctions ψ_i as solutions to the time-independent Schrödinger equation $H\psi_i = E_i\psi_i$, and then by building the time dependence in as a superposition of time-independent eigenfunctions.

Consider then the nature of the time-independent eigenfunctions. We focus attention on the system at a fixed energy E in the continuum where, as we shall show, the essence of controlling reactions is manifest. The system requires, at such an energy, an independent wavefunction to describe each of the possible outcomes that can be observed in the product regions [21]. As a consequence, one expects substantial degeneracy at energy E. Further, the fact that this set of degenerate wavefunctions of the separated products exists, implies [22] that a related set of degenerate eigenfunctions of the total Hamiltonian exists.

This requirement, that total system eigenfunctions correlate with specific asymptotic product state eigenfunctions, may be included as a boundary condition on the total system wavefunctions and serves to considerably simplify the understanding of the dynamics of unimolecular decay, which will serve subsequently as our primary example. Specifically, say we distinguish the different possible chemical product arrangements of the decay of ABC by the numerical value of an index q, (e.g., $q = 1$ denotes A + BC in Equation 6.1, etc.) and m denotes all additional identifying state labels (e.g., j', v', scattering angle, etc.). Then we can define the set of Hamiltonian continuum eigenstates $|E,m,q^-\rangle$ via the Schrödinger equation $H|E,m,q^-\rangle = HE|E,m,q^-\rangle$ and via the requirement that this eigenstate describes, at large distances, the state of the separated products, denoted $|E,m,q^0\rangle$, which is of energy E, arrangement q, and remaining quantum numbers m. The "minus" superscript serves to indicate this choice of boundary condition.

Imposition of such boundary conditions and the description of the system in terms of $|E,m,q^-\rangle$ has a number of important simplifying consequences. For example, if one sets up (either experimentally or conceptually) a state at $t = 0$, at energy E, consisting of $|E,m,1^-\rangle$ then the probability of observing the product in the $q = 1$ arrangement, and with quantum numbers m, is unity since $|E,m,1^-\rangle$ uniquely correlates with that particular product state. Similarly, if we set up the system in a linear combination of states:

$$|\psi(0)\rangle = \sum_m [c_{1m}|E,m,1^-\rangle + c_{2m}|E,m,2^-\rangle] \tag{6.4}$$

Then the probability of observing one of the arrangments (e.g., $q = 1$) is

$$\sum_m |c_{1m}|^2$$

This apparently simple discussion allows us to make a few crucial statements:

1. The product yield (i.e., the probability of obtaining a particular chemical product at long-time, is solely determined by the state created at $t = 0$). Furthermore, our choice of "minus states" $|E,m,q^-\rangle$ allows expression of this fact in a relatively simple way (i.e., the coefficients in the $t = 0$ superposition state are identical and equal to the coefficients at long-time whose squares are the product probabilities). The fact that the long-time state is predetermined by the initially created state is, admittedly, intuitively obvious. However, consequences of this feature are often misunderstood. For example it makes clear that arguments such as "intramolecular energy scrambling makes reaction control difficult", are misleading. Viewed from the proper perspective (i.e., the minus states) there simply is no time-evolving scrambling on the way to product. We have discussed these issues and their role in understanding control of reactions in detail in [15].
2. Since product probabilities are predetermined by the composition of the prepared ($t = 0$) superposition state, the route to controlling a chemical reaction is to control the content of the initially prepared superposition state.
3. The branching of the reaction probabilities into various product channels occurs at a fixed energy E. As such, a continuum wave packet built of states over a range of energies and its associated time dependence need not be introduced in order to consider control over reaction yields.

Next, we demonstrate that the key to laser control of chemical reactions is to use the lasers to alter the nature of the prepared superposition state and hence to alter the product probabilities. That this strategy is the essence of controlling chemical reactions should be clear from the above discussion. First, however, we discuss preparation of states from the viewpoint of perturbation theory.

B. Perturbation Theory, System Preparation, and Coherence

In preparation for a discussion of laser induced unimolecular dissociation consider the effect of an electric field on a molecule. Consider an isolated molecule with Hamiltonian H_M in an eigenstate $|\varphi_g\rangle$ which is subjected to a perturbing incident radiation field. The overall Hamiltonian is then given by:

$$H = H_M - \mathbf{d}[\bar{\varepsilon}(t) + \bar{\varepsilon}^*(t)] \tag{6.5}$$

where H_M is the molecular Hamiltonian and **d** is the component of the dipole moment along the electric field, $\bar{\varepsilon}(t)$.

To ascertain the effect of the field on the molecule requires that we solve the time-dependent Schrödinger equation for the given Hamiltonian, including the perturbation. To do so we invoke time-dependent perturbation theory and expand the solution $\Psi(t)$ in solutions to the problem in the absence of the perturbation

$$H_M\varphi_i = E_i\varphi_i \tag{6.6}$$

Specifically, writing the full wavefunction $\Psi(t)$ as:

$$\Psi(t) = \sum_i c_i(t)\varphi_i e^{-iE_it/\hbar} \tag{6.7}$$

with the $c_i(t)$ as yet to be determined expresses the solution precisely in terms we want physically (i.e., $|c_i(t)|^2$ gives the probability of being in the molecular state $|\varphi_i>$ at time t).

Inserting Equation 6.7 into the Schrödinger Equation yields a set of ordinary differential equations for $c_i(t)$ that may be solved numerically [23]. For weak fields this is not necessary and a simple perturbation theory solution for the long-time (i.e., when the electric field is off) behavior can be obtained as:

$$c_i(t \rightarrow \infty) = \frac{\sqrt{2\pi}}{i\hbar}\,\varepsilon(\omega_{E_i,E_g})<\varphi_i|\mathbf{d}|\varphi_g> \tag{6.8}$$

with

$$\varepsilon(\omega) = \frac{1}{\sqrt{2\pi}}\int_{-\infty}^{\infty} e^{i\omega t}\,\bar{\varepsilon}(t)\,dt \tag{6.9}$$

In this case

$$\omega = \omega_{E_i,E_g} = (E_i - E_g)/\hbar$$

Note that the object which is created by the effect of the incident electric field on the bound state is a pure state (i.e., it can be described by a wavefunction $\Psi(t)$). The fact that a well defined electric field produces a pure state (and hence a state which is phase-coherent) is essential to the subsequent discussion.

Consider now laser-induced photodissociation where we excite a molecule in an eigenstate $|\varphi_g\rangle$ with an electric field which provides enough energy to dissociate the molecule. Our interest is in ascertaining the probability of forming particular products. With the product of the electric field and dipole moment assumed to be small enough to allow the use of first-order perturbation theory [24] we proceed in the standard fashion [25] and expand the wavefunction in eigenstates of the molecular Hamiltonian. Since the photon lifts the system into the continuum, our expansion is in terms of the eigenstates $|E,m,q^-\rangle$,

$$|\Psi(t)\rangle = c_g|\varphi_g\rangle e^{-iE_gt/\hbar} + \sum_{m,q} \int dE\; c_{E,m,q}(t)|E,m,q^-\rangle e^{-iEt/\hbar} \tag{6.10}$$

Following through with standard perturbation theory gives, for the probability $P(E,q)$ of forming product in arrangement q:

$$P(E,q) = \lim_{t\to\infty} \sum_m |c_{E,m,q}(t)|^2 = \frac{2\pi}{\hbar^2} \sum_m |\varepsilon(\omega_{E,E_g})\langle\varphi_g|\mathbf{d}|E,m,q^-\rangle|^2 \tag{6.11}$$

The ratio $R(1,2;E)$ of products in channel $q = 1$ to $q = 2$ at energy E is:

$$R(1,2;E) = \frac{\sum_m |\langle\varphi_g|\mathbf{d}|E,m,1^-\rangle|^2}{\sum_m |\langle\varphi_g|\mathbf{d}|E,m,2^-\rangle|^2} \tag{6.12}$$

An understanding of the qualitative structure of this yield ratio (see Section 6.2.C) is crucial to recognizing the difficulties associated with experimental attempts to alter the yield in a traditional unimolecular decay experiment. It also motivates the specific control approach which we advocate.

C. Coherent Radiative Control of Chemical Reactions

Our primary goal is to experimentally alter the yield ratio R so as to control the product distribution. Equation (6.12) makes clear that this can not be achieved, for example, by moderately altering the laser power, which cancels out in forming the ratio R. Hence this, and any quantity which appears in a similar form in both the numerator and denominator, can not serve as a handle on yield control. An alternate possibility for controlling the reaction yield is to vary the frequency of excitation $[(E - E_g)/\hbar]$ and see the effect on the ratio. However, such a procedure is not systematic and its success, if any, is based solely upon a *chance* occurence of a desirable result with variations in the laser frequency.

There is, however, an alternate possibility. Specifically, note the form of Equation 6.12, which has the square of an amplitude in both the numerator and the denominator. If we could manage to experimentally alter the quantity within the square then the effect on the numerator and the denominator might differ and we would have experimental control over the ratio. Our coherent radiative control approach reflects this philosophy and is coupled with the recognition that quantum interference phenomena alter the amplitude within the square in a particularly useful fashion.

As a pedagogical example, consider starting with a molecule prepared in a superposition $c_1|\varphi_1\rangle + c_2|\varphi_2\rangle$ of molecular eigenstates $|\varphi_i\rangle$ (Figure 6.1). The c_1 and c_2 coefficients, which are determined by the method of preparation, have phases and magnitudes which are functions of the *experimentally controllable* parameters.

This bound superposition state is subjected to an electric field which contains frequency components that can independently transform both of these states to a state of energy E; for example, excitation with two CW sources

$$\bar{\varepsilon}(t) = \varepsilon_1 e^{-i\omega_1 t + i\chi_1} + \varepsilon_2 e^{-i\omega_2 t + i\chi_2}$$

where $\hbar\omega_i = E - E_i$. We then ask for the yield ratio R under these circumstances. A straightforward computation [1] gives the result:

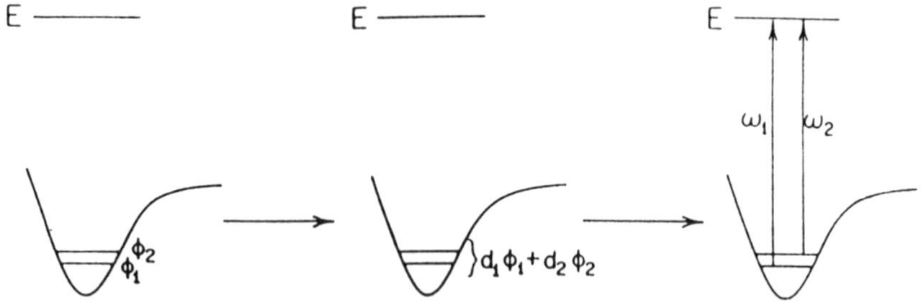

Figure 6.1 A general two step scheme for inducing controllable quantum interference effects into the continuum state at energy E. The two bound states φ_1, φ_2 belong to a lower electronic state whereas the level at energy E is that of an excited electronic state. Coherence introduced in the first step is carried into the continuum. *Source*: from [12].

$$R(1,2;E) = \frac{\sum_m |\langle \tilde{\varepsilon}_1 c_1 \varphi_1 + \tilde{\varepsilon}_2 c_2 \varphi_2 | \mathbf{d} | E,m,1^- \rangle|^2}{\sum_m |\langle \tilde{\varepsilon}_1 c_1 \varphi_1 + \tilde{\varepsilon}_2 c_2 \varphi_2 | \mathbf{d} | E,m,2^- \rangle|^2} \tag{6.13a}$$

where $\tilde{\varepsilon}_i = \varepsilon_i \exp(i\chi_i)$. Expanding the square gives:

$$R(1,2;E) =$$

$$\frac{\sum_m [|\tilde{\varepsilon}_1 c_1|^2 |\langle \varphi_1 | \mathbf{d} | E,m,1^- \rangle|^2 + |\tilde{\varepsilon}_2 c_2|^2 |\langle \varphi_2 | \mathbf{d} | E,m,1^- \rangle|^2 + 2\mathrm{Re}[c_1 c_2^* \tilde{\varepsilon}_1 \tilde{\varepsilon}_2^* \langle E,m,1^- | \mathbf{d} | \varphi_2 \rangle \langle \varphi_1 | \mathbf{d} | E,m,1^- \rangle]}{\sum_m [|\tilde{\varepsilon}_1 c_1|^2 |\langle \varphi_1 | \mathbf{d} | E,m,2^- \rangle|^2 + |\tilde{\varepsilon}_2 c_2|^2 |\langle \varphi_2 | \mathbf{d} | E,m,2^- \rangle|^2 + 2\mathrm{Re}[c_1 c_2^* \tilde{\varepsilon}_1 \tilde{\varepsilon}_2^* \langle E,m,2^- | \mathbf{d} | \varphi_2 \rangle \langle \varphi_1 | \mathbf{d} | E,m,2^- \rangle]} \tag{6.13b}$$

The result of this computation is qualitatively straightforward. Specifically, the superposition state $[c_1\varphi_1 + c_2\varphi_2]$ has replaced the initial state φ_g of Equation 6.12 and the two electric fields, which now remain in this expression, are those which raise each of the individual levels to the excited state.

The structure of each of the numerator and denominator of Equation 6.13 is clearly of the type desired (i.e., each has two terms associated with the two different independent excitations of levels at energies E_1 and E_2 and set of terms corresponding to the interference between these two processes). The interference term can either constructively enhance or destructively cancel out contributions to either product channel. What makes Equation 6.13 so important in practice is that the interference terms have coefficients whose magnitude and sign depend upon *experimentally controllable* parameters. Thus the experimentalist can manipulate laboratory parameters and, in doing so, directly alter the reaction product yield by varying the magnitude of the interference term. In the case of Equation 6.13 the experimental parameters which alter the yield [1] are the absolute magnitude and phase of the quantity

$$A = \frac{\tilde{\varepsilon}_2 c_2}{\tilde{\varepsilon}_1 c_1}$$

Results of a specific computational example based upon Equation 6.13 are shown in Figure 6.2. Here we consider control over the relative probability of forming $^2P_{3/2}$ versus $^2P_{1/2}$ atomic iodine, denoted I and I^*, in the dissociation of methyl iodide:

$$\begin{aligned} CH_3I &\rightarrow CH_3 + I \\ &\rightarrow CH_3 + I^* \end{aligned} \tag{6.14}$$

Although this reaction is an example of electronic branching of products, the same principles of control apply. Note that computations were carried out with realistic potential surfaces within the framework of a fully quantum computational photodissociation method.

Figure 6.2 shows a typical plot of the yield of I^* as a function of A, where

$$S = \frac{A^2}{(1 + A^2)}$$

and

$$\theta_1 - \theta_2 = arg(A)$$

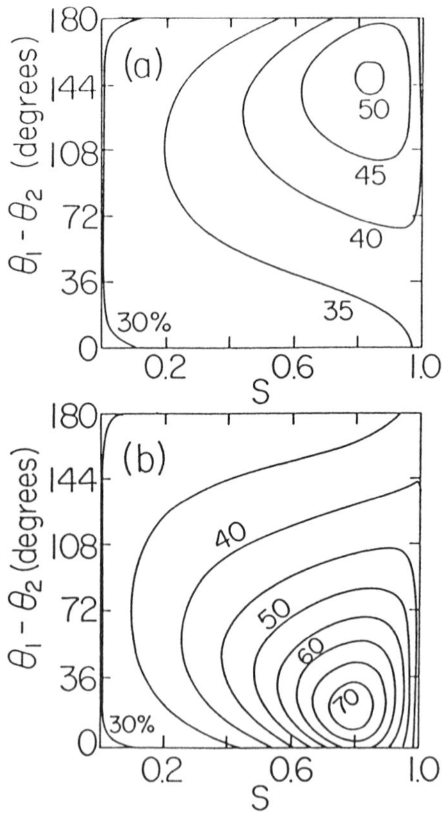

Figure 6.2 Contour plot of the yield of I^* (i.e., fraction of I^* as product) in the photodissociation of CH_3I from a superposition state comprised of (a) $(v_1, J_1, M_1) = (0,0,0) + (v_2, J_2, M_2) = (0,1,0)$ and (b) $(0,0,0) + (0,2,0)$. Here v_i, J_i, M_i are the vibrational, rotational and rotational projection quantum numbers of the ith bound state. *Source*: from [1].

With this choice, $S = 0$ corresponds to $\tilde{\varepsilon}_1 = 0$ (i.e., laser 1 off), whereas $S = 1$ corresponds to $\tilde{\varepsilon}_2 = 0$ (i.e., laser 2 off). Note the large range of control with variation in S and $\theta_1 - \theta_2$ (i.e., the yield varies from 30–70% I). Higher and lower ratios can also be achieved [27] with different choices of the initial pair of states $|\varphi_1\rangle$ and $|\varphi_2\rangle$.

This scenario assumes starting with a superposition of two bound states. Experimental possibilities for creating such an initial state include pulsed excitation with an electromagnetic field whose frequency width spans the two levels [9–11] or stimulated emission pumping through an intermediate electronic state [19]. A coherent control scenario, in which the superposition state preparation is followed by a second pulse, as distinct from photodissociation with CW sources, is discussed in detail in Section 6.2.B below. Furthermore, a theoretical scheme, in which there are N product channels and the initial superposition is comprised of N levels, is discussed elsewhere [16]. In this case we have shown that under certain conditions one can control the product yields *completely*. That is, one may specify a desired product distribution and analytically determine the N electromagnetic fields required to attain this product distribution.

D. The Essential Principle

The two step approach of Figures 6.1 and 6.2 is but one particular implementation of the principle of coherent control. That is, the essence of control is in the content of the superposition state at energy E where dissociation into various products is possible. Numerous other scenarios may be designed which rely upon the same essential principle: that coherently driving a pure state through multiple optical excitation routes to the same final state allows for the possibility of control. This procedure has a well-known analogy: the interference between paths as a beam of either particles or of light passes through a double slit. In that case interference between two coherent beams leads to interference, manifest as patterns of enhanced or reduced probabilities on an observation screen. In the case of coherent control the overall coherence of a pure state plus laser source allows for the constructive or destructive manipulation of probabilities in product channels.

The coherence of the laser and the knowledge of the molecular phase are essential elements of control. It is easy to see that, for example, total laser incoherence leads to loss of control. That is, laser incoherence implies that the phases of $\tilde{\varepsilon}_1$ and $\tilde{\varepsilon}_2$ in Equation 6.13 are random. Doing an ensemble average over these phases results in the disappearance of the interference term. Nonetheless, we have also shown that control does persist in the presence of *partial* laser incoherence [14] and when the initial state is described by a (nondiagonal) density matrix [7], so some degree of phase incoherence can be tolerated without total loss of phase control.

6.2 REPRESENTATIVE CONTROL SCENARIOS

A. Interference Between *N*-Photon and *M*-Photon Routes

As a second example of a useful experimental implementation of coherent control consider Equation 6.12 once again. There we introduced quantum interference by modifying the initial state prior to excitation into the continuum. An alternative method readily suggests itself. Specifically, consider adding a second, multiphoton, optical route to energy E. The simultaneous one and multiphoton excitation then lead to controllable interference contributions and thus to yield control.

As the simplest example, we examine one photon plus three photon absorption. (The approach generalizes in a staightforward fashion to the case of two or more excitation routes with N and M photons as long as the choice of fields obeys requisite selection rules; see Section 6.3.A). Let H_g and H_e be the nuclear Hamiltonians for the ground and excited states, respectively, and $|E_i\rangle$ be the ground eigenstate (i.e., $H_g|E_i\rangle = E_i|E_i\rangle$). The kets $|E,n,q^-\rangle$ are continuum eigenstates of H_e with incoming boundary conditions as discussed earlier.

The molecule, initially in $|E_i\rangle$, is subjected to two electric fields (Figure 6.3) given by

$$\varepsilon(t) = \varepsilon_1 \cos(\omega_1 t + \boldsymbol{k}_1 \cdot \boldsymbol{R} + \theta_1) + \varepsilon_3 \cos(\omega_3 t + \boldsymbol{k}_3 \cdot \boldsymbol{R} + \theta_3) \tag{6.15}$$

Here $\omega_3 = 3\omega_1$, $\varepsilon_l = \varepsilon_l \hat{\varepsilon}_l$, $l = 1, 3$; ε_l is the magnitude and $\hat{\varepsilon}_l$ is the polarization of the electric fields. The two fields are chosen parallel, with $\boldsymbol{k}_3 = 3\boldsymbol{k}_1$. The probability $P(E,q;E_i)$ of producing product with energy E in arrangement q from a state $|E_i\rangle$ is given by

$$P(E,q;E_i) = P_3(E,q;E_i) + P_{13}(E,q;E_i) + P_1(E,q;E_i) \tag{6.16}$$

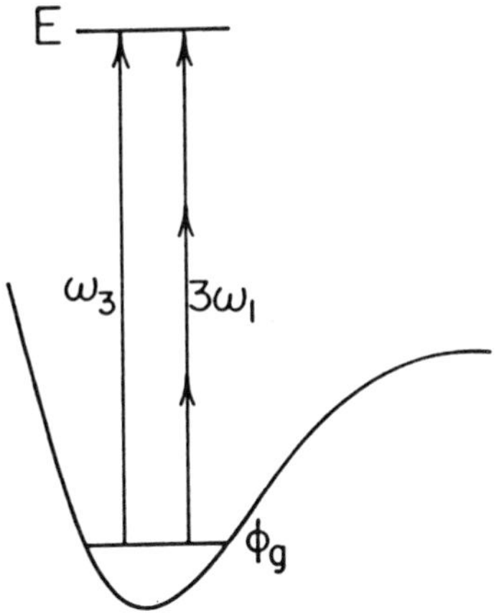

Figure 6.3 A multiple optical-route scheme to inducing controllable quantum interference effects into the continuum state at energy E. Here the level φ_g is a bound state of a lower electronic state and that at E is a continuum state of the excited electronic state. Simultaneous application of frequencies ω_1 and $\omega_3 = 3\omega_1$ leads to interference in the continuum state. *Source*: from [12].

where $P_i(E,q;E_1)$ and $P_3(E,q;E_i)$ are the probabilities of dissociation due to the ω_1 and ω_3 excitation, and $P_{13}(E,q;E_i)$ is the term due to interference between the two excitation routes.

In the weak field limit, $P_3(E,q;E_i)$ is given by

$$P_3(E,q;E_i) = \left(\frac{\pi}{h}\right)^2 \varepsilon_3^2 \sum_n |\langle E,n,q^-|(\hat{\varepsilon}_3 \cdot \mathbf{d})_{e,g}|E_i\rangle|^2 \tag{6.17}$$

Here $\mathbf{d}$ is the electric dipole operator and

$$(\hat{\varepsilon}_3 \cdot \mathbf{d})_{e,g} = \langle e|\hat{\varepsilon}_3 \cdot \mathbf{d}|g\rangle \tag{6.18}$$

where $|g\rangle$ and $|e\rangle$ are the ground and excited electronic state wavefunctions, respectively. Assuming also that $E_i + 2\hbar\omega_1$ is below the dissociation threshold, with dissociation occuring from the excited electronic state, $P_1(E,q;E_i)$ is given in third-order perturbation theory by [6]

$$P_1(E,q;E_i) = \left(\frac{\pi}{\hbar}\right)^2 \varepsilon_1^6 \sum_n |\langle E,n,q^-|T|E_i\rangle|^2 \tag{6.19}$$

with

$$T = (\hat{\varepsilon}_1 \cdot \mathbf{d})_{e,g}(E_i - H_g + 2\hbar\omega_1)^{-1}(\hat{\varepsilon}_1 \cdot \mathbf{d})_{g,e}(E_i - H_e + \hbar\omega_1)^{-1}(\hat{\varepsilon}_1 \cdot \mathbf{d})_{e,g} \tag{6.20}$$

A similar derivation [6] gives the cross-term in Equation 6.16 as

$$P_{13}(E,q;E_i) = -2\left(\frac{\pi}{\hbar}\right)^2 \varepsilon_3\varepsilon_1^3 \cos(\theta_3 - 3\theta_1 + \delta_{13}^{(q)})|F_{13}^{(q)}| \tag{6.21}$$

with the amplitude $|F_{13}^{(q)}|$ and phase $\delta_{13}^{(q)}$ defined by

$$|F_{13}^{(q)}| \exp(i\delta_{13}^{(q)}) = \sum_n \langle E_i|T|E,n,q^-\rangle\langle E,n,q^-|(\hat{\varepsilon}_3 \cdot \mathbf{d})_{e,g}|E_i\rangle \tag{6.22}$$

The branching ratio $R_{qq'}$ for channels q and q', can then be written as

$$R_{qq'} = \frac{P(E,q;E_i)}{P(E,q';E_i)}$$

$$= \frac{\varepsilon_3^2 F_3^{(q)} - 2\varepsilon_3\varepsilon_1^3 \cos(\theta_3 - 3\theta_1 + \delta_{13}^{(q)})|F_{13}^{(q)}| + \varepsilon_1^6 F_1^{(q)}}{\varepsilon_3^2 F_3^{(q')} - 2\varepsilon_3\varepsilon_1^3 \cos(\theta_3 - 3\theta_1 + \delta_{13}^{(q')})|F_{13}^{(q')}| + \varepsilon_1^6 F_1^{(q')}} \tag{6.23}$$

where

$$F_3^{(q)} = \left(\frac{\hbar}{\pi\varepsilon_3}\right)^2 P_3(E,q;E_i)$$

$$F_1^{(q)} = \left(\frac{\hbar}{\pi\varepsilon_1^3}\right)^2 P_1(E,q;E_i) \tag{6.24}$$

with $F_3^{(q')}$ and $F_1^{(q')}$ defined similarly. Next, we rewrite Equation 6.23 in a more convenient form. We define a dimensionless parameter $\bar{\varepsilon}_i$ and a parameter x as follows:

$$\varepsilon_l = \bar{\varepsilon}_l \varepsilon_0; \quad x = \frac{\bar{\varepsilon}_1^3}{\bar{\varepsilon}_3} \tag{6.25}$$

for $l = 1, 3$. The quantity ε_0 essentially carries the unit for the electric fields; variations of the magnitude of ε_0 can also be used to account for unknown transition dipole moments. Utilizing these parameters, Equation 6.23 becomes

$$R_{qq'} = \frac{F_3^{(q)} - 2x \cos (\theta_3 - 3\theta_1 + \delta_{13}^{(q)})\varepsilon_0^2|F_{13}^{(q)}| + x^2\varepsilon_0^4 F_1^{(q)}}{F_3^{(q')} - 2x \cos (\theta_3 - 3\theta_1 + \delta_{13}^{(q')})\varepsilon_0^2|F_{13}^{(q')}| + x^2\varepsilon_0^4 F_1^{(q')}} \tag{6.26}$$

The numerator and denominator of Equation 6.26 each display what may be regarded as the canonical form for coherent control: of independent contributions from more than one route, modulated by an interference term. Since the interference term is controllable through variation of laboratory parameters, so too is the product ratio $R_{qq'}$. Thus the principle upon which this control scenario is based is the same as in the first example but the interference is introduced in an entirely different way.

Experimental control over $R_{qq'}$ is obtained by varying the difference $(\theta_3 - 3\theta_1)$ and the parameter x. The former is the phase difference between the ω_3 and the ω_1 laser fields and the latter, via Equation 6.25, incorporates the ratio of the two lasers amplitudes. Experimentally, one envisons using tripling to produce ω_3 from ω_1, the subsequent variation of the phase of one of these beams provides a straightforward method of altering $\theta_3 - 3\theta_1$. Indeed, generating ω_3 from ω_1 allows for compensation of any phase jumps in the two laser sources. Thus the relative phase $\omega_3 - 3\omega$, is well defined. Were it not (e.g., as in the case of incoherent sources), then the cross-term and control would vanish.

With the qualitative principle of interfering pathways established, it remains to determine the quantitative extent to which coherent control alters the yield ratio in a realistic system. To this end we consider an application to one photon versus three photon photodissociation of IBr. In particular, we focus on the energy regime where IBr dissociates to both $I(^2P_{3/2})$ +

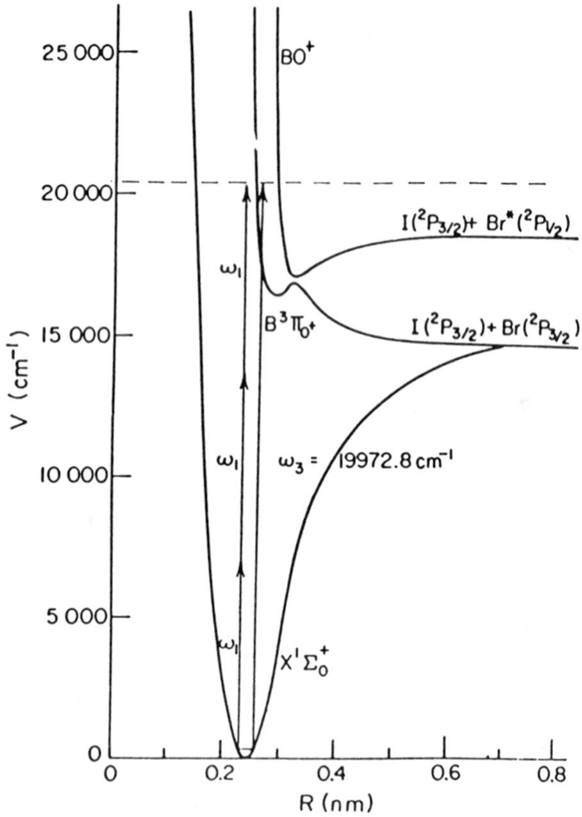

Figure 6.4 IBr potential curves relevant in the one-plus-three photon induced dissociation. *Source*: from [13].

$Br(^2P_{3/2})$ and $I(^2P_{3/2}) + Br^*(^2P_{1/2})$. The IBr potential curves used in the calculation are shown in Figure 6.4.

A complete IBr computation requires inclusion of angular momentum. Thus the notation $|E_i\rangle$ is replaced by $|E_i,J_i,M_i\rangle$, where J_i is the angular momentum, M_i is its z projection, and the ket is of energy E_i, whose value incorporates specification of the vibrational quantum number v. Where no confusion arises, we continue to use $|E_i\rangle$ for simplicity.

The primary quantity required in the control calculation is the photodissociation amplitude $\langle E,n,q^-|(\hat{\varepsilon}_l \cdot \mathbf{d})_{e,g}|E_i,J_i,M_i\rangle$, $\hat{\varepsilon}_l = \hat{\varepsilon}_1, \hat{\varepsilon}_3$ where the quantum number n in the continuum state $|E,n,q^-\rangle$ denotes the scattering angles $\hat{\mathbf{k}} = \varphi_k, \theta_k$ and $q = 1, 2$ labels either the $Br(^2P_{3/2})$ or $Br^*(^2P_{1/2})$ channel. In terms of this notation, the one-photon photodissociation amplitude is given by

$$\langle E,\hat{\mathbf{k}},q^-|(\hat{\varepsilon}_l \cdot \mathbf{d})_{e,g}|E_i,J_i,M_i\rangle =$$

$$\frac{(2\mu k)^{1/2}}{\hbar} \sum_J (2J+1)^{1/2} \begin{pmatrix} J & 1 & J_i \\ -M_1 & 0 & M_i \end{pmatrix} D^J_{0,M_i}(\varphi_k, \theta_k, -\theta_k)\, t(E,J,q|E_i,J_i) \tag{6.27}$$

Here μ is the reduced mass of IBr, $\boldsymbol{k}$ is the relative momentum of the dissociated particles, D^J_{0,M_i} is the rotation matrix element, and

$$\begin{pmatrix} j_1 & j_2 & j_3 \\ M_i & M_2 & M_3 \end{pmatrix}$$

is the Wigner $3-j$ symbol and $t(E,J,q|E_i,J_i)$ is the (M_i-independent) reduced amplitude, containing the essential dynamics of the photodissociation process [28]. Equation (27) follows from the usual procedure of expanding $\langle E,\hat{\mathbf{k}},q^-|\varphi_k,\theta_k,r\rangle$, where r is the I–Br distance, in partial waves labeled by angular momentum J and its z projection M. The angular integration of the resulting expression follows from the Wigner-Eckart theorem and the r integration is incorporated in $t(E,J,q|E_i,J_i)$. We have calculated these $t(E,J,q|E_i,J_i)$ exactly [13] using the potential curves and coupling strengths given by Child [26].

As shown here, the required P_i and P_{13} (Equation 6.16) are conveniently expressed in terms of the primary quantities $\mathbf{d}^{(q)}(E_j,J_j,M_j;E_i,J_i,M_i;E)$, where

$$\mathbf{d}^{(q)}(E_j,J_j,M_j;E_i,J_i,M_i;E)$$
$$= \int d\hat{\mathbf{k}} <E_j,J_j,M_j|(\hat{\varepsilon}_l \cdot \mathbf{d})_{g,e}|E,\hat{\mathbf{k}},q^-><E,\hat{\mathbf{k}},q^-|(\hat{\varepsilon}_l \cdot \mathbf{d})_{e,g}|E_i,J_i,M_i>$$
$$= \frac{8\pi\mu k}{\hbar^2} \sum_J \begin{pmatrix} J & 1 & J_i \\ -M_i & 0 & M_i \end{pmatrix} \begin{pmatrix} J & 1 & J_j \\ -M_j & 0 & M_j \end{pmatrix} \delta_{M_i M_j} t^*(E,J,q|E_j,J_j) t(E,J,q|E_i,J_i) \tag{6.28}$$

The integration over the scattering angles $\hat{\mathbf{k}}$ corresponds to the n-summation in Equations 6.17 and 6.19 and the Kronecker delta $\delta_{M_i M_j}$ arises from the fact that the laser-molecule interaction does not depend on the azimuthal angle for the case of linearly polarized light.

The probability P_3 is given, from the definition (Equation 6.17) of $\mathbf{d}^{(q)}$, by

$$P_3(E,q;E_i,J_i,M_i) = \left(\frac{\pi}{\hbar}\right)^2 \varepsilon_3^2 \mathbf{d}^{(q)}(E_i,J_i,M_i;E_i,J_i,M_i;E) \tag{6.29}$$

The terms P_{13} and P_1 can also be written in terms of the $\mathbf{d}^{(q)}$. To do so we express $<E,n,q^-|T|E_i>$ in Equation 6.20 explicitly in terms of $<E,n,q^-|(\hat{\varepsilon}_1 \cdot \mathbf{d})_{e,g}|E_i>$ by inserting appropriate resolutions of the identity:

$$<E,n,q^-|T|E_i> = \sum_{j,n',q'} \int dE' \times$$
$$\frac{<E,n,q^-|(\hat{\varepsilon}_1 \cdot \mathbf{d})_{e,g}|E_j><E_j|(\hat{\varepsilon}_1 \cdot \mathbf{d})_{g,e}|E',n',q'^-><E',n',q'^-|(\hat{\varepsilon}_1 \cdot \mathbf{d})_{e,g}|E_i>}{(E_j - E_i - 2\hbar\omega_1)(E' - E_i - \hbar\omega_1)} \tag{6.30}$$

Here, as noted previously, $|E_i>$ denotes all of $|E_i,J_i,M_i>$ and the j-summation indicates a sum over all bound states $|E_j>$ of the ground $X^1\Sigma_0^+$ potential surface. Computationally, of all the bound eigenstates of $X^1\Sigma_0^+$, the contribution to $<E,n,q^-|T|E_i>$ is dominated by those states with energy E_j satisfying the near two-photon resonance condition

$$E_j \approx E_i + 2\hbar\omega_1 \tag{6.31}$$

From Equation 6.30 and the definition of $\mathbf{d}^{(q)}$, $|F_{13}^{(q)}| \exp(i\delta_{13}^{(q)})$ in the cross-term P_{13}, is given by

$$|F_{13}^{(q)}| \exp(i\delta_{13}^{(q)}) = \sum_{E_j,J_j,q'} \int dE' \times$$

$$\frac{\mathbf{d}^{(q)}(E_j,J_j,M_i;E_i,J_i,M_i;E)\,\mathbf{d}^{(q')*}(E_j,J_j,M_i;E_i,J_i,M_i;E')}{(E_j - E_i - 2\hbar\omega_1)(E' - E_i - \hbar\omega_1)} \tag{6.32}$$

where the J_j and the E_j summation indicates a summation over all bound eigenstates of the $X^1\Sigma_0^+$ state.

Similarly, using Equations 6.19 and 6.30, the probability for the three-photon process is given by

$$P_1(E,q;E_i,J_i,M_i) = \sum_{E_j,E_l,J_j,J_l,\bar{q},q'} \int \mathrm{d}E' \int d\bar{E} \times$$

$$\frac{\mathbf{d}^{(q)}(E_l,J_l,M_i;E_j,J_j,M_i;E)\,\mathbf{d}^{(q')}(E_j,J_j,M_i;E_i,J_i,M_i;E')\,\mathbf{d}^{(\bar{q})*}(E_l,J_l,M_i;E_i,J_i,M_i;\bar{E})}{(E_j - E_i - 2\hbar\omega_1)(E_l - E_i - 2\hbar\omega_1)(E' - E_i - \hbar\omega_1)(\bar{E} - E_i - \hbar\omega_1)} \tag{6.33}$$

These expressions are quite complex and simple qualitative rules for tabulating the terms are provided elsewhere [13]. Given these results, the quantities $F_3^{(q)}$ and $F_1^{(q)}$ in the branching ratio ($R_{qq'}$) expression of Equation 6.26 are then written easily in terms of $\mathbf{d}^{(q)}$ using Equations 6.24, 6.29, 6.32, and 6.33.

Computational results were obtained using this scenario for the case of IBr photodissociation:

$$\mathrm{I}(^2P_{3/2}) + \mathrm{Br}^*(^2P_{1/2}) \leftarrow \mathrm{IBr} \rightarrow \mathrm{I}(^2P_{3/2}) + \mathrm{Br}(^2P_{3/2})$$

Two different cases were examined, those corresponding to fixed initial M_i values and those corresponding to averaging over a random distribution of M_i, for fixed J_i. Results typical of those obtained are shown in Figures 6.5 and 6.6, where we provide a contour plot of the yield of $\mathrm{Br}^*(^2P_{1/2})$ for the case of excitation from $J_i = 1$, $M_i = 0$, and $J_i = 42$ with an average over M_i,

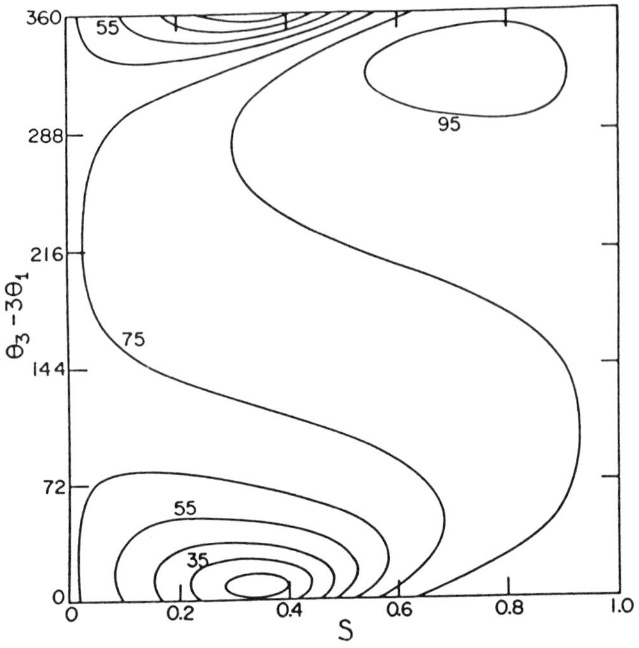

Figure 6.5 Contour plot of the yield of $Br^*(^2P_{1/2})$ (percentage of Br^* as product) in the photodissociation of IBr from an initial bound state in $X^1\Sigma_0^+$ with $v = 0$, $J_i = 1$, $M_i = 0$. Results arise from simultaneous (ω_1, ω_3) excitation $(\omega_3 = 3\omega_1)$, with $\omega_1 = 6657.5\ cm^{-1}$ *Source*: from [13].

as a function of laser control parameters (relative intensity and phase). The range of control in each case is vast, with no loss of control with averaging over M_j. A related, high field study of two-photon + four-photon control in the photodissociation of Cl_2 has been carried out by Bandrauk et al. [31].

This three-photon + one-photon scenario has now been experimentally implemented [32, 33], in studies of Hg and HCl ionization through a resonant-bound Rydberg state. Specifically, in the work of Gordon et al. [33], HCl is excited to a selected rotational state in the $^3\Sigma^-(\omega^+)$ manifold using $\omega_1 = 336$ nm; ω_3 is obtained by third harmonic generation in a Krypton gas cell. The relative phase of the light fields was then varied by passing the beams through a second gas cell and varying the gas pressure. The population of the resultant Rydberg state was interrogated by ionizing to HCl^+ with an

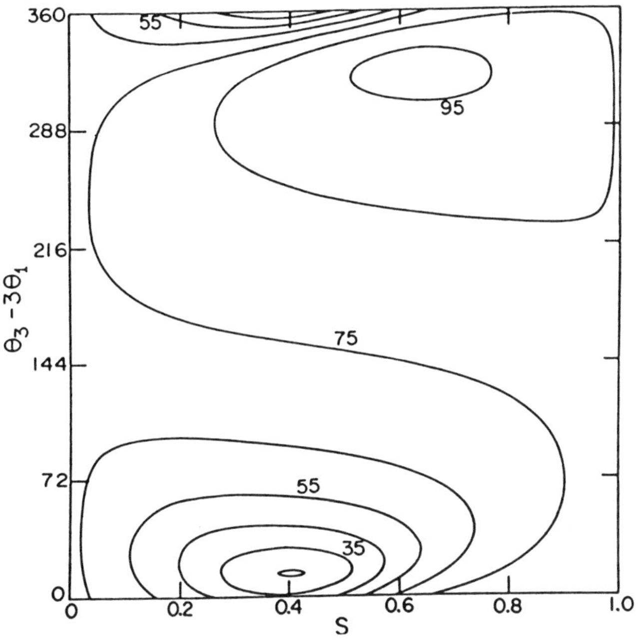

Figure 6.6 As in Figure 6.5 but for $v = 0$, $J_i = 42$, $\omega_1 = 6635.0$ cm^{-1} and *M*-averaged ($\varepsilon_0 = 1/8$). *Source*: from [13].

additional photon. This REMPI-type experiment showed that the HCl^+ ion probability depended upon both the relative phase and intensity of the two exciting lasers, in accord with the theory described here. A similar phase control experiment has been performed on atoms [32], in the simultaneous three-photon + five-photon ionization of Hg. Although the effect of the relative laser intensity was not studied, the Hg^+ ionization probability was shown to be a function of relative phase of the two lasers.

In another laboratory study, photocurrent directionality, which we predicted to be achievable in semiconductors using coherent control techniques with no bias voltage (see Section 6.4.B) has been demonstrated using one-photon + two-photon interferences [34].

These experiments clearly show that coherent control of simple molecular processes through quantum interference of multiple optical excitation routes is both feasible and experimentally observable. Further experimental studies designed to show control over processes with more than one product channel are in progress by a number of experimental groups.

B. Pump-Dump Scenarios

An important generalization of the scenario outlined in Section 6.2.A arises when the initial superposition of bound states is prepared with a laser pulse and is subsequently dissociated with a laser pulse. The scenario is shown qualitatively in Figure 6.7.

The pump and dump steps are assumed to be temporally separated, with a time delay τ_d between their temporal centers. The analysis here shows that under these circumstances the convenient control parameters are the central frequency of the pump pulse and the time delay between the pulses.

Consider a molecule, initially ($t = 0$) in eigenstate $|E_g\rangle$ of Hamiltonian H_M, which is subjected to two sequential transform-limited light pulses. The total Hamiltonian is of the form:

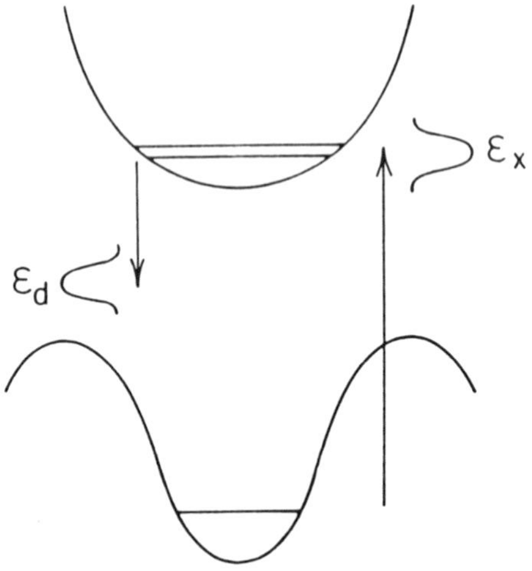

Figure 6.7 Coherent radiative control via a picosecond pulse scheme. In this case a single level is excited with a laser pulse to produce a superposition of two bound states in an excited electronic state. Subsequent deexcitation of this state to the continuum of the ground state allows control over the reaction on the ground state surface. *Source*: from [12].

$$H = H_M + V = H_M - \mathbf{d}[\overline{\varepsilon}(t) + \overline{\varepsilon}^*(t)] \tag{6.34}$$

where **d** is the dipole operator along the electric field direction. The field $\overline{\varepsilon}(t)$ consists of two temporally separated pulses $\overline{\varepsilon}(t) = \overline{\varepsilon}_x(t) + \overline{\varepsilon}_d(t)$, with the Fourier transform of $\overline{\varepsilon}_x(t)$ denoted $\varepsilon_x(\omega)$, etc. For convenience, these are chosen as gaussian pulses peaking at $t = t_x$ and t_d respectively. The $\overline{\varepsilon}_x(t)$ pulse induces a transition to a linear combination of two excited, bound, electronic states with nuclear eigenfunctions $|E_1\rangle$ and $|E_2\rangle$, and the $\overline{\varepsilon}_d(t)$ pulse dissociates the molecule by further exciting it to the continuous part of the spectrum. Both fields are chosen to be sufficiently weak so that perturbation theory is valid [24].

The superposition state prepared by the $\overline{\varepsilon}_x(t)$ pulse, whose width is chosen to encompass just the two E_1 and E_2 levels, is given in first-order perturbation theory as,

$$|\varphi(t)\rangle = |E_g\rangle e^{-iE_g t/\hbar} + c_1|E_1\rangle e^{-iE_1 t/\hbar} + c_2|E_2\rangle e^{-iE_2 t/\hbar}$$

where

$$c_k = \frac{\sqrt{2\pi}}{i\hbar} \langle E_k \, \mathbf{d} | E_g\rangle \varepsilon_x(\omega_{kg}), \qquad k = 1, 2, \tag{6.35}$$

with

$$\omega_{kg} \equiv \frac{E_k - E_g}{\hbar}.$$

After a time delay $\tau \, (= \tau_d - \tau_x)$ the system is subjected to the $\overline{\varepsilon}_d(t)$ pulse. Following that pulse the system wavefunction, expanded in a complete set of continuum eigenstates $|E,n,q^-\rangle$, labeled by energy E, arrangement channel label q and remaining labels n, is then,

$$|\psi(t)\rangle = |\varphi(t)\rangle + \sum_{n,q} \int dE \, B(E,n,q|t)|E,n,q^-\rangle e^{-iEt/\hbar} \tag{6.36}$$

Here we are interested in the probability $P(E,q)$ of forming product in arrangement channel q at energy E. However, we require $P(E,m_j,q)$ (i.e., the probability of forming product in q, E, and total fragment angular momentum projection m_j along the space fixed axis; see Section 6.4.A). The two probabilities are related by $P(E,q) = \Sigma_{m_j} P(E,m_j,q)$. Hence we first produce $P(E,m_j,q)$ and subsequently sum to obtain P(E,q).

Using first-order perturbation theory and the rotating wave approximation in conjunction with Equation 6.36, gives:

$$P(E,m_j,q) = \sum_n {}' |B(E,n,q|t=\infty)|^2$$

$$= \left(\frac{2\pi}{\hbar^2}\right) \sum_n {}' \left|\sum_{k=1,2} c_k <E,n,q^-|\mathbf{d}|E_k> \varepsilon_d(\omega_{EE_k})\right|^2 \qquad (6.37)$$

where $\omega_{EE_k} = (E - E_k)/\hbar$, c_k is given by Equation 6.35 and where the prime denotes summation over all quantum numbers n other than m_j.

Expanding the square and using the gaussian pulse shape gives:

$$P(E,m_j,q) =$$

$$\left(\frac{2\pi}{\hbar^2}\right) [|c_1|^2 \mathbf{d}_{1,1}^{(q)} \varepsilon_1^2 + |c_2|^2 \mathbf{d}_{2,2}^{(q)} \varepsilon_2^2 + 2|c_1 c_2 \varepsilon_1 \varepsilon_2 \mathbf{d}_{1,2}^{(q)}| \cos(\omega_{2,1}(t_d - t_x) + \alpha_{1,2}^{(q)}(E) + \varphi)] \qquad (6.38)$$

where $\varepsilon_i = |\varepsilon_d(\omega_{EE_i})|$, $\omega_{2,1} = (E_2 - E_1)/\hbar$ and the phases φ, $\alpha_{1,2}^{(q)}(E)$ are defined by

$$<E_1|\mathbf{d}|E_g><E_g|\mathbf{d}|E_2> \equiv |<E_1|\mathbf{d}|E_g><E_g|\mathbf{d}|E_2>| e^{i\varphi}$$

$$\mathbf{d}_{i,k}^{(q)}(E) \equiv |\mathbf{d}_{i,k}^{(q)}(E)| e^{i\alpha_{i,k}^{(q)}(E)} = \sum_n {}' <E,n,q^-|\mathbf{d}|E_i><E_k|\mathbf{d}|E,n,q^-> \qquad (6.39)$$

Integrating over E to encompass the width of the second pulse, assumed sufficiently small so that $\mathbf{d}_{i,k}^{(q)}(E)$ can be assumed constant, and forming the ratio

$$Y = \frac{\sum_{m_j} P(q,m_j)}{\sum_{m_j,q} P(q,m_j)} \tag{6.40}$$

gives the ratio of products in each of the two arrangement channels (i.e., the quantity we wish to control). Once again it is the sum of two direct photodissociation contributions plus an interference term.

Examination of Equation 6.38 makes clear that the product ratio Y can be varied by changing the delay time $\tau = (t_d - t_x)$ or ratio $x = |c_1/c_2|$; the latter is most conveniently done by detuning the initial excitation pulse.

It is enlightening to consider this scenario as applied [9] to a model collinear reaction with masses of D and H,

$$H + HD \leftarrow DH_2 \rightarrow D + H_2$$

Typical results (see also [9]) for control are shown in Figure 6.8 where the yield is seen to vary from 16–72% as the time delay and tuning of the initial excitation pulse are varied. This is an extreme range of control, especially in light of the fact that the two product channels differ only by mass factors.

It is highly instructive to examine the nature of the superposition state prepared in the initial excitation (Equation 6.35) and its time evolution during the delay between pulses. An example of such a state is shown in Figure 6.9 where we plot the wavefunction for a collinear model of reaction (40). Specifically, the coordinates are the reaction coordinate S and its orthogonal conjugate x. The wavefunction is shown evolving over one half of its total possible period. Examination of Figure 6.9 shows that de-exciting this superposition state during frame (b) would result in a substantially different product yield than de-exciting at the time of frame (e). However, there is clearly no particular preference of the wavefunction for large positive or large negative S at these particular times, which would be the case if the reaction control were a result of some spatial characteristics of the wavefunction. Rather, the essential control characteristics of the wavefunction are carried in the quantum amplitude and phase of the created superposition state.

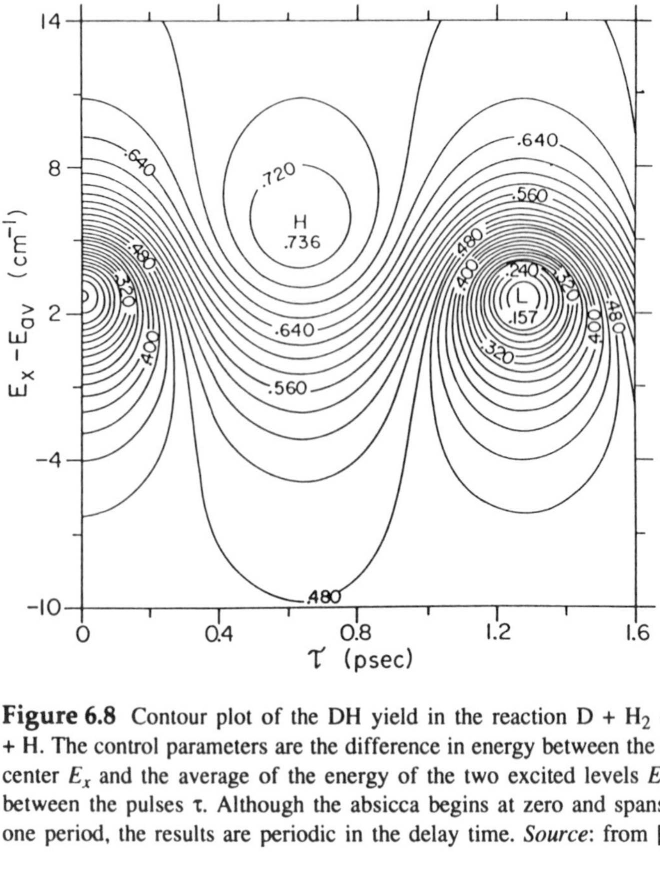

Figure 6.8 Contour plot of the DH yield in the reaction D + H_2 ← DH_2 → DH + H. The control parameters are the difference in energy between the excitation pulse center E_x and the average of the energy of the two excited levels E_{av} and the time between the pulses τ. Although the absicca begins at zero and spans approximately one period, the results are periodic in the delay time. *Source*: from [9].

A second example of pump-dump control [11] is provided by IBr photodissociation. Specifically, we showed that it is possible to control the Br* versus Br yield in this process by using two conveniently chosen picosecond pulses in the following way. The first pulse prepares a linear superposition of two bound states which arise from mixing of the *X* and *A* states. A subsequent pulse pumps this superposition to dissociation where the relative yields of Br and Br* are examined. Results typical of those obtained are shown in Figure 6.10 where the relative yield is shown as a function of the delay between pulses and the detuning of the pump pulse from the energetic center of the two bound states in the initial superposition. The

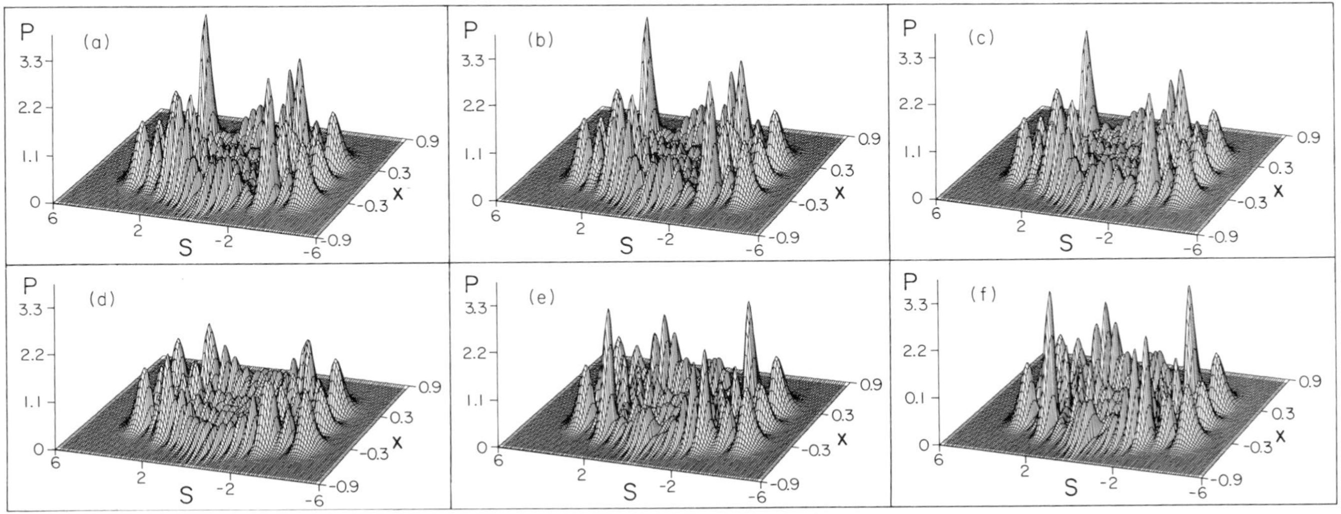

Figure 6.9 Time evolution of the square of the wavefunction for a superposition state comprised of levels 56 and 57 of the G1 surface of H_3. The probability is shown as a function of the reaction coordinate S and orthogonal distance x at times (a) 0, (b) 0.0825 psec, (c) 0.165 psec, (d) 0.33 psec, (e) 0.495 psec, (f) 0.66 psec, which correspond to equal fraction of one half the period $2\pi/\omega_{2,1}$. *Source*: from [9].

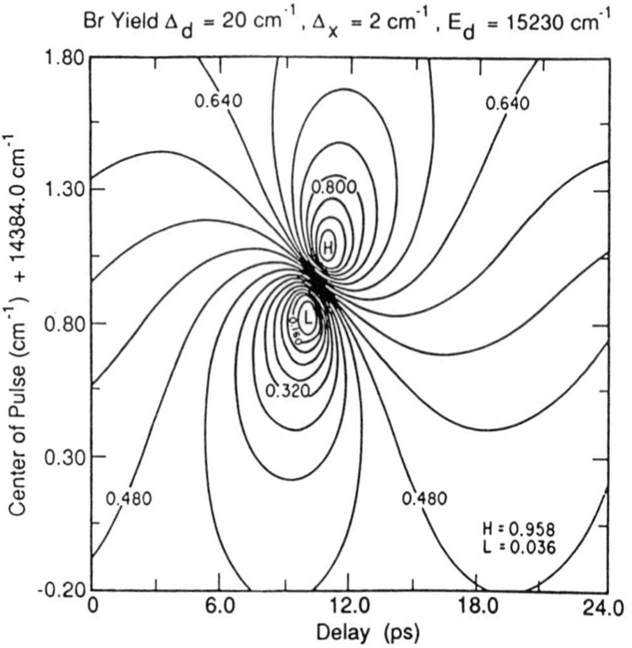

Figure 6.10 Computed control over the Br yield as a function of E_x—the excitation pulse detuning and τ—the time delay between the pulses. Parameters are of Figure 6.6, Ref. [11]. *Source*: from [11].

results show the vast range of control which is possible with this relatively simple experimental setup. Once again it is worth noting that both the potential energy surfaces and quantum photodissociation computations are state-of-the-art, so that the results should be representative of results expected in the laboratory.

6.3 DESIGNING CONTROL SCENARIOS

Given the general principle established above, one can design numerous scenarios to obtain control over reactions in the laboratory. Such scenarios must, however, properly account for a number of factors which reduce or eliminate control. In this section we briefly note four such features and subsequently discuss two in greater detail.

1. In any proposed scenario it is necessary to minimize extraneous uncontrolled satellites; that is, photoexcitation often results in contributions which are uncontrolled, in addition to the desired controlled contributions. An example is the $J = 3$ state created by three photon absorption in the above scenario. Since there is only one route to $J = 3$ (unlike $J = 1$) photodissociation from the $J = 3$ component is insensitive to the relative phase and intensities of the one and three photon routes. Similar uncontrolled contributions arise in other scenarios as well. Effective control schemes must insure that such contributions are small compared to the controlled component.
2. A successful laboratory scenario must insure properly treated laser spatial dependence and phase jitter. Since the ability to accurately manipulate the laser relative phase is crucial to coherent control, one must account for all laboratory features which affect the phase. For example, scenarios must be designed so as to eliminate effects due to the $k \cdot R$ spatial dependence of the laser phase. Not doing so results in the reduction of control resulting from the variation of this term over the molecule beam-laser beam intersection volume.
3. The laboratory scenario must be designed to compete effectively against effects which tend to destroy coherence (e.g., collisions and laser phase instabilities). The collision and laser incoherence studies [7, 17] (discussed in Section 6.3.A) indicate that control can survive moderate levels of such phase destructive processes or that clever scenarios can overcome their effects.
4. Selection rules play a fundamental role in the nature of the processes which can be controlled. Control requires that the interference term (e.g., $F_{13}^{(q)}$) between optical routes is nonzero. In general, this is the case only if the various optical excitation routes satisfy specific rules regarding conserved integrals of motion. The importance of the issue of selection rules in designing control scenarios leads us to discuss them in detail.

A. Selection Rules

So far we have not considered the question of *selection rules* in coherent control. In this section we show that indeed there are some strict limitations on the type of processes which can be used to control integral attributes. The term *integral*, in contrast to *differential*, is used here to describe a quantity in which averaging over angles and/or final polarizations takes place in the detection process. It turns out that most of the limitations imposed on the control of integral quantities do not apply to the control of differential attributes. Naturally, if our main interest is in the *total yield* of a chemical reaction, we must consider processes which control integral quantities. There are, however, many important applications, such as the control of current directionality [8] (see Section 6.4.B), and angular distributions of photofragments [5], in which we aim to control the differential attributes.

For example, in order to maintain control over integral attributes in the scenario described in Section 6.1.C we must ensure that the interference term of Equation 6.13b does not vanish when we perform the averaging which defines the particular integral quantity of interest. Thus, if we wish to control integral *cross-sections*, the interference term should survive integration over scattering angles; if we wish to control a *nonpolarized* but otherwise differential quantity we must make sure that the interference term survives the averaging over the appropriate magnetic quantum-numbers.

The discussion below can be summarized in terms of two general selection rules for integral control:

1. The (two) interfering pathways must be able to access continuum states with the same magnetic quantum numbers. This rule holds even when the initial states are M-polarized.
2. If the initial state is not M-polarized, integral control can only be achieved via interference between continuum states of equal parity. Two pathways which generate states of opposite parity, such as a one-photon and a two-photon absorption, cannot lead to integral control of unpolarized initial states. However, these pathways can [5, 8] and do [34] lead to differential control of unpolarized states or integral control of polarized beams (subject to selection rule 1).

In order to see how these selection rules come about, we use a symmetric-top molecule as a working example and the the superposition-state control scenario outlined in Section 6.1.C. To be more specific, we take CH_3I as an example. The symmetric-top CH_3I is in many ways equivalent to a linear triatomic molecule [36], since in both the ground and the first few excited states the I, the C, and the H_3 (C.M.) do not deviate significantly from the collinear configuration.

As mentioned in Section 6.1.C, CH_3I breaks apart to yield $CH_3(v) + I^*(^2P_{1/2})$ and $CH_3(v) + I(^2P_{3/2})$. The relevant quantum numbers for the bound states $|\varphi_i\rangle$, $|\varphi_j\rangle$ which make up the initial superposition state are E_i—the energy and J_i and M_i—the total angular momentum and its z-projection, respectively; hence we denote the states as $|E_i,J_i,M_i>$, etc. The products prime label m is composed of v—the final CH_3 (umbrella) vibrational state, $\hat{\mathbf{k}}$ $(= \varphi_k, \theta_k)$—the CH_3 scattering angles relative to the polarization direction of the photolysis laser, and q = 1, 2—the products prime electronic state index.

The molecule is a symmetric top rather than a simple rotator due to the presence of λ, the projection of the *total* angular momentum on the CH_3–I axis. In general λ is a projection of both the electronic angular momentum and the rotation of the CH_3 group about the C–I axis. In the present discussion we ignore the nuclear component of λ and concentrate on the electronic component.

In the photoexcitation to the first (A) continuum of CH_3I, λ assumes the values 0 (the ground and the the $^3Q_0$ states) and $\pm$ 1 (the $^1Q_1$ state). In the diabatic representation [28], λ = 0 correlates with q = 1—the CH_3 + $I^*(^2P_{1/2})$ fragment-channel and $\lambda = \pm 1$ with q = 2—the CH_3 + $I(^2P_{3/2})$ channel. We therefore use λ and q interchangeably in describing the products prime electronic states.

For a symmetric-top molecule the three dimensional photodissociation amplitude of Equation 6.27 can be written as [28], where **d** is now along the electric field

$$<E,\hat{\mathbf{k}},v,\lambda^-|\mathbf{d}|E_i,J_i,M_i> = \frac{(2\mu k_{v\lambda})^{1/2}}{\hbar} \cdot$$

$$\sum_J \begin{pmatrix} J & 1 & J_i \\ -M_i & 0 & M_i \end{pmatrix} (2J+1)^{1/2} D^J_{\lambda,M_i}(\varphi_k,\theta_k,-\varphi_k)\, t(E,J,\lambda,v|E_i,J_i) \tag{6.41}$$

Here μ is the reduced-mass of the CH_3,I pair, $k_{v\lambda}$ is the magnitude of the $CH_3(v)$–I(λ) relative wave-vector, and $t(E,J,\lambda,v|E_i,J_i)$ are the (M_i-independent) reduced amplitudes, containing the essential dynamics of the photodissociation process [28].

With the use of Equation 6.11, the integral attributes which enter the general coherent control expression (Equation 6.13b), are,

$$\mathbf{d}^{(\lambda)}(E_i,J_i,M_i;E_j,J_j,M_j;E) = \sum_v \int d\hat{\mathbf{k}} <E_i,J_i,M_i|\mathbf{d}_\varepsilon|E,\hat{\mathbf{k}},v,\lambda^-><E,\hat{\mathbf{k}},v,\lambda^-|\mathbf{d}_\varepsilon|E_j,J_j,M_j>$$

$$= \frac{8\pi\mu}{\hbar^2}\,\delta_{M_i,M_j}\sum_{v,J} k_{v\lambda}\begin{pmatrix} J & 1 & J_i \\ -M_i & 0 & M_i\end{pmatrix}\begin{pmatrix} J & 1 & J_j \\ -M_j & 0 & M_j\end{pmatrix}\times$$

$$t(E,J,\lambda,v|E_i,J_i)t^*(E,J,\lambda,v|E_j,J_j) \tag{6.42}$$

The δ_{M_i,M_j} factor arises from the angular integration and the orthogonality of the Wigner D functions [37]. We see immediately that even for the $i \neq j$ interference term, $M_i = M_j$. Since coherent control vanishes if the interference term vanishes, we conclude that for the superposition-state scenario, control with linearly polarized light is possible only when the states that make up the superposition state have *equal* magnetic quantum numbers.

This result is actually due to the fact that the two excitation pathways must be able to access continuum states with the same M quantum number. In the case of linear polarization the photoexcitation process cannot change M, hence the requirement that $M_i = M_j$ in the initial superposition state. For circular polarization M changes by $\pm$ 1 and the selection rule is that $M_i = M_j \pm 1$. These two different selection rules immediately preclude the use of a *single* bound state with two *different* polarizations, (linear + circular or two circular polarizations of opposite sense) for *integral* control. However there are no limitations on differential control [5] since two states of different M quantum numbers can interfere.

We now proceed to explore two cases of interest; the first, where the initial state is M-selected and the second where no such selection is assumed. Each of these cases is treated in turn.

M-Polarized Initial States

We consider exciting with linearly polarized light a superposition of two bound states: $|E_1,J_1,M_1\rangle$ and $|E_2,J_2,M_2 = M_1\rangle$. The choice $M_2 = M_1$ is a result of the earlier discussion and the use of linearly polarized light in the excitation step. The excitation by radiation with two colors raises the system to energy E as described above. Equation 6.13b, in conjunction with Equation 6.42, is now directly applicable.

The symmetry properties of the 3-j symbols [37] imply that,

$$\mathbf{d}^{(\lambda)}(E_i,J_i,M_i;E_j,J_j,M_j;E) = (-1)^{(J_i+J_j)}\,\mathbf{d}^{(\lambda)}(E_i,J_i,-M_i;E_j,J_j,-M_j;E) \tag{6.43}$$

Therefore, the relative product yield $R(1,2;E)$ is identical for the case of $|M_1|$ and $-|M_1|$ if $(J_1 + J_2)$ is even. In the case of odd $(J_1 + J_2)$, the interference term changes sign when going from $|M_1|$ to $-|M_1|$. The control map (i.e., yield versus S) and $(\theta_1 - \theta_2)$ of the M-polarized case is identical for the $|M_1|$ and $-|M_1|$ case, except for a shift in the relative phase $(\theta_1 - \theta_2)$ of π. For the *unpolarized* case, this result is shown below to lead to cancellation of the interference term for states of different parity.

Figures 6.11(a) and 6.12(a) display the yield of $I^*(^2P_{1/2})$ for two different M-selected initial bound-state superpositions. Results are shown at $\omega_1 = 39638\ \text{cm}^{-1}$, which is near the absorption maximum. For the present discussion, the main feature worth noting is that the equal-parity case of Figure 6.12(a), where $J_1 = J_2 = 1$, is strikingly different from the unequal parity case of Figure 6.11(a), where $J_1 = 1$ and $J_2 = 2$. The equal-parity maps show a wider range of control as compared with the unequal parity results.

In addition to these results, the actual value assumed by M of the initial beam is of importance. This is most noticeable in the unequal-parity case, where the $M = 1$ case of Figure 6.11(a), is drastically different than the $M = 0$ case of Figure 6.13 which shows *no phase control*. This loss of control follows from the properties of the 3-j symbols of Equation 6.42 which are zero whenever $M_1 = 0$ and $(J_1 + J_2)$ is odd [37].

M-Averaged Initial States

In this case the initial state is defined by the density matrix

$$\rho_0 = \frac{1}{J_1 + 1} \sum_{M_1} c_1 |E_1,J_1,M_1\rangle\langle E_1,J_1,M_1| + c_2 |E_2,J_2,M_1\rangle\langle E_2,J_2,M_1|$$

Each of the superposition states which make up our initial density matrix may be treated independently in the subsequent two-color irradiation which lifts the system to E. The resultant probability of observing product channel q at energy E, $P(q,E)$, is obtained as an average over the $(2J_1 + 1)$ superpositions,

$$P(q,E) = \frac{(\pi/\hbar)^2}{2J_1 + 1} \sum_{i=1,2} \sum_{j=1,2} \sum_{M_1} F_{i,j} \mathbf{d}^{(\lambda)}(E_i,J_i,M_1;E_j,J_j,M_1;E) \qquad (6.44)$$

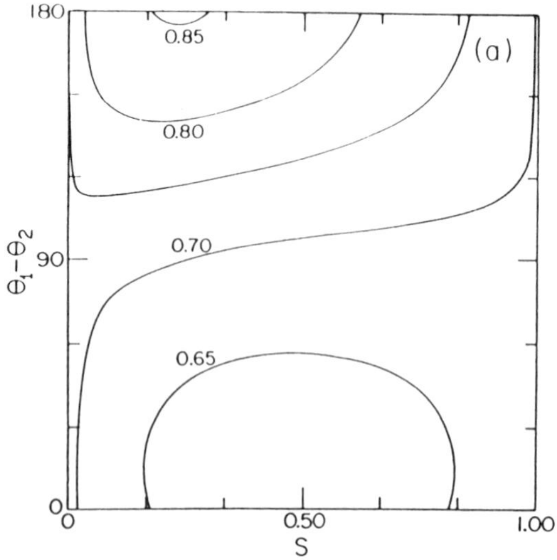

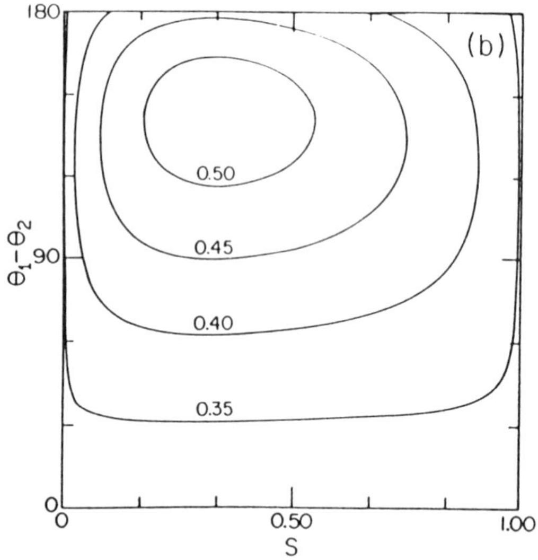

Figure 6.11 Contour plot of the yield of $I^*(^2P_{1/2})$ (i.e., percentage of I^* as product) in the photodissociation of CH_3I from a polarized superposition state composed of $v_1 = 0$, $J_1 = 1$, and $v_2 = 0$, $J_2 = 2$, where $M_1 = M_2 = 1$, at (a) $\omega_{E_1} = 39638\ cm^{-1}$ and (b) $\omega_{E_1} = 42367\ cm^{-1}$. $v = 0$ denotes the ground vibrational state of CH_3I. The abscissa is labelled by $S = x^2/(1 + x^2)$, where x is the ratio of laser field intensities and the ordinate by the relative phase parameter $\theta = \theta_1 - \theta_2$. After Ref. [3].

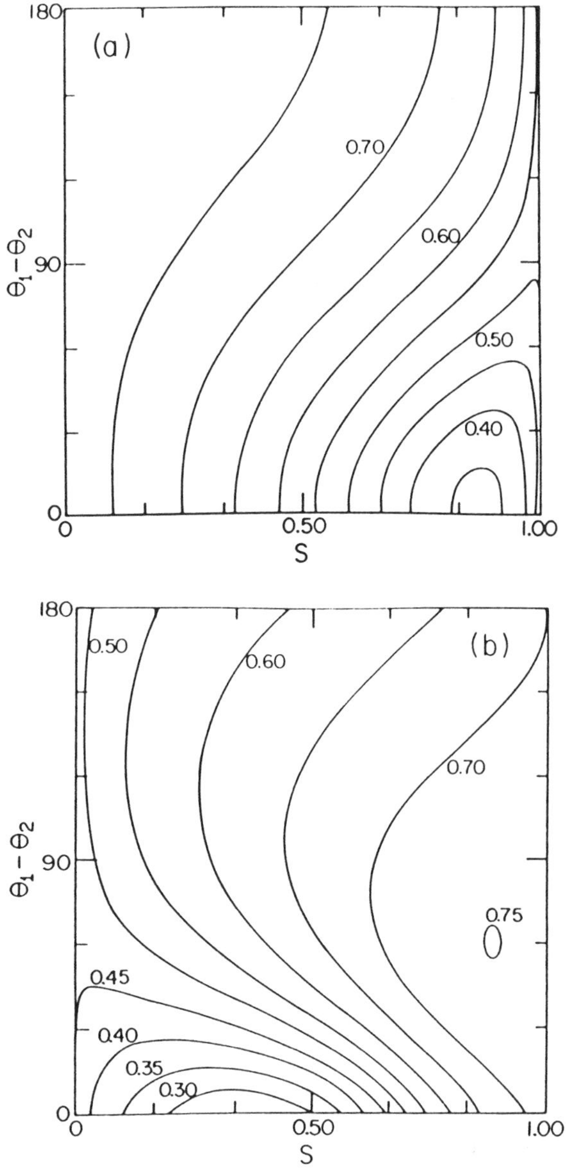

Figure 6.12 As in Figure 6.11 but $v_1 = 0$, $J_1 = 2$, $v_2 = 1$, $J_2 = 2$, $M_1 = M_2 = 0$, at (a) $\omega_{E_1} = 39638$ cm^{-1} and (b) $\omega_{E_1} = 42367$ cm^{-1}. $v = 1$ denotes the first vibrationally excited state of CH_3I (essentially one quantum of the C–I stretch). After Ref. [3].

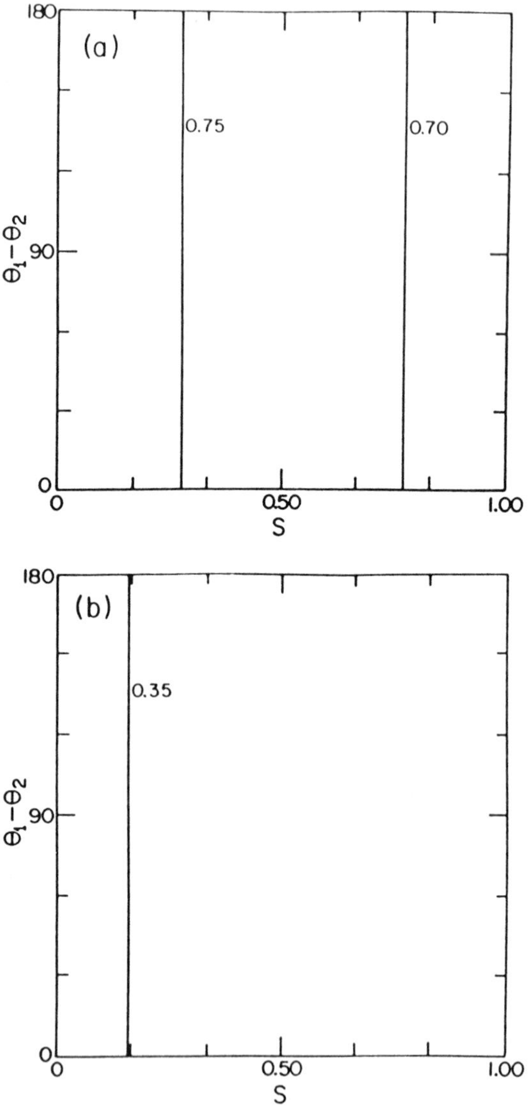

Figure 6.13 As in Figure 6.11 but $v_1 = 0$, $J_1 = 1$, $v_2 = 0$, $J_2 = 2$, $M_1 = M_2 = 0$. No phase control is seen since $J_1 + J_2$ is odd and $M_1 = 0$. (a) $\omega_{E_1} = 39638$ cm^{-1} and (b) $\omega_{E_1} = 42367$ cm^{-1}. After Ref. [3].

where $F_{i,j} \equiv c_i c_j^* \tilde{\varepsilon}_i \tilde{\varepsilon}_j^*$. From Equation 6.42 it follows that the M_1-dependence of $\mathbf{d}^{(\lambda)}(E_i,J_i,M_1;E_j,J_j,M_1;E)$ is entirely contained in the 3-j product

$$\begin{pmatrix} J & 1 & J_i \\ -M_1 & 0 & M_1 \end{pmatrix} \begin{pmatrix} J & 1 & J_j \\ -M_1 & 0 & M_1 \end{pmatrix}$$

Hence, the M_1 summation can be performed separately. Defining,

$$C_{i,j}(J) \equiv \frac{1}{2J_1+1} \sum_{M_1} \begin{pmatrix} J & 1 & J_i \\ -M_1 & 0 & M_1 \end{pmatrix} \begin{pmatrix} J & 1 & J_j \\ -M_1 & 0 & M_1 \end{pmatrix} F_{i,j} \tag{6.45}$$

we have that

$$P(q,E) = \left(\frac{\pi}{\hbar}\right)^2 \sum_{i=1,2} \sum_{j=1,2} t^{(q)}(E_i,J_i;E_j,J_j;E) \tag{6.46}$$

where

$$t^{(q)}(E_i,J_i;E_j,J_j;E) \equiv \sum_{J,v} C_{i,j}(J) t(E,J,\lambda,v|E_i,J_i) t^*(E,J,\lambda,vE_j,J_j) \tag{6.47}$$

It follows immediately from the symmetry properties of the 3-j symbols [37] and Equation 6.45 that $t^{(q)}(E_1,J_1;E_2,J_2;E)$ is zero if $(J_1 + J_2)$ is odd; that is, yield control in M-averaged situations requires J_1 and J_2 of equal parity. Another way of reaching the same conclusion is to note that for odd $J_1 + J_2$, when we perform the M_1 summation, the positive M_1 terms cancel out the negative M_1 terms, and the $M_1 = 0$ term is identically zero. We conclude that for nonpolarized initial states, only two states of equal parity can be made to interfere allowing control of integral cross-sections.

The expression for the yield now follows directly as the ratio of $P(1,E)/\Sigma_q P(q,E)$ of Equation 6.46. Figure 6.14 shows the result for coherent radiative control of an initial M-averaged pair of states of *equal* parity at two different values of ω_1. The range of control demonstrated is very wide: at the peak of the absorption, (Figure 6.14(a)), the $I^*(^2P_{1/2})$ quantum yield changes from 30%, for $S = 0.9$ and $\theta_1 - \theta_2 = 0^\circ$, to 75% for $S = 0.2$ and $\theta_1 - \theta_2 =$

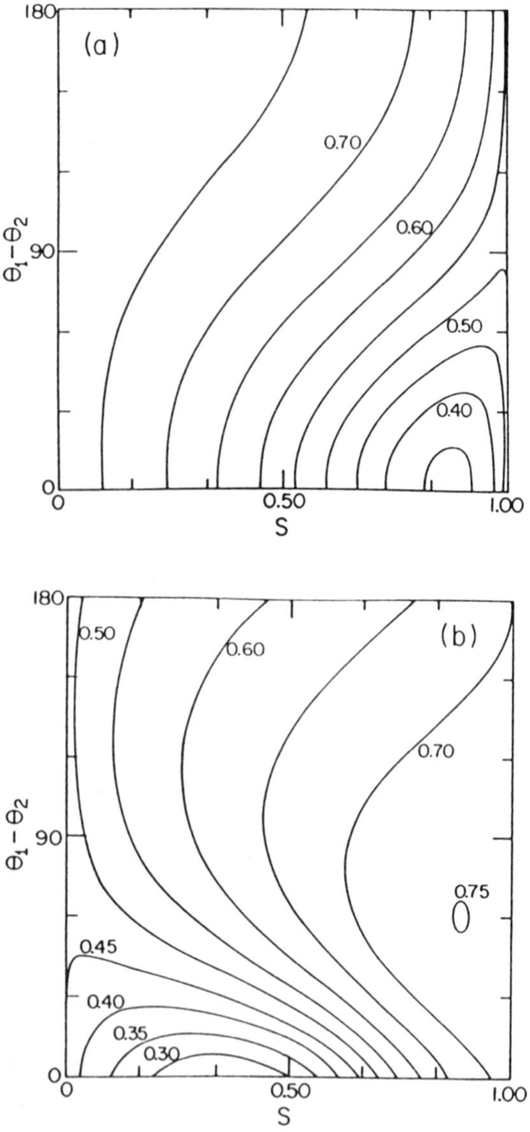

Figure 6.14 $I^*(^2P_{1/2})$ yield in the photodissociation of CH_3I starting from an M-averaged (unpolarized) ensemble of superposition states. The J and v are as in Figure 6.12. (a) $\omega_{E_1} = 39639$ cm^{-1} and (b) $\omega_{E_1} = 42367$ cm^{-1}. After Ref. [3].

140°. A comparison with the even $J_1 + J_2$ polarized case (Figure 6.12) shows that the range of control degrades only slightly with M-averaging. This is to be contrasted (Figure 6.11 versus Figure 6.13) with the odd $J_1 + J_2$ case.

B. Control in the Presence of Incoherence Effects

In practice there are a number of sources of incoherence which tend to diminish control. Prominent amongst these are effects due to an initial thermal distribution of states and effects due to partial coherence of the laser source. Hence, we describe one approach, based upon resonant two-photon photodissociation, which deals effectively with both problems and results in a viable scheme to control coherence in a thermal environment. An alternative method in which coherence is retained in the presence of collisions is discussed elsewhere [7].

Here we discuss a method [17] where it is possible to maintain control in a molecular system that is initially in thermal equilibrium by interfering two independent resonant two-photon routes to photodissociation. The resonant character of the excitations insure that only a selected state from the molecular thermal distribution participates. The proposed control scenario also provides a method for overcoming destructive interference loss due to phase jitter in the laser source and allows the reduction of contributions from uncontrolled ancillary photodissociation routes. Computational results support the feasibility and broad range of control afforded by this approach.

The specific scheme we advocate is depicted, for the particular case of Na_2 photodissociation, in Figure 6.15. Here the molecule is lifted from an initial bound state $|E_i,J_i,M_i\rangle$ to energy E via two independent two-photon routes. To introduce notation, first consider a single such two-photon route. Absorption of the first photon of frequency ω_1 lifts the system to a region close to an intermediate bound state $|E_m J_m M_m\rangle$, and a second photon of frequency ω_2 carries the system to the dissociating states $|E,\hat{\mathbf{k}},q^-\rangle$, where the scattering angles are specified by $\hat{\mathbf{k}} = (\theta_k,\varphi_k)$. Here J is the angular momentum, M is the projection along the z-axis, and the values of energy, E_i and E_m, include specification of the vibrational quantum numbers. If we denote the phases of the coherent states by φ_1 and φ_2, the wavevectors by $\boldsymbol{k}_1$ and $\boldsymbol{k}_2$ with overall phases $\theta_i = \boldsymbol{k}_i \cdot \boldsymbol{R} + \varphi_i$ $(i = 1, 2)$, and the electric field amplitudes by ε_1 and ε_2, then the probability amplitude for resonant two-photon $(\omega_1 + \omega_2)$ photodissociation is given [35] by

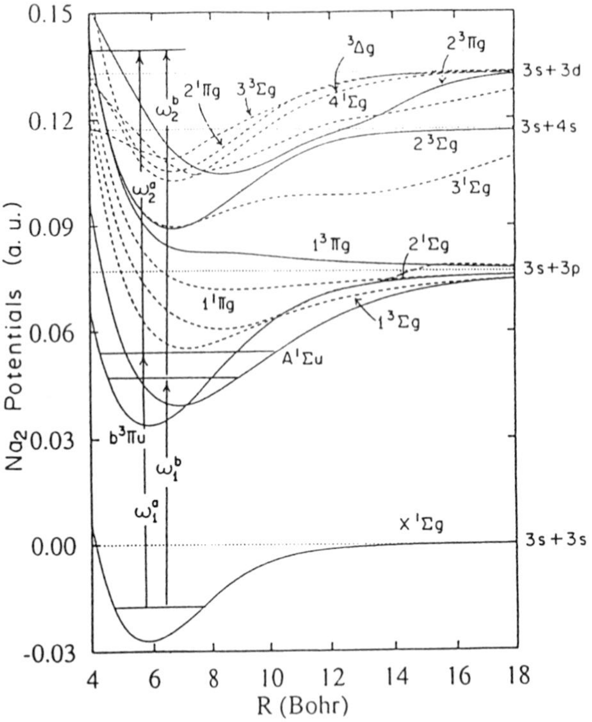

Figure 6.15 Two resonant two-photon paths in the photodissociation of Na_2. *Source*: from [17].

$$T_{\boldsymbol{k}q,i}(E,E_iJ_iM_i,\omega_2,\omega_1)$$

$$= \sum_{E_m J_m} \frac{\langle E,\boldsymbol{k},q^-|\mathbf{d}_2\varepsilon_2|E_mJ_mM_i\rangle\langle E_mJ_mM_i|\mathbf{d}_1\varepsilon_1|E_iJ_iM_i\rangle}{\omega_1-(E_m+\delta_m-E_i)+i\Gamma_m} \exp\,[i(\theta_1+\theta_2)]$$

$$= \frac{\sqrt{2\mu k_q}}{\hbar}\sum_{J,p,\lambda\geq 0}\sum_{E_m J_m}\begin{pmatrix} J & 1 & J_m \\ -M_i & 0 & M_i\end{pmatrix}\begin{pmatrix} J_m & 1 & J_i \\ -M_i & 0 & M_i\end{pmatrix}\times$$

$$\sqrt{2J+1}\, D^{Jp}_{\lambda,M_i}(\theta_k,\varphi_k,0)t(E,E_iJ_i,\omega_2,\omega_1,q|Jp\lambda,E_mJ_m)\exp\,[i(\theta_1+\theta_2)] \tag{6.48}$$

Here $\mathbf{d}_i$ is the component of the dipole moment along the electric-field vector of the ith laser mode, $E = E_i + (\omega_1 + \omega_2)$, δ_m and Γ_m are respectively the radiative shift and width of the intermediate state, μ the reduced mass, and k_q is the relative momentum of the dissociated product in q-channel. The D^{Jp}_{λ,M_i} is the parity-adapted rotation matrix [17] with λ the magnitude of the projection on the internuclear axis of the electronic angular momentum and $(-1)^J p$ the parity of the rotation matrix. We have set $\hbar \equiv 1$ and assumed for simplicity lasers which are linearly polarized and with parallel electric-field vectors. Note that the T-matrix element in Equation 6.48 is a complex quantity whose phase is the sum of the laser phase $\theta_1 + \theta_2$ and the molecular phase (i.e., the phase of $\mathbf{t}$).

The probability of producing the fragments in q-channel is obtained by integrating the square of Equation 6.48 over the scattering angles $\hat{\mathbf{k}}$, with the result:

$$P^{(q)}(E,E_i,J_i,M_i,\omega_2,\omega_1) = \int d\hat{\mathbf{k}}\, |T_{\mathbf{k}q,i}(E,E_i,J_i,M_i,\omega_2,\omega_1)|^2$$

$$= \frac{8\pi\mu k_q}{\hbar^2} \sum_{J,p,\lambda \geq 0} \Big| \sum_{E_m,J_m} \begin{pmatrix} J & 1 & J_m \\ -M_i & 0 & M_i \end{pmatrix} \begin{pmatrix} J_m & 1 & J_i \\ -M_i & 0 & M_i \end{pmatrix} \times$$

$$t(E,E_i,J_i,\omega_2,\omega_1,q|Jp\lambda,E_m,J_m)|^2 \tag{6.49}$$

Because the t-matrix element contains a factor of $[\omega_1 - (E_m + \delta_m - E_i) + i\Gamma_m]^{-1}$ the probability is greatly enhanced by the approximate inverse square of the detuning $\delta = \omega_1 - (E_m + \delta_m - E_i)$ as long as the line width Γ_m is less than δ. Hence only the levels closest to the resonance d = 0 contribute significantly to the dissociation probability. This allows us to selectively photodissociate molecules from a thermal bath, thereby reestablishing coherence necessary for quantum-interference-based control and overcoming dephasing effects due to collisions.

Consider then the following coherent control scenario. A molecule is irradiated with three interrelated frequencies, ω_0, ω_+, ω_- where photodissociation occurs at $E = E_i + 2\omega_0 = E_i + (\omega_+ + \omega_-)$ and where ω_0 and ω_+ are chosen resonant with intermediate bound state levels. The probability of photodissociation at energy E into arrangement channel q is then given by the square of the sum of the T-matrix elements from pathway a $(\omega_0 + \omega_0)$ and pathway b $(\omega_+ + \omega_-)$. That is, the probability into channel q

$$P_q(E,E_i J_i M_i;\omega_0,\omega_+,\omega_-) \equiv \int dk \left| T_{kq,i}(E,E_i J_i M_i,\omega_0,\omega_0) + T_{kq,i}(E,E_i J_i M_i,\omega_+,\omega_-) \right|^2$$

$$\equiv P^{(q)}(a) + P^{(q)}(b) + P^{(q)}(ab) \qquad (6.50)$$

Here $P^{(q)}(a)$ and $P^{(q)}(b)$ are the independent photodissociation probabilities associated with routes a and b respectively and $P^{(q)}(ab)$ is the interference term between them. Note that the two T-matrix elements in Equation 6.50 are associated with different lasers and as such contain different laser phases. Specifically, the overall phase of the three laser fields are

$$\theta_0 = \boldsymbol{k}_0 \cdot \boldsymbol{R} + \varphi_0,\ \theta_+ = \boldsymbol{k}_+ \cdot \boldsymbol{R} + \varphi_+ \text{ and } \theta_- = \boldsymbol{k}_- \cdot \boldsymbol{R} + \varphi_-$$

where φ_0, φ_+, and φ_- are the photon phases, and $\boldsymbol{k}_0$, $\boldsymbol{k}_+$, and $\boldsymbol{k}_-$ are the wavevectors of the laser modes ω_0, ω_+, and ω_- whose electric field strengths are ε_0, ε_+, ε_- and intensities I_0, I_+, I_-.

The optical path-path interference term $P^{(q)}(ab)$ is given by

$$P^{(q)}(ab) = 2|F^{(q)}(ab)| \cos(\alpha_a^q - \alpha_b^q) \qquad (6.51)$$

with relative phase

$$\alpha_a^q - \alpha_b^q = (\delta_a^q - \delta b^q) + (2\theta_0 - \theta_+ - \theta_-) \qquad (6.52)$$

where the amplitude $|F^{(q)}(ab)|$ and the molecular phase difference $(\delta_a^q - \delta_b^q)$ are defined by

$$|F^{(q)}(ab)| \exp[i(\delta_a^q - \delta_b^q)] \equiv$$

$$\frac{8\pi\mu k_q}{\hbar^2} \sum_{J,p,\lambda \geq 0} \sum_{E_m, J_m} \sum_{E'_m, J'_m} \begin{pmatrix} J & 1 & J_m \\ -M_i & 0 & M_i \end{pmatrix} \begin{pmatrix} J_m & 1 & J_i \\ -M_i & 0 & M_i \end{pmatrix} \begin{pmatrix} J & 1 & J'_i \\ -M_i & 0 & M_i \end{pmatrix} \times$$

$$\begin{pmatrix} J_m' & 1 & J_i \\ -M_i & 0 & M_i \end{pmatrix} t(E,E_i,J_i,\omega_0,\omega_0,q|Jp\lambda,E_m,J_m)t^*(E,E_i,J_i,\omega_-,\omega_+,q|Jp\lambda,E_m'J_m') \tag{6.53}$$

Consider now the quantity of interest $R_{qq'}$, the branching ratio of the product in q-channel to that in q'-channel. Noting that in the weak field case $P^{(q)}(a)$ is proportional to ε_0^4, $P^{(q)}(b)$ to $\varepsilon_+^2\varepsilon_-^2$, and $P^{(q)}(ab)$ to $\varepsilon_0^2\varepsilon + \varepsilon_-$ we can write

$$R_{qq'} = \frac{\mu_{aa}^{(q)} + x^2\mu_{bb}^{(q)} + 2x|\mu_{ab}^{(q)}| \cos(\alpha_a^q - \alpha_b^q) + (B^{(q)}/\varepsilon_0^4)}{\mu_{aa}^{(q')} + x^2\mu_{bb}^{(q')} + 2x|\mu_{ab}^{(q')}| \cos(\alpha_a^{q'} - \alpha_b^{q'}) + (B^{(q')}/\varepsilon_0^4)} \tag{6.54}$$

where

$$\mu_{aa}^{(q)} = \frac{P^{(q)}(a)}{\varepsilon_0^4}$$

$$\mu_{bb}^{(q)} = \frac{P^{(q)}(b)}{(\varepsilon_+^2\varepsilon_-^2)}$$

$$|\mu_{ab}^{(q)}| = \frac{|F^{(q)}(ab)|}{(\varepsilon_0^2\varepsilon_+\varepsilon_-)}$$

and

$$x = \frac{\varepsilon_+\varepsilon_-}{\varepsilon_0^2} = \frac{\sqrt{I_+I_-}}{I_0}$$

The terms with $B^{(q)}$, $B^{(q')}$ correspond to resonant photodissociation routes to energies other than $E = E_i + 2\hbar\omega_0$ and hence [4] to terms which do not coherently interfere with the a and b pathways. Minimization of these terms, which result from absorption of $(\omega_0 + \omega_-)$, $(\omega_0 + \omega_+)$, $(\omega_+ + \omega_0)$, or $(\omega_+ + \omega_+)$, is discussed elsewhere [17]. Here we emphasize that the product ratio in Equation 6.54 depends upon both the laser intensities and relative laser phase. Hence manipulating these laboratory parameters allows for control over the relative cross-section between channels.

The proposed scenario embodied in Equation 6.54 also provides a means by which control can be improved by eliminating effects due to laser jitter. Specifically, the term $2\varphi_0 - \varphi_+ - \varphi_-$ contained in the relative phase $\alpha_a^q - \alpha_b^q$ can be subject to the phase fluctuations arising from laser instabilities. If such fluctuations are sufficiently large then the interference term in Equation 6.54, and hence control, disappears [14]. The following experimentally desirable implementation of the two-photon plus two-photon scenario readily compensates for this problem. Specifically, consider generating $\omega_+ = \omega_0 + \delta$ and $\omega_- = \omega_0 - \delta$ in a parametric process by passing a beam of frequency $2\omega_0$ with phase $2\varphi_0$ through a nonlinear crystal. This latter beam is assumed generated by second harmonic generation from the laser ω_0 with the phase φ_0. Then the quantity $2\varphi_0 - \varphi_+ - \varphi_-$ in the phase difference between the $(\omega_0 + \omega_0)$ and $(\omega_+ + \omega_-)$ routes is a constant. That is, in this particular scenario, fluctuations in φ_0 cancel and have no effect on the relative phase $\alpha_a^q - \alpha_b^q$. Thus the two-photon plus two-photon scenario is insensitive to the laser jitter of the incident laser fields.

To examine the range of control afforded by this scheme consider the photodissociation of Na_2 in the regime below the Na(3*d*) threshold, where dissociation is to two product channels Na(3*s*) + Na(3*p*) and Na(3*s*) + Na(4*s*). Two-photon dissociation of Na_2 from a bound state of the ${}^1\Sigma_g^+$ state occurs [35] in this region by initial excitation to an excited intermediate bound state $|E_m J_m M_m\rangle$. The latter is a superposition of states of the $A^1\Sigma_u^+$ and $b^3\Pi_u$ electronic curves, a consequence of spin-orbit coupling. That is, the two-photon photodissociation can be viewed [35] as occuring via intersystem crossing subsequent to absorption of the first photon. The continuum states reached in the excitation can be either of singlet or triplet character but, despite the multitude of electronic states involved in the computation, the predominant contributions to the products Na(3*p*) and Na(4*s*) are found to come from the ${}^3\Pi_g$ and ${}^3\Sigma_g^+$ states, respectively. Methods for computing the required photodissociation amplitude, which involves eleven electronic states, are discussed elsewhere [35]. Since the resonant character of the two-photon excitation allows us to select a single initial state from a thermal ensemble we consider here the specific case of $v_i = J_i = 0$ without loss of generality, where v_i, J_i denote the vibrational and rotational quantum numbers of the initial state.

The ratio $R_{qq'}$ depends on a number of laboratory control parameters including the ratio of laser intensities x, relative laser phase, and the ratio of ε_+ and ε_- via η. In addition, the relative cross-sections can be altered by modifying the detuning. Typical control results are shown in Figure 6.16 which provides contour plots of the Na(3*p*) yield (i.e., the ratio of the probability of observing Na(3*p*) to the sum of the probabilities to form Na(3*p*) plus Na(4*s*)). The figure axes are the ratio of the laser amplitudes x and the relative laser phase $\delta\theta = 2\theta_0 - \theta_+ - \theta_-$. Here $\omega_0 = 631.899$, $\omega_+ = 562.833$, and $\omega_- = 720.284$ nm and control is seem to be large, ranging from 30% Na(3*p*) to 90% as $\delta\theta$ and x are varied.

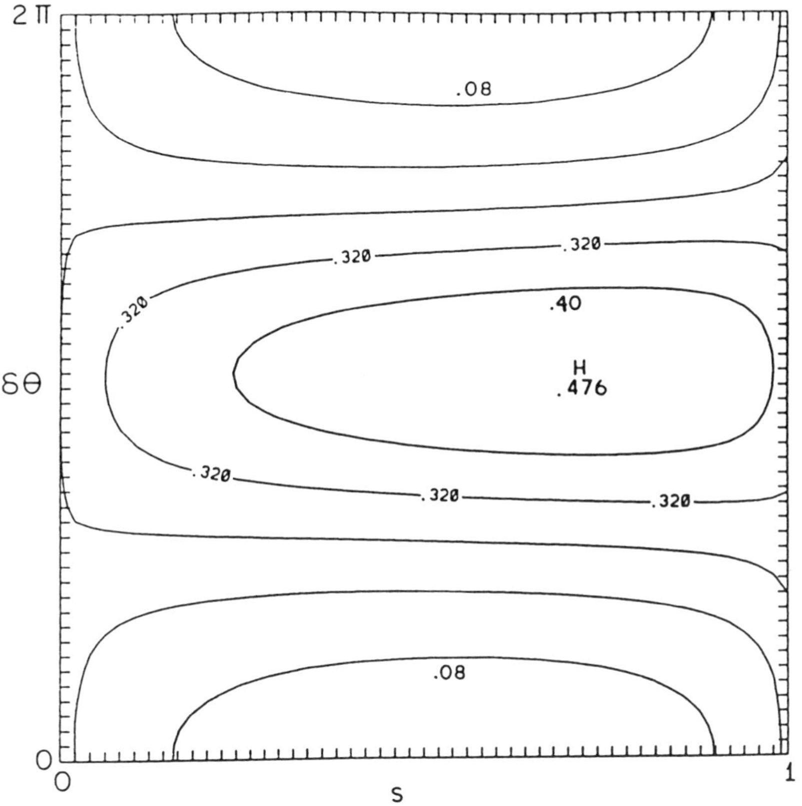

Figure 6.16 Contours of equal Na(3*p*) yield. Ordinate is the relative laser phase and the abscissa is $S = x^2/(1 + x^2)$ where x is the field intensity ratio. Here $\omega_0 = 627.584$, $\omega_+ = 611.207$, $\omega_- = 644.863$ nm, and $\eta = 0.5$. See [17] for a discussion of η which can be used to minimize background contributions. *Source*: from [35].

Note that the proposed approach is not limited to the specific frequency scheme discussed previously. Essentially all that is required is that the two resonant photodissociation routes lead to interference and that the cumulative laser phases of the two routes be independent of laser jitter. As one sample extension, consider the case where paths a and b are composed of totally different photons, $\omega_+^{(a)}$ and $\omega_-^{(a)}$ and $\omega_+^{(b)}$ and $\omega_-^{(b)}$, with $\omega_+^{(a)} + \omega_-^{(a)} = \omega_+^{(b)} + \omega_-^{(b)}$. Both of these sets of frequencies can be generated, for example, by passing $2\omega_0$ light through nonlinear crystals, hence yielding two pathways whose relative phase is independent of laser jitter in the initial 2 ω_0 source. Given these four frequencies we now have an additional degree of freedom in order to optimize control, although the experiment is considerably more complicated than in the three frequency case. Typical four-frequency results for Na_2 are provided elsewhere [17]. Note also that the control is not limited to two product channels. Computations [17] on higher energy Na_2 photodissociation, where more product arrangment channels are available, show equally large ranges of control for the three channel case.

6.4 NOVEL APPLICATIONS OF CONTROL

A. Control of Symmetry Breaking

The pump-dump scenario has been extended to include applications to symmetry breaking in chemical reactions [18]. We briefly review these developments in this section.

The possibility of controlling symmetry-breaking processes using coherent control techniques constitutes an exciting application of this field. In general, symmetry breaking occurs whenever a system executes a transition to a *nonsymmetric* eigenstate of the Hamiltonian. Strictly speaking, nonsymmetric eigenstates (i.e., states which do not belong to any of the symmetry-group representations) can occur if there exist several degenerate eigenstates, each belonging to a different irreducible representation. Any linear combination of such eigenstates is nonsymmetric.

In practice, symmetry breaking also occurs even if the degeneracy is only approximate, as in the problem of a symmetric double-well potential. If the barrier between the two wells is such that tunneling is miniscule, then the ground state of the system is composed, to all intents and purposes, of a doublet of *symmetric* and *antisymmetric* degenerate states. States localized at either well can then result by taking the ± linear combinations of this doublet. Because of the near degeneracy, these nonsymmetric localized states are essentially eigenstates of the Hamiltonian insofar as their time evolution can be immeasureably slow.

Nonsymmetric eigenstates of a symmetric Hamiltonian also occur in the continuous spectrum of a *BAB* type molecule. It is clear that the $|E,n,R^-\rangle$ state, which correlates asymptotically with the dissociation of the right *B* group, must be degenerate with the $|E,n,L^-\rangle$ state, which gives rise to the departure of the left *B* group. It is also possible to form a symmetric $|E,n,s^-\rangle$ and an antisymmetric $|E,n,a^-\rangle$ eigenstate of the same Hamiltonian by taking the ± combination of these states. There is an important physical distinction between the nonsymmetric states and states which are symmetric/antisymmetric: Any experiment performed in the asymptotic *B–AB* or *BA–B* regions must, by necessity, measure the probability of populating a nonsymmetric state. This follows because when the *B–AB* distance or the *BA–B* distance is large, a given group *B* is either far away from or close to group *A*. Thus symmetric and antisymmetric states are not directly observable in the asymptotic regime.

We conclude that the very act of observation of the dissociated *BAB* molecule entails the collapse of the system to one of the nonsymmetric states. As long as the probability of collapse to the $|E,n,R^-\rangle$ state is equal to the probability of collapse to the $|E,n,L^-\rangle$ state, the collapse to a nonsymmetric state does not lead to a preference of *R* over *L* in an *ensemble* of molecules. This is the case when the above collapse (symmetry breaking) is *spontaneous* (i.e., occurring due to some (random) factors not in our control). Coherent control techniques allow us to influence these probabilities. In this case symmetry breaking is stimulated rather than spontaneous. This has far reaching physical and practical significance.

One of the most important cases of symmetry breaking arises when the two *B* groups (now denoted as *B* and *B'*) are not identical but are enantiomers of one another. (Two groups of atoms are said to be *enantiomers* of one another if one is the mirror image of the other). If these groups are also *chiral* (i.e., they lack a center of inversion symmetry) then the two enantiomers are distinguishable and can be detected through the distinctive direction of rotation of linearly polarized light.

The existence and role of enantiomers is recognized as one of the fundamental broken symmetries in nature [38]. It has motivated a long-standing interest in asymmetric synthesis (i.e., a process which preferentially produces a specific chiral species). Contrary to the prevailing belief [39] that asymmetric synthesis must necessarily involve either chiral reactants, or chiral external system conditions such as chiral crystalline surfaces, we show below that preferential production of a chiral photofragment can occur even though the parent molecule is not chiral. In particular two results are demonstrated:

1. Ordinary photodissociation, using linearly polarized light, of a BAB' "prochiral" molecule may yield different cross-sections for the production of right-handed (B) and left-handed (B') products, when the projection of the angular momentum (m_j) of the products is selected; and
2. That this natural symmetry breaking may be enhanced and controlled using coherent lasers.

To treat this problem we return to the formulation of the pump-dump scenario with attention focused on control of the relative yield of two product arrangement channels but with angular momentum projection m_j fixed (Equation 6.40).

Explicitly considering the dissociation of BAB' into right- (R) and left-hand (L) products we have:

$$Y = \frac{P(L,m_j)}{P(L,m_j) + P(R,m_j)}$$

The yield Y is a function of the delay time $\tau = (t_d - t_x)$ and ratio $x = |c_1/c_2|$, the latter by detuning the initial excitation pulse. Active control over the products $B + AB'$ versus $B' + AB$ (i.e., a variation of Y with τ and x) and hence control over left- versus right-handed products, will result only if $P(R,m_j)$ and $P(L,m_j)$ have different functional dependences on x and τ.

We now show that $P(R,m_j)$ may be different from $P(L,m_j)$ for the $B'AB$ case. We first note that this molecule belongs to the C_s point group which is a group possesing only one symmetry plane. This plane, denoted as σ, is defined as the collection of the C_{2v} points (i.e., points satisfying the B–A = A–B' condition), where B–A designates the distance between the B and A groups. In order to do that we choose the intermediate state $|E_1>$ to be symmetric and the state $|E_2>$ to be antisymmetric with respect to reflection in the σ plane. Furthermore, we shall focus upon transitions between electronic states of the same representations (e.g., A' to A' or A'' to A'', where A' denotes the symmetric representation and A'' the antisymmetric representation of the C_s group). We further assume that the ground vibronic state belongs to the A' representation.

The first thing to demonstrate is that it is possible to excite simultaneously, by optical means, both the symmetric $|E_1>$ and antisymmetric $|E_2>$ states. Using Equation 6.35 we see that this requires the existence of both a symmetric $\mathbf{d}$ component, denoted as $\mathbf{d}_s$, and an antisymmetric $\mathbf{d}$ component, denoted $\mathbf{d}_a$, because, by the symmetry properties of $|E_1>$ and $|E_2>$,

$$<E_1|\mathbf{d}|E_g> = <E_1|\mathbf{d}_s|E_g>, \quad <E_2|\mathbf{d}|E_g> = <E_2|\mathbf{d}_a|E_g> \tag{6.55}$$

The existence of both dipole-moment components occurs in $A' \rightarrow A'$ electronic transitions whenever a bent B'–A–B molecule deviates considerably from the equidistance C_{2v} geometries (where $\mathbf{d}_a = 0$). The effect is non-Franck-Condon in nature because we no longer assume that the dipole-moment does not vary with the nuclear configurations. (In the theory of vibronic-transitions terminology the existence of both $\mathbf{d}_s$ and $\mathbf{d}_a$ is due to a Herzberg-Teller intensity borrowing [42] mechanism).

We conclude that the excitation pulse can create a $|E_1>$, $|E_2>$ superposition consisting of two states of different reflection symmetry which is therefore nonsymmetric. We now wish to show that this nonsymmetry established by exciting *nondegenerate bound* states translates to a nonsymmetry in the probability of populating the two *degenerate* $|E,n,R^->$, $|E,n,L^->$ *continuum* states. We proceed by examining the properties of the bound-free transition matrix elements of Equation 6.39 entering the probability expression of Equation 6.38.

Although the continuum states of interest $|E,n,q^->$ are nonsymmetric, they satisfy a closure relation since $\sigma|E,n,R^-> = |E,n,L^->$ and vice-versa. Working with the symmetric and antisymmetric continuum eigenfunctions,

$$|E,n,R^-> \equiv \frac{|E,n,s^-> + |E,n,a^->}{\sqrt{2}}$$

$$|E,n,L^-> \equiv \frac{|E,n,s^-> - |E,n,a^->}{\sqrt{2}} \tag{6.56}$$

using the fact that $|E_1>$ is symmetric and $|E_2>$ antisymmetric, and adopting the notation $A_{s2} \equiv <E,n,s^-|\mathbf{d}_a|E_2>$, $S_{a1} \equiv <E,n,a^-|\mathbf{d}_s|E_1>$, etc., we have,

$$\mathbf{d}_{11}^{(q)} = \sum{}' [|S_{s1}|^2 + |A_{a1}|^2 \pm 2Re(A_{a1}S^*_{s1})]$$

$$\mathbf{d}_{22}^{(q)} = \sum{}' [|A_{s2}|^2 + |S_{a2}|^2 \pm 2Re(A_{s2}S^*_{a2})]$$

$$\mathbf{d}_{12}^{(q)} = \sum{}' [S_{s1}A^*_{s2} + A_{a1}S^*_{a2} \pm S_{s1}S^*_{a2} \pm A_{a1}A^*_{s2}] \tag{6.57}$$

where the plus sign applies for $q = R$ and the minus sign for $q = L$.

Equation 6.57 displays two noteworthy features:

1. $\mathbf{d}_{kk}^{(R)} \neq \mathbf{d}_{kk}^{(L)}$, $k = 1, 2$. That is, the system displays *natural symmetry breaking* in photodissociation from state $|E_1>$ or state $|E_2>$, with right- and left-handed product probabilities differing by $4\Sigma' Re(S^*_{s1}A_{a1})$ for excitation from $|E_1>$ and $4\Sigma' Re(A_{s2}S^*_{a2})$ for excitation from $|E_2>$. Note that these symmetry-breaking terms may be relatively small since they rely upon non Franck-Condon contributions.
2. However, even in the Franck-Condon approximation: $\mathbf{d}_{12}^{(R)} \neq \mathbf{d}_{12}^{(L)}$. Thus laser controlled symmetry breaking, which depends upon $\mathbf{d}_{12}^{(q)}$ in accordance with Equation 6.38, is therefore possible, allowing enhancement of the enantiomer ratio for the m_j polarized product.

To demonstrate the extent of expected control, as well as the effect of m_j summation, we considered a model of the enantiomer selectivity (i.e., HOH photodissociation in three dimensions) where the two hydrogens are assumed distinguishable. The computation is done using the formulation and computational methodology of Segev et al. [43]. We briefly summarize the angular momentum algebra and some other details involved in performing three dimensional quantum calculations of triatomic photodissociation [44].

We first specify the relevant n and i quantum numbers which enter the bound-free matrix elements of Equation 6.39. For the continuum states, $n = \hat{\mathbf{k}},v,j,m_j$ where $\hat{\mathbf{k}}$ is the scattering direction, v and j are the vibrational and rotational product quantum numbers, and m_j is the space-fixed z-projection of j. For the bound states $i = \{E_i,M_i,J_i,p_i\}$, where J_i, M_i, and p_i are, respectively, the bound state angular momentum, its space-fixed z-projection, and its parity. The full (6-dimensional) bound-free matrix element can be written as a product of analytic functions involving $\hat{\mathbf{k}}$ and (three-dimensional) radial matrix elements,

$$<E,\hat{\mathbf{k}},v,j,m_j,q^-|\mathbf{d}|E_i,M_i,J_i,p_i> =$$

$$\left(\frac{\mu k_{vj}}{2\pi^2\hbar^2}\right)^{1/2} \sum_{J\lambda} (2J+1)^{1/2}(-1)^{M_i-j-m_j} D^J_{\lambda M_i}(\varphi_k,\theta_k,0) D^j_{-\lambda-m_j}(\varphi_k,\theta_k,0) t^{(q)}(E,J,v,j,\lambda|E_i,J_i,p_i) \tag{6.58}$$

where D are the rotation matrices, μ is the reduced mass, k_{vj} is the momentum of the products and $t^{(q)}(E,J,v,j,\lambda|E_i,J_i,p_i)$ is proportional to the radial partial wave matrix element [44] $<E,J,M,p,v,j,\lambda,q^-|\mathbf{d}|E_i,M_i,J_i,p_i>$. Here λ is the projection of J along the body-fixed axis of the H–OH (c.m.) product separation.

The product of the bound-free matrix elements of Equation 6.58, which enter Equation 6.39, integrated over scattering angles and averaged over the initial [45] M_k $(= M_i)$ quantum numbers, is [46]

$$(2J_i+1)^{-1} \sum_{M_i} \int d\hat{\mathbf{k}} <E_k,M_i,J_k,p_k|\mathbf{d}|E,\hat{\mathbf{k}},v,j,m_j,q'^-><E,\hat{\mathbf{k}},v,j,m_j,q^-|\mathbf{d}|E_i,M_i,J_i,p_i>$$

$$= (-1)^{m_j} \frac{16\pi^2 v\mu}{\hbar^2(2J_i+1)} \sum_{vj} k_{vj} \sum_{J\lambda J'\lambda'} [(2J+1)(2J'+1)]^{1/2}(-1)^{\{\lambda-\lambda'+J+J'+J_i\}}$$

$$\sum_{\ell=0,2} (2\ell+1) \begin{pmatrix} J & J' & \ell \\ \lambda & -\lambda' & \lambda'-\lambda \end{pmatrix} \begin{pmatrix} j & j & \ell \\ -\lambda & \lambda' & \lambda-\lambda' \end{pmatrix} \begin{pmatrix} 1 & 1 & \ell \\ 0 & 0 & 0 \end{pmatrix}$$

$$\begin{pmatrix} j & j & \ell \\ -m_j & m_j & 0 \end{pmatrix} \begin{Bmatrix} 1 & 1 & \ell \\ J & J' & J_i \end{Bmatrix} t^{(q)}(EJvj\lambda p|E_iJ_ip_i) t^{(q')*}(EJ'vj\lambda' p|E_kJ_kp_k) \tag{6.59}$$

Here $J_i = J_k$ has been assumed for simplicity.

These equations, in conjunction with the artificial channel method [44] for computing the T-matrix elements, were used to compute the yield Y of the HO + H (as distinct from the H + OH) product in a fixed m_j state. Specifically, Figure 6.17 shows the result of first exciting the superposition of symmetric plus asymmetric vibrational modes [(1,0,0) + (0,0,1)] with $J_i = J_k = 0$ in the ground electronic state, followed by dissociation at 70,700 cm^{-1} to the B state using a pulse width of 200 cm^{-1}. Results show that

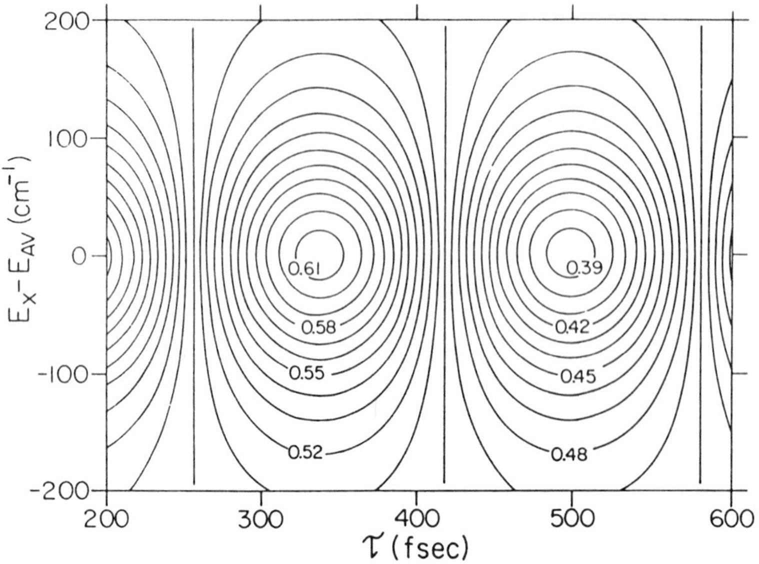

Figure 6.17 Contour plot of percent HO + H (as distinct from H + OH) in HOH. Ordinate is the detuning from $E_{av} = (E_2 - E_1)/2$, abscissa is time between pulses. *Source*: from [18].

varying the time delay between pulses allows for controlled variation of Y from 61–39%!

Finally we sketch the effect of a summation over product m_j states on symmetry breaking and chirality control. In this regard the three body model is particularly informative. Specifically, note that Equation 6.57 provides $\mathbf{d}_{i,k}^{(q)}$ in terms of products of matrix elements involving $|E,n,a^-\rangle$ and $|E,n,s^-\rangle$. Focus attention on those products which involve both wavefunctions (e.g., $A_{a1}S_{s1}^*$). These matrix element products can be written in the form of Equation 6.59 where q and q' now refer to the antisymmetric or symmetric continuum states, rather than channels 1 and 2. Thus, for example, $A_{a1}S_{s1}^*$ results from using $|E,n,a^-\rangle$ in Equation 6.58 to form A_{a1} and $|E,n,s^-\rangle$ to form S_{s1}^*. The resultant $A_{a1}S_{s1}^*$ has the form of Equation 6.59 with $t^{(q)}$ and $t^{(q')}$ associated with the symmetric and antisymmetric continuum wavefunctions, respectively. Consider now the effect of summing over m_j. Standard formulae [43, 44] imply that this summation introduces a $\delta_{\ell,0}$ which, in turn forces $\lambda = \lambda'$ via the first and second $3j$ symbol in Equation 6.59. However, a rather involved argument [47] shows that t-matrix elements associated with symmetric continuum eigenfunctions and those associated with antisymmetric continuum eigenfunctions must have λ of different parities. Hence summing over m_j eliminates all contributions to Equation 6.57 which involve both $|E,n,a^{-}\rangle$ and $|E,n,s^{-}\rangle$. Specifically, we find [47] after m_j summation:

$$\mathbf{d}_{ii}^{(1)} = \mathbf{d}_{ii}^{(2)} = \sum [|S_{si}|^2 + |A_{ai}|^2] \tag{6.60}$$

$$\mathbf{d}_{12}^{(1)} = \mathbf{d}_{12}^{(2)} = \sum [S_{si}A_{sj}^{*} + A_{ai}S_{aj}^{*}] \tag{6.61}$$

That is, natural symmetry breaking is lost upon m_j summation, both channels $q = R$ and $q = L$ having equal photodissociation probabilities, and control over the enantiomer ratio is lost since the interference terms no longer distinguish the $q = R$ and $q = L$ channels [48].

Having thus demonstrated the principle of m_j-selected enantiomer control it remains to determine the extent to which realistic systems can be controlled. Such studies are in progress [47].

B. Control of Photocurrent Directionality

The possibility of controlling product channels applies not only to chemically or electronically distinct products but also to control over the distribution of internal states (e.g., rotation, vibration) of the product as well [2]. Thus, for example, one may use coherent pathways to produce inversion amongst product states. In this section we demonstrate that it is possible to control the product angular distribution using the scheme outlined in Section 6.1.C. Our specific application is to the case of photoionization so that the result is control over the direction of the photocurrent induced by the interference. An alternative method of controlling differential cross-sections by varying the degree of elliptic polarization of the light source is described elsewhere [5].

Properties of a photocurrent generated in a semiconductor are controlled by a bias voltage for a given concentration and spatial distribution of charge carriers [49]. The role of this voltage is to give *thermodynamic* preference to the flow of photoelectrons in one direction (the forward or backward direction in a p–n junction.) In a p-type or n-type semiconductor the probability of carrier photoemission (from a single impurity) without an external voltage is anisotropic only inasmuch as the crystal possesses mass or dielectric constant anisotropies, but the probabilities of emission backward and forward along a given crystal axis are equal. Although photocurrents are commonly produced by laser illumination, the laser coherence does not affect the process.

Here we review our scheme [8] for generating and controlling photocurrents *without bias voltage*, relying instead on the coherence of the illuminating source. The method is an application of the scenario of Section 6.1.C to the photoionization of bound states of donors. Specifically, a superposition of two bound donor (or exciton) states is photoionized by two mutually phase-locked lasers at slightly different frequencies with the same polarization axis. The result is a current along the direction of polarization. The realization of the scheme is discussed for shallow-level donors in semiconductors.

Consider a semiconductor doped with shallow-level donors. The bound state wavefunction of such a donor is described by the hydrogenic effective-mass theory [50] with wavefunction:

$$\chi_n(\mathbf{r}) = \langle \mathbf{r}|n\rangle = V^{-1/2}\int_{-\infty}^{\infty} B_{n,\mathbf{k}}u_{\mathbf{k}}(\mathbf{r})e^{i\mathbf{k}\cdot\mathbf{r}}d\mathbf{k} \tag{6.62}$$

Here $u_{\mathbf{k}}(\mathbf{r})$ is the conduction band Bloch state correlated to the asymptotic free-electron momentum $\hbar\mathbf{k}$, V is the normalization volume and $B_{n,\mathbf{k}}$ is the corresponding Fourier component of the hydrogenic wavefunction envelope χ_n. For semiconductors with effective-mass anisotropy, the χ_n are evaluated variationally [51–54]. Although the theory described below holds for any superposition of bound donor states, a superposition of $|1s\rangle$ and $|2p_o\rangle$ states will be considered explicitly. For these cases a simple variational procedure [54], whose results agree reasonably well with those of more refined procedures [51–55], yields

$$\chi_{1s} = \pi^{1/3}\exp\left\{-\left[\frac{x^2+y^2}{a^2+z^2/b^2}\right]^{1/2}\right\}$$

$$\chi_{2p_o} = \sqrt{2}\pi^{3/4}b^{-1}z\exp\left\{-\left[\frac{(x^2+y^2)}{a^2+z^2/b^2}\right]^{1/2}\right\} \tag{6.63}$$

Here the coordinates [normalized to the effective Bohr radius $a^* = \hbar^2/(m_\perp e^2)$] coincide with the main axes of the cubic crystal. Depending on the ratio $\gamma = m_\perp/m_\parallel$ (the parallel direction coinciding with z), the a and b parameters vary between $a = b = 1$ for nearly isotropic materials with $\gamma = 1$ (e.g., GaAs, GaSb, InAs) and $a \approx \frac{4}{3}\pi$; $b \approx (\frac{1}{3})(4/\pi)^{2/3}\gamma^{1/3}$ for highly anisotropic materials (e.g., Si or Ge) with $\gamma << 1$.

Let a superposition of the $|1s\rangle$ and $|2p_o\rangle$ states be prepared by some coherent process. As pointed out previously, this can be achieved by a short coherent laser pulse or various other means. It is possible to discriminate against the excitation of the $|2p_{\pm 1}\rangle$ states either by frequency tuning, (e.g., the $2p_{\pm 1} - 2p_o$ splitting is ~5meV in Si) or by linearly polarizing the laser along the z-axis. Consider now the simultaneous excitation of this superposition state to a kinetic energy level E_k in the conduction band continuum by two z-polarized infrared or visible lasers with frequencies $\omega_{1s}, \omega_{2p_o}$; the former lifts the $|1s\rangle$ state to E_k and the latter lifts the $|2p_o\rangle$ state to E_k. These excitations involve the energy conservation relation:

$$E_k = \frac{\hbar^2 k_\perp^2}{2m_\perp} + \frac{\hbar^2 k_z^2}{2m_\parallel} = \hbar\omega_n - |E_n| - \sum_p p\hbar\omega \tag{6.64}$$

Here the **n**-state energy is measured from the conduction-band edge and the last term accounts for the emission ($p > 0$) or absorption ($p < 0$) of p phonons of frequency ω. For the sake of simplicity, we shall use the zero phonon-frequency line [55, 56]; hence, $\hbar\omega_{1s} = E_k + |E_{1s}|, \hbar\omega_{2p_o} = E_k + |E_{2p_o}|$.

In what follows we consider only electric-dipole induced optical transitions with the electric field along the z axis. The electric dipole transition amplitudes from an impurity state $|\boldsymbol{n}\rangle$ to the asymptotic (far from impurity) plane wave $\langle \boldsymbol{r}|\boldsymbol{k}\rangle = V^{-1/2} e^{i\boldsymbol{k}\cdot\boldsymbol{r}} u_{\boldsymbol{k}}(\boldsymbol{r})$ is

$$\langle \boldsymbol{k}|\mu_z|\boldsymbol{n}\rangle = \frac{-ie\hbar}{m_\parallel(E_{\boldsymbol{k}} + |E_{\boldsymbol{n}}|)} \left\langle \boldsymbol{k} \left| \frac{-i\hbar\partial}{\partial z} \right| \boldsymbol{n} \right\rangle \tag{6.65}$$

The last factor is, using Equation 6.62, simply given as

$$\left\langle \boldsymbol{k} \right| \frac{-i\hbar\partial}{\partial z} \boldsymbol{n}\rangle = \hbar k_z \langle \boldsymbol{k}|\boldsymbol{n}\rangle = \hbar k_z B_{\boldsymbol{n},\boldsymbol{k}} \tag{6.66}$$

We now consider the photoionization of the superposition state,

$$|\psi\rangle = c_1|1\rangle + c_2|2\rangle \tag{6.67}$$

where 1 denotes the $1s$ state and 2 the $2p_o$ state. We let a z-polarized two-color source, whose electric field is given as,

$$\varepsilon_z(t) = \varepsilon_1 \cos(\omega_1 t + \varphi_1) + \varepsilon_2 \cos(\omega_2 t + \varphi_2) \tag{6.68}$$

act on this superposition state. The rate (probability per unit time and unit solid angle) of photoemission to a conduction state with momentum $\hbar\boldsymbol{k}$ resulting from this action is,

$$P(\cos\theta) = \left(\frac{2\pi}{\hbar}\right)\rho(k)\left|\sum_{n=1,2} e^{-i\varphi_n}\varepsilon_n c_n\langle\boldsymbol{k}|\mu_z|\boldsymbol{n}\rangle\right|^2 \tag{6.69}$$

Here,

$$\cos\theta = \frac{k_z}{k};\ \sin\theta = \frac{k_\perp}{\gamma^{1/2}k}$$

$$k = \frac{(2m_\parallel E_k)^{1/2}}{\hbar}$$

$$\rho(k) = \left(\frac{m_\perp V}{8\pi^3\hbar^2}\right)k \tag{6.70}$$

and ρ(k) is the density of final states. The Franck-Condon factor for the zero-phonon line has been set here to unity.

Denoting $c_n = |c_n| \exp(i\alpha_n)$ and using Equations 6.65 and 6.66 in Equation 6.69 gives the form:

$$P(\cos\theta) =$$

$$[A_1|B_{1s,\boldsymbol{k}}|^2 + A_2|B_{2p_o,\boldsymbol{k}}|^2 + A_{12}\cos(\alpha_1 - \alpha_2 - \varphi_1 + \varphi_2 + \alpha_{12})|B_{1s,\boldsymbol{k}}B_{2p_o,\boldsymbol{k}}|]\cos^2\theta \tag{6.71}$$

where

$$A_n = \frac{2\pi e^2\hbar^3 k^2 \rho(k)|\varepsilon_n c_n|^2}{m_{\|}^2 (E_k + E_n)^2} \qquad (n = 1, 2)$$

$$A_{12} = \frac{4\pi e^2\hbar^3 k^2 \rho(\kappa)|\varepsilon_1 \varepsilon_2 c_1 c_2|}{m_{\|}^2 (E_k + E_1)(E_k + E_2)} \tag{6.72}$$

Here α_{12} is defined by $B_{1s,k}B^*_{2p_o,k} = |B_{1s,k}B_{2p_o,k}| \exp(i\alpha_{12})$ and $E_1 = |E_{1s}|, E_2 = |E_{2p_o}|$.

The evaluation of $P(\cos\theta)$ requires the Fourier components $B_{n,k}$. For the present choice of impurity states and z axis these components are obtained from Equation 6.63 as,

$$B_{1s,k} = \frac{8\pi^{4/3} a^2 b V^{-1/2}}{G^2}$$

$$B_{2p_o,k} = \frac{-i\sqrt{2}(32)a^2 b^2 \pi^{7/4} V^{-1/2} a^* k_z}{G^3} \tag{6.73}$$

with

$$G = G(\cos^2\theta) = [1 + \gamma(a^* a k)^2 + (b^2 - a^2\gamma)(a^* k)^2 \cos^2\theta]$$

It is clear from Equation 6.73 that $\alpha_{12} = \pi/2$.

Given the above expression, the net current flowing in the z-direction is given as

$$\Gamma_z^+ = \left(\frac{eNV\hbar}{m_{\|}}\right)\tau \times F \int_o^{2\pi}\int_o^{\pi} d\omega P(\cos\theta) k \cos\theta$$

$$= 256\left(\frac{eNV\hbar^4 k^5}{m_{\|}^3}\right)\tau \times F a^4 b^3 \pi^{25/12} \frac{|\varepsilon_1 \varepsilon_2 c_1 c_2|}{(E_k + E_1)(E_k + E_2)}$$

$$\cos\left(\alpha_1 - \alpha_2 - \varphi_1 + \varphi_2 + \frac{\pi}{2}\right)\int_{-1}^{+1} dx \frac{x^4}{[G(x^2)]^5} \tag{6.74}$$

where τ is the free electron collisional relaxation time, N is the donor concentration in cm^{-3}, and F is the x–y cross-sectional area of the sample.

We note that contributions from the diagonal A_1 and A_2 terms are odd in $\cos\theta$ and that they have vanished, whereas the interference term induces a directional current flow! Thus coherent interference contributions result in a controlled directional current flow.

Several additional remarks are in order. First, the phases φ_1 and φ_2 of Equation 6.68 contain the spatial phase factors $\exp[i\mathbf{k}\cdot\mathbf{R}]$, where $\mathbf{k}$ is the light wave vector. The difference in the spatial phases can be exactly offset by the phase difference $\alpha_1 - \alpha_2$ in the preparation step (e.g., in a Raman preparation of $|\psi\rangle$) or eliminated by phase matching. Second, there are substantial experimental simplications associated with applying the photodissociating lasers at the same time as initiating the preparation of the superposition state. Third, two-color light also causes excitation (via ω_{2p_o}) of the $|1s\rangle$ level to the state at $[E_k + |E_{2p_o}| - |E_{1s}|]$ and of the $|2p_o\rangle$ level (via ω_{1s}) to the state at $[E_k + |E_{1s}| - |E_{po}|]$ (i.e., the uncontrolled satellite contributions discussed previously). In this case, however, these terms contribute to the A_1 and A_2 terms in Equation 6.71 and hence do not contribute to degrade the controlled current I_z^+.

The magnitude and sign of the current is controllable for a given host material and superposition state parameters via: a) the optical phase difference $\varphi_1 - \varphi_2$, b) the donor number N, and/or, c) the ionizing field strengths ε_1 and ε_2 and their frequencies ω_1 and ω_2. To estimate a typical current, consider the I_z resulting from the following parameters: $\varepsilon_1 = \varepsilon_2 = 0.1$ Volts/cm, $k = 5 \times 10^7\ cm^{-1}$, $|c_1c_2| = 0.25$, and $\tau = 10^{-14}$ to 10^{-13} sec. The latter corresponds to a mean free path ($\hbar k\tau/m$) of 100–1000Å, a typical value for the ballistic electrons at the cited k value. Furthermore, $N(Si)V = 10^{18}$ cm^{-3}V *where V is the effective interaction volume. For a sample of 0.1 micron × 10 micron × 10 micron,* $V = 10^{-11}$ cm^3. Utilizing Equation 6.74 and these parameter values, we obtain a current $I_z = 10$–100 mA. Thus, sizeable currents may be readily produced due to the high quantum efficiency of the silicon photoionization.

Equations 6.72–6.74 apply, evidently, to photoionization of other $|ns\rangle - |n'p_o\rangle$ superpositions, where $|n - n'| = 1$, upon substituting the appropriate Fourier coefficients $B_{ns,k}$ and $B_{n'p_o,k}$. It may turn out to be more practical to use other states than those discussed earlier.

6.5 COHERENCE CHEMISTRY

Our discussion makes clear that the characteristic features which we invoke in order to control chemical reactions are purely quantum in nature. There is, for example, little classical about the time-dependent picture where the ultimate outcome of the de-excitation (i.e., product H + HD or H_2 + D) depends entirely upon the phase and amplitude characteristics of the wavefunction. Indeed, as repeatedly emphasized, if, for example, collisional effects are sufficiently strong so as to randomize the phases then reaction control is lost. Hence reaction dynamics are intimately linked to the wavefunction phases which are controllable through coherent optical phase excitation.

These results must be viewed in light of the history of molecular reaction dynamics. Possibly the most useful result of the reaction dynamics research effort has been the recognition that the vast majority of qualitatively important phenomena in reaction dynamics are well described by classical mechanics. Quantum and semiclassical mechanics were viewed as necessary only insofar as they correct quantitative failures of classical mechanics for unusual circumstances and/or for the dynamics of very light particles. Therefore, it appeared to be correct to considering reaction dynamics in traditional chemistry to be essentially classical in character for the vast majority of naturally occuring molecular processes; coherence played no role. The approach which we have introduced above makes clear, however, that coherence phenomena have great potential for application. The quantum phase is always present and can be used to our advantage, even though it is irrelevant to traditional chemistry. By calling attention to the extreme importance of coherence phenomena to controlled chemistry we herald the introduction of a new focus in atomic and molecular science (i.e., introducing coherence in controlled environments to modify molecular processes) thus defining the area of coherence chemistry.

ACKNOWLEDGMENTS

We acknowledge support for this research by the U.S. Office of Naval Research.

REFERENCES

1. P. Brumer and M. Shapiro, *Chem. Phys. Lett.*, **126**, 541 (1986).
2. M. Shapiro and P. Brumer, *J. Chem. Phys.*, **84**, 4103 (1986).
3. P. Brumer and M. Shapiro, *Faraday Disc. Chem. Soc.*, **82**, 177 (1986).
4. M. Shapiro and P. Brumer, *J. Chem. Phys.*, **84**, 4103 (1986).
5. C. Asaro, P. Brumer and M. Shapiro, *Phys. Rev. Lett.*, **60**, 1634 (1988).
6. M. Shapiro, J. Hepburn and P. Brumer, *Chem. Phys. Lett.*, **149**, 451 (1988).
7. P. Brumer and M. Shapiro, *J. Chem. Phys.*, **90**, 6179 (1989).
8. G. Kurizki, M. Shapiro and P. Brumer, *Phys. Rev. B*, **39**, 3435 (1989).
9. T. Seideman, M. Shapiro and P. Brumer, *J. Chem. Phys.*, **90**, 7136 (1989).
10. J. Krause, M. Shapiro and P. Brumer, *J. Chem. Phys.*, **92**, 1126 (1990).
11. I. Levy, M. Shapiro and P. Brumer *J. Chem. Phys.*, **93**, 2493 (1990).
12. P. Brumer and M. Shapiro, *Accounts Chem. Res.*, **22**, 407 (1989).
13. C. K. Chan, P. Brumer and M. Shapiro, *J. Chem. Phys.*, **94**, 2688 (1991).
14. X-P. Jiang, P. Brumer and M. Shapiro (to be published).
15. P. Brumer and M. Shapiro, *Chem. Phys.*, **139**, 221 (1989).
16. M. Shapiro and P. Brumer, *J. Chem. Phys.*, **97**, 6259 (1992)
17. X. Jiang, P. Brumer and M. Shapiro, (in preparation).
18. M. Shapiro and P. Brumer, *J. Chem. Phys.*, **95**, 8658 (1991).
19. J. Dods, M. Shapiro and P. Brumer, *Can. J. Chem.* (submitted); J. Dods, M.Sc. Dissertation, University of Toronto (1992).
20. For a discussion of the basic principles of coherence, quantum interference, time dependence, which are fundamental to coherent control; see, for example, J. D. Macomber, *The Dynamics of Spectroscopic Transitions*, J. Wiley, New York, 1976.
21. To see this requirement note that it should be possible to set up initial reactant conditions (i.e., relative velocities, internal rotation, and vibration states, etc.) so as to produce a wide variety of possible reaction outcomes (e.g., probabilities of observing products with different relative velocities, internal rotational and vibrational states, scattering angles, etc.). To properly describe these product states requires then a host of wavefunctions, all at energy E, which can be added arbitrarily together so as to yield this wide variety of allowed final states.
22. This the Asymptotic Condition of Scattering Theory [see J. R. Taylor, *Scattering Theory*, J. Wiley, New York, 1972].
23. See, for example, R. D. Taylor and P. Brumer, *Disc. Faraday Society*, **75**, 17 (1983); P. Brumer and M. Shapiro (to be published).
24. Contrary to popular expectation, perturbation theory does not imply a small total photodissociation yield. Computational results (P. Brumer and M. Shapiro) indicate that perturbation theory is quantitatively correct for dissociation probabilities as large as 0.2.
25. R. Bersohn and M. Shapiro, *Ann. Rev. Phys. Chem.*, **33**, 409 (1982); M. Shapiro, *J. Chem. Phys.*, **56**, 2582 (1972).
26. M. S. Child, *Mol. Phys.*, **32**, 495 (1976).

27. M. Shapiro and P. Brumer, in *Methods of Laser Spectroscopy*, (A. Prior, A. Ben-Reuven and M. Rosenbluh, Eds.) Plenum, New York, 1986.
28. M. Shapiro, *J. Phys. Chem.*, **90**, 3644 (1986).
29. D. J. Tannor and S. A. Rice, *J. Chem. Phys.*, **83**, 5013 (1985); D. J. Tannor, R. Kosloff and S. A. Rice, *J. Chem. Phys.*, **85**, 5805 (1986).
30. See, for example, R. B. Bernstein and R. D. Levine, *Molecular Reaction Dynamics*, Oxford University Press, New York, 1987.
31. A. D. Bandrauk, J. M. Gauthier, J. F. McCann, *Chem. Phys. Lett.*, **200**, 399 (1992)
32. C. Chen, Y-Y. Yin, and D. S. Elliott, *Phys. Rev. Lett.*, **64**, 507 (1990); C. Chen, Y-Y. Yin, and D. S. Elliott, *Phys. Rev. Lett.*, **65**, 1737 (1990).
33. S. M. Park, S-P. Lu, and R. J. Gordon, *J. Chem. Phys.*, **94**, 8622 (1991); S-P. Lu, S. M. Park, Y. Xie, R. J. Gordon, *J. Chem. Phys.*, **96**, 6613 (1992).
34. B. A. Baranova, A. N. Chudinov, and B. Ya Zel prime dovitch, *Opt. Comm.*, **79**, 116 (1990).
35. Z. Chen, M. Shapiro and P. Brumer, *J. Chem. Phys.*, **98**, 6843 (1993).
36. M. Shapiro and R. Bersohn, *J. Chem. Phys.*, **73**, 3810 (1980).
37. A. R. Edmonds, *Angular Momentum in Quantum Mechanics*, 2nd edition, Princeton University Press, Princeton, 1960.
38. L. D. Barron, *Molecular Light Scattering and Optical Activity*, Cambridge University Press, Cambridge, 1982; R.G. Woolley, *Adv. Phys.*, **25**, 27 (1975); *Origins of Optical Activity in Nature*, (D. C. Walker, Ed.), Elsevier, Amsterdam, 1979.
39. For a discussion see L. D. Barron, *Chem. Soc. Rev.*, **15**, 189 (1986). For historical examples see J. A. Bel, *Bull. Soc. Chim. Fr.*, **22**, 337 (1874); J. H. Van't Hoff, *Die Lagerung der Atome und Raume*, 2nd. Ed., 1894, p. 30.
40. An example, although requiring a slight extension of the BAB prime notation, is the Norrish type II reaction: $D(CH_2)_3CO(CH_2)_3D$ prime dissociating to $DCHCH_2$ + D prime $(CH_2)_3COCH_3$ and D prime $CHCH_2$ + $D(CH_2)_3COCH_3$ where D and D prime are enantiomers.
41. See T. Seideman, M. Shapiro and P. Brumer, *J. Chem. Phys.*, **90**, 7136 (1989); J. Krause, M. Shapiro and P. Brumer, *J. Chem. Phys.*, **92**, 1126 (1990); C. K. Chan , P. Brumer and M. Shapiro, *J. Chem. Phys.*, **94**, 2688 (1991) and references therein.
42. J. M. Hollas, *High Resolution Spectroscopy*, Butterworths, London, 1982.
43. E. Segev and M. Shapiro, *J. Chem. Phys.*, **77**, 5604 (1982).
44. G.G. Balint-Kurti and M. Shapiro, *Chem. Phys.*, **61**, 137 (1981).
45. $M_i = M_k$ since both $|E_i\rangle$ and $|E_k\rangle$ arise by excitation, with linearly polarized light, from a common eigenstate.
46. Equation 6.59 is a generalized version of Equation 6.48 [44].
47. M. Shapiro and P. Brumer, (to be published).
48. As an aside we note that control is not possible in collinear models since, in that case, $\mathbf{d}_a$ and $\mathbf{d}_s$ can not both couple to the same electronically excited state.
49. K. Seeger, *Semiconductor Physics*, Springer Verlag, Berlin, 1973.

50. S. T. Pantelides, *Rev. Mod. Phys.*, **50**, 797 (1978).
51. R. A. Faulkner, *Phys. Rev.*, **184**, 713 (1969).
52. A. Kasami, *J. Phys. Soc. Japan*, **24**, 551 (1968).
53. A. Baldereschi and M. G. Diaz, *Nuov. Cim.*, **68B**, 217 (1970).
54. W. Kohn and J. M. Luttinger, *Phys. Rev.*, **98**, 915 (1955).
55. B. K. Ridley, *J. Phys.*, **C13**, 2015 (1980).
56. K. Huang and A. Rhys, *Proc. Roy. Soc. A*, **04**, 406 (1950).
57. D. Hsu and J. L. Skinner, *J. Chem. Phys.*, **81**, 1604 (1984); **81**, 5471 (1984).
58. M. J. M. Ziman, *The Principles of the Theory of Solids*, Cambridge University Press, 1965, Equation 7.16.

7

Coherent Optical Spectroscopic Studies of Collisional Dynamics

Mark A. Banash[*] and Warren S. Warren

Princeton University, Princeton, New Jersey

7.1 INTRODUCTION

Over the last decade, the effects of controlled laser fields (amplitude and frequency modulated shaped laser pulses and interpulse phase control) have been widely explored, both theoretically and experimentally. Much of the interest in these effects stems from the belief that specific tailored pulse shapes (or sequences of shaped laser pulses) will ultimately lead to laser control over chemical reactions. For example, calculations on model systems have confirmed that such waveforms can force molecules up anharmonic ladders, trap energy in strong bonds to rupture them, prepare rotationally oriented molecules, or turn off intramolecular vibrational redistribution in a variety of applications. However, these dramatic theoretical results generally demand extraordinarily complicated waveforms (peak powers, modulation

* *Current affiliation:* Grintel Optics, Newton, Pennsylvania

rates, and bandwidths which are well beyond current experimental capabilities), unrealistic assumptions (such as neglect of rotational levels or inhomogeneous variations in field strength), or essentially perfect knowledge of the molecular Hamiltonian. Experiments on a picosecond or subpicosecond timescale have succeeded in preparing wavepackets, generating narrowband excitation with broadband pulses, enhancing impulsive stimulated Raman signals, or modulating fluorescence; while these applications are not as dramatic as the theoretical potential, they nonetheless also show the utility of the technique.

In fact, shaped laser pulses and pulse sequences are capable of extracting additional information from molecules even when the technology is far from the present upper limits of speed or power. In particular, nanosecond pulses and long delays are often the best choice for understanding energy transfer and collisional dynamics in simple molecular gases. Long ago, common fluorescence techniques analyzed and determined energy transfer rates that involve quantum state changes (such as rotational energy transfer) and this has led to ever more sophisticated levels of theory. However, in the case of translational energy transfer, past experimental studies have been far from complete. The best efforts yield only averaged or indirect data. The purpose of this chapter has been to develop and refine an experimental technique that can overcome this obstacle and directly observe such a process in a molecular gas.

Past characterizations of gases are based largely on thermodynamic results and have led to limited models. For example, the use of second virial coefficients often simply determines whether a hard-sphere approximation is valid [1]. Many parameters necessary to develop more extensive models, such as the intermolecular potential between a ground state molecule and an identical molecule in an electronically excited state, are not known. Attempts to evaluate them using approximate combinatory rules yield analyses that are of questionable value when applied to the results of molecular energy transfer studies [2].

Optical coherent transient experiments (including most notably the photon echo) have become a common technique for exploring gas-phase collisional dynamics [3]. The optical dephasing rate (T_2) measured by a photon echo generally reflects contributions from a variety of processes. In the limit of noninteracting molecules, where excited state relaxation is assumed to occur only by spontaneous emission, it is expected that $2T_1 = T_2$, where T_1 is the excited state lifetime. However, intermolecular interactions can change this relationship dramatically; for example T_1 and T_2 are known to be pressure-dependent. The simplest way to resolve these different contributions (used in essentially all the gas phase work to date) is to measure the pressure dependence of both T_2 and the fluorescence lifetime τ_{flu} and then define a "pure dephasing rate" T_2' by:

$$\frac{1}{T_2'} = \frac{1}{T_2} - \frac{1}{2\tau_{flu}'} \tag{7.1}$$

which in principle removes lifetime effects. This pure dephasing rate can generally be fit to a Sterm-Vollmer plot and translated into some sort of dephasing cross-section.

There are good reasons to be skeptical of this kind of approach. For example, even in the limit of two ideal but coupled two-level systems, it can be shown that Equation (7.1) is wrong [4]. The limitations of information that can be gleaned by this approach are perhaps best seen by comparing photon echo data on the $X^1\Sigma_g \rightarrow B^3\Pi_u$ transition in molecular iodine from a variety of research groups (see Table 7.1). All of these experiments used laser pulses with lengths in the nanosecond–microsecond range. The only consistency is an extremely large (and variable) echo decay cross-section (100–1000 sq. Å). Even if consistency was not a problem, this would still be a reason to examine the dephasing process more closely. At the upper end of this range, the cross-section corresponds to a classical intermolecular separation of 35 Å—enormously larger than what would be compatible with the normal concepts of molecular size, particularly since this molecule lacks a permanent dipole moment to cause such strong intermolecular attraction and it has only a weak transition dipole moment. Much of the problem comes from imposing language appropriate for a two-level system on a true collisional process. In a fundamental sense, measurements of parameters like T_2 and τ_{flu} only produce a few highly averaged weightings of the many possible and distinct collisional processes and thus can give, at best, only vague information.

In this chapter we describe the results of our measurements on a variety of rotational and vibrational levels of the $X^1\Sigma_g \rightarrow B^3\Pi_u$ transition in I_2, using different pulse sequences to clarify the relative importance of different relaxation mechanisms [5]. Comparisons are made between fluorescence life-

Table 7.1 Summary of Previous Photon Echo Work

Wavelength (nm)	Transition	Cross-section (sq. 143)	Reference
589.748	R59 (15-2)?	780 (phase interrupt)	Brewer and Genack *Phys. Rev. Lett.*, **36**(16):959 (1976).
589.75	P114 (14-1)? R59 (15-2)? P40 (17-3)?	968 (phase interrupt)	Orlowski, Jones and Zewail *Chem. Phys. Lett.*, **50**(1):45 (1977).
589.75	P114 (14-1)? R59 (15-2)? P40 (17-3)	Observed decay at single pressure. No cross-section	Orlowski, Jones, and Zewail, *Chem. Phys. Lett.*, **54**(2):197 (1978)
589.75	P114 (14-1)? R59 (15-2)? P40 (17-3)?	Merely observes it No cross-section	Warren and Zewail, *J. Chem. Phys.*, **78**(5):2279 (1982).
589.7	P107 (16-2)? P53 (15-2)?	590 (phase interrupt)	Sleva and Zewail, *Chem. Phys. Lett.*, **110**(6):582 (1984).
589.746	R59 (15-2)?	145 (phase interrupt)	Brewer and Kano, Springer Series in *Chem. Phys.*, vol. 6, p. 54.1980.

times, two-pulse phase-incoherent single frequency measurements, two-pulse phase-incoherent dual frequency measurements, and phase-sensitive three-pulse photon echoes. In all cases, the largest part of the photon echo decay is shown to come from phase-interrupting collisions that perturb the velocity only slightly (≤ 6 m/sec, compared to a mean thermal velocity of 100 m/sec). Additionally, substantial and reproducible variations in the dephasing rates are observed with surprising sensitivity to the ground state vibrational level v'' and to the exact position in the inhomogeneous lineshape. This explains the large variations in previously reported measurements and leads to a picture of collisional dephasing as a far more subtle process than was appreciated previously.

7.2 THEORETICAL BACKGROUND

Traditionally, the pure dephasing T_2' in Equation 7.1 measures processes that generate phase shifts to scramble the macroscopic sum of the polarization vectors induced by the initial pulse, in close analogy to the magnetic resonance case. In fact many other processes can contribute to the decay of a photon echo, particularly when selective pulses are used (as will be the case for all the experiments we describe). For example, a collision that changes the rotational state will not alter the fluorescence lifetime significantly but it will change the molecular resonance frequencies by an amount much larger than the pulse bandwidth. Therefore it will contribute to the decay of a photon echo though normally this is not called pure dephasing. Additionally, since our studies use laser pulses that excite only a small fraction of the Doppler profile, velocity changing collisions that move molecules out of the excitation bandwidth but do not affect polarization will also contribute to T_2.

The density matrix formalism for dealing with the effects of velocity changing collisions can be constructed using collisional kernels [6]. Separate kernels describe the probability densities for the diagonal (i.e., population) and off–diagonal (i.e., coherence) elements. These have been calculated by solving for the equations of transport and assuming a hard-sphere model for scattering. According to Berman, the pertinent mechanics of molecular scattering can be separated into two sections—classical and quantum [6]. The uncertainty principle is used to determine the regions where each type of scattering is valid. For a collision with an impact parameter b scattering into angle θ, classical mechanics are valid in the region where

$$\Delta b < b, \quad \Delta\theta < \theta \tag{7.2}$$

where Δb and $\Delta\theta$ are the respective uncertainties. The uncertainty principle requires $2\pi\Delta b\Delta p_t \geq h$ where Δp_t is the uncertainty in the transverse component of the excited molecule's momentum. Since $\Delta p_t \equiv m\Delta v_t \approx mv\Delta\theta/2\pi$ (m is the excited molecule's mass and $k \equiv 2\pi mv/h$), this can be rewritten as

$$k\Delta b\,\Delta\theta \geq 1 \tag{7.3}$$

For a classical scattering treatment to be valid, it follows

$$\theta >> \frac{1}{kb} \tag{7.4}$$

If b_i is some characteristic range for scattering for a given perturbing molecule to affect molecules in state i, it follows that there is a corresponding characteristic range $\theta_i^d = (kb)^{-1}$ that demarcates the classical regions from the quantum, and if $\theta < \theta_i^d$ then the classical picture of scattering is not appropriate. For a molecule like iodine with a most probable relative velocity of ≈ 200 m/sec at 296°K, an estimate of 10Å to the critical impact parameter gives $\theta_i^d \approx 0.001$. This range is important because of a fundamental difference in the effect on coherence between a classical and quantum collision. During a classical collision the interaction between the molecules causes the populations of the two molecular states to follow two separate and distinct trajectories. Since the populations scatter independently relative to space, the density matrix elements ρ_{12} (that depend on state overlap) disappear immediately after such an event. Only quantum mechanical effects (that give rise to diffractive scattering) can produce nonzero off-diagonal elements after certain types of collisions. These collisions are limited in I_2 to those possessing $\theta \lesssim 0.001$. Both classical scattering (that changes the velocities of the collision partners significantly) and diffractive scattering (that does not) can be measured separately in the laboratory by appropriate pulse sequences. A variety of two-pulse sequences we discuss later use combinations of selective pulses with center frequencies to monitor velocity changes; at the opposite extreme, a three-pulse photon echo with full interpulse phase control gives a signal only from molecules that have not undergone either classical or diffractive scattering.

7.3 DERIVATION OF PROBABILITY EQUATIONS

However, while θ_i^d gives a limit, it still does not establish the distribution of possible deflection angles and thus does not state whether velocity-changing collisions affect coherence decay in I_2. To provide a better understanding, the density matrix formalism for dealing with the effects of velocity changing collisions can be constructed using collisional kernels [6]. The diagonal elements of the density matrix (populations of the two resonant levels) and the off-diagonal elements (coherences) can experience different collisional effects; these are reflected in the equations of transport. The population kernel W_{ii} must satisfy the equation of transport standard in dealing with velocity transport [6]

$$\frac{\partial \rho_{ii}(\vec{v},t)}{\partial t} = -\Gamma_i(v)\rho_{ii}(\vec{v},t) + \int W_{ii}(\vec{v}' - \vec{v})\rho_{ii}(\vec{v},t)\, d\vec{v}' \tag{7.5}$$

The first term on the right hand side of Equation 7.5 is the loss rate $\Gamma_i(v)$ of population density $\rho_{ii}(\vec{v},t)$ while the second term give the increase of $\rho_{ii}(\vec{v},t)$ resulting from collisions which change molecular velocity from $\vec{v}'$ to $\vec{v}$. The collision kernel $W_{ii}(\vec{v}' - v)$ gives the probability density per unit time that a collision changes the molecular velocity to $\vec{v}'$ from $\vec{v}$. It is related to the differential scattering cross–section by [6]

$$W_{ii}(\vec{v}' - \vec{v}) = Nv \mid f_i(v,\theta) \mid^2 v^{-2}\, \delta(v - v') \tag{7.6}$$

where N is the perturber density and θ is the angle between $\vec{v}'$ and $\vec{v}$. The elastic scattering amplitude for a molecule in state $j, f(\theta)$, generally can be determined using the method of partial waves [7]

$$f_\varphi(\theta) = \frac{1}{2ik} \left| \sum_{l=0}^{\infty} (2l+1)\,(e^{2i\eta_l^j} - 1)\, P_l(\cos\theta) \right| \tag{7.7}$$

where the η_l^j terms are the phase shifts (this is in the center of mass frame). For hard-sphere scattering these are equal to [6]

$$\eta_l^j = \tan^{-1}\left[\frac{j_l(kb_j)}{n_l(kb_j)}\right] \tag{7.8}$$

where j_l and n_l are the spherical Bessel and Neumann functions (here b_j is the radius of an impenetrable sphere determined by the state j scattering potential). The collision rate $\Gamma_i(v)$ is given by

$$\Gamma_i(v) = \int W_{ii}\,(\vec{\mathbf{v}}' - \vec{\mathbf{v}})\,d\vec{\mathbf{v}}' \tag{7.9}$$

where $\Gamma_i(v) = Nv\sigma_i(v)$ and the cross-section $\sigma_i(v) = |\,f_\iota(\theta)\,|^2\,d\Omega_v$. For hard-sphere scattering the method of stationary phase yields the following expression for the collisional kernel [5]

$$W_{ii}(\vec{\mathbf{v}}' - \vec{\mathbf{v}}) \cong N\,v^{-1}\,\delta(v - v')\begin{cases} \dfrac{b_i^2}{4},\ \theta >> (kb_i)^{-1} \\ \left(\dfrac{k^2b_i^4}{4}\right)\exp\left(\dfrac{1}{4}k^2b_i^2\theta^2\right),\ \theta < (kb_i)^{-1} \end{cases} \tag{7.10}$$

The first instance corresponds to classical scattering while the second refers to quantum contributions. The collision cross-section is simply $\sigma_i(v) = 2\pi b_i^2$, where the classical and quantum scattering each add πb_i^2 to the cross-section.

To develop the coherence kernel involves solving an equation of transport similar to Equation 7.8 [6]

$$\frac{\partial\rho_{12}(\vec{\mathbf{v}},t)}{\partial t} = -\frac{1}{2}[\Gamma_1(v) + \Gamma_2^*(v)]\,\rho_{12}\,(\vec{\mathbf{v}}\,,t) + \int W_{12}(\vec{\mathbf{v}}' - \vec{\mathbf{v}})\,\rho_{12}(\vec{\mathbf{v}},t)\,d\vec{\mathbf{v}}' \tag{7.11}$$

The terms on the right hand side of Equation 7.11 are defined analogously with those in Equation 7.5. For hard-sphere scattering in the classical limit, the collision kernel $W_{12}(\vec{\mathbf{v}}' - \vec{\mathbf{v}})$ is well approximated by [5]

$$W_{12}(\vec{v}' - \vec{v}) = \frac{1}{4} N v^{-1} \delta(v - v')\, b_1 b_2 \exp\left[-2ik(b_2 - b_1) \sin\left(\frac{\chi}{2} \right) \right] \tag{7.12}$$

In the quantum mechanical case, the collision kernel is

$$W_{12}(\vec{v}' - \vec{v}) = \frac{1}{4} N v^{-1} \delta(v - v')\, k^2 b_1 b_2 \exp\left[\frac{1}{8} k^2 (b_1^2 + b_2^2)\, \chi \right] \tag{7.13}$$

Observe that in the classical limit $k(b_2 - b_1) >> 1$ that the collisional kernel is approximately zero.

Further refinement of theory leads to the case where molecular motion must be considered when the system is exposed to a laser pulse. Assuming that the electric field of the laser propagates in the $\pm z$ direction, the density matrix element $\rho_{12}(\vec{v}_t)$ may be factored into two components

$$\rho_{12}(\vec{v},t) = \rho_{12}(\vec{v}_t)\, \rho_{12}(v_z,t) \tag{7.14}$$

where $(\vec{v}_t)$ is a velocity transverse to the z axis. Since it will not be affected by the EM field, the transverse component of $\rho_{12}(\vec{v}_t)$ may be taken as constant in time. Assuming this to be approximately true even in the presence of collisions, one further assumes that $\rho_{12}(\vec{v}_t)$ is described as a simple thermal distribution

$$\rho_{12}(\vec{v}_t) = \frac{1}{\pi u^2} \exp\left(\frac{-v_t^2}{u^2} \right) \tag{7.15}$$

where u is the most probable molecular speed. One can also develop expressions for the time dependence of $\rho_{12}(\vec{v}_t)$ as well as for the collisional kernel and total collision rate. The final expression developed for the kernel $W_{12}(v_z' \rightarrow v_z)\; W_{12}(v_z' \rightarrow v_z)$ is [6]

$$W_{12}(v_z' \to v_z) =$$

$$N\sigma_{12}^{vc}\,\theta_0^{-1}\left[\frac{1}{2} + \frac{|(v_z' - v_z)v_z'|}{\theta_0 u^2} + \frac{v_z'^2}{u^2}\right]\exp\left[\frac{-(v_z' - v_z)^2}{\theta_0^2 u^2}\right]\exp\left[\frac{-2|(v_z' - v_z)v_z'|}{\theta_0 u^2}\right] \tag{7.16}$$

where σ_{12}^{vc} is the velocity-changing coherence cross-section

$$\sigma_{12}^{vc} = \frac{2\pi b_1^2 b_2^2}{(b_1^2 + \ b_2^2)} \tag{7.17}$$

with corresponding rate $\Gamma_{12}^{vc}(v)$. We can also define a total cross-section

$$\sigma_{12}^t = \pi(b_1^2 + b_2^2) \tag{7.18}$$

with rate $\Gamma_{12}^t(v)$ as well as the phase-interrupting cross-section

$$\sigma_{12}^{ph} = \ = \sigma_{12}^t - \sigma_{12}^{vc} = \frac{\pi(b_1^4 + b_2^4)}{(b_1^2 + b_2^2)} \tag{7.19}$$

with rate $\Gamma_{12}^{ph}(v)$. When a laser pulse is applied to the system, a coherence $\rho_{12}(\vec{v},t)$ is created that follows the field $(\rho_{12}(\vec{v},t) = \tilde{\rho}_{12}(\vec{v},t)e^{+i\Omega t}$ where $\tilde{\rho}_{12}(\vec{v},t)$ is a slowly varying function of $\vec{v}$ and t). The frequency Ω' seen in the molecular rest frame is $\Omega' = \Omega - Kv_z$ where K is a propagation vector. Therefore $\rho_{12}(\vec{v},t)$ really varies as

$$\rho_{12}(\vec{v},t) = \tilde{\rho}_{12}(\vec{v},t)e^{+i\Omega t}\, e^{-iKv_z t} \tag{7.20}$$

which leads to the following equation of motion

$$\frac{\partial\tilde{\rho}_{12}(\vec{v},t)}{\partial t} = -\,\Gamma_{12}^t\,(v_z)\,\tilde{\rho}_{12}(\vec{v},t) + \int W_{12}\,(v_z' \to v_z)\,\tilde{\rho}_{12}(v_z',t)\,dv_z' \tag{7.21}$$

Notice now that the coherence kernel limits $|v_z' - v_z|$ to values less than or equal to $u\theta_o$ or

$$|v_z' - v_z| \leq u\theta_o \equiv \delta u << u$$

where

$$\delta u = \sqrt{\frac{2h^2}{\pi\sigma_{12}' m^2}} \tag{7.22}$$

For a molecule of I_2's mass, δu is approximately 62 cm/sec using a typical echo decay cross-section of 400 sq. Å—velocity changes larger than this would not contribute significantly to W_{12}. That is, collisions which change molecular velocity by more than this small amount have approximately zero probability of transferring coherence. Photon echo decay is therefore sensitive to the presence of velocity-changing collisions.

However, the photon echo rate has other components such as, for example, velocity-changing collisions that remove the molecule from the pulse bandwidth will also contribute to echo decay regardless of whether the polarization is scrambled. For this reason it is useful to look at a purely classical treatment of molecular translational energy redistribution to establish the magnitude of an average velocity change. For two molecules in a classical encounter where nothing but kinetic energy transfer occurs, the angle of deflection for an elastic collision is given in the center-of-mass frame as [8]

$$\chi(b,v_t) = \pi - 2b\int_{r_m}^{\infty} \frac{dr}{r^2}\left[1 - \left(\frac{2\mathbf{u}(r)}{\mu v_r^2}\right) - \left(\frac{b^2}{r^2}\right)\right]^{1/2} \tag{7.23}$$

where b is the impact parameter (the distance upon impact between the molecular centers), $\mathbf{u}(r)$ is the intermolecular potential, v_r is the relative velocity between the two particles and r_m is found by setting the denominator of the above integrand equal to zero and solving for r. Figure 7.1 illustrates these quantities. χ is then related to the classical cross-section for the relative velocity collision by the relation

$$\sigma(v_r) = 2\pi\int_0^{\infty} (1 - \cos\chi)\, b\, db \tag{7.24}$$

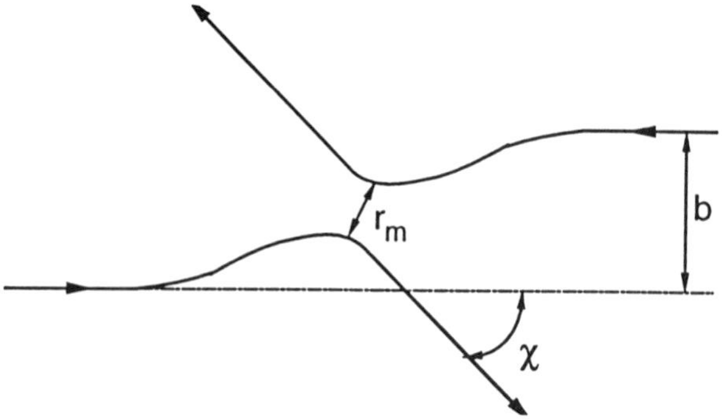

Figure 7.1 The center-of-mass frame illustrating listed terms.

The mean square change in velocity is given by

$$<(\Delta v_1)^{2}> = \int (\Delta v_1)^2 \, 2\pi b \, db \tag{7.25}$$

where v_1 is the velocity of our target molecule in the laboratory frame. The fraction of collisions of molecules traveling between v_r and $v_r + dv_r$ is

$$p(v_r)\, dv_r = \frac{1}{2}\left(\frac{\mu}{kT}\right) v_r^3 \exp\left(\frac{-\mu v_r^2}{2kT}\right) dv_i \tag{7.26}$$

Integrate over v_r to get $<(\Delta v_1)^2>$

$$<(\Delta v_1)^2> = \left(\frac{\mu}{kT}\right)^2 \int_0^\infty v_r^5 (1 - \cos\chi) \exp\left(\frac{-\mu v_r^2}{2} kT\right) dv_r \tag{7.27}$$

From the definition of $\sigma(v_r)$ it follows

$$<(\Delta v_1)^2> = \left(\frac{\mu}{kT}\right)^2 \int_0^\infty v_r^5 \, \sigma(v_r) \exp\left(\frac{-\mu v_r^2}{2} kT\right) dv_r \tag{7.28}$$

Equation 7.28 is the expression for the mean-square change in velocity. For all but the hard-sphere intermolecular potential exact analytical expressions for $\sigma(v_r)$ are not possible. However, when $\mathbf{u}(r)$ has the form of an inverse power potential such as $\mathbf{u}(r) = Kr^{-\delta}$ it is possible to simplify the above equations. The equation for deflection angle in the center of mass frame reduces to

$$\chi(y_0) = \pi - 2 \int_0^{y_m(y_0)} \left[1 - y^2 - \frac{1}{\delta}\left(\frac{y}{y_0}\right)^{-\delta} \right]^{-1/2} dy \tag{7.29}$$

where

$$y = \frac{b}{r} \quad y_m = \frac{b}{r_m} \quad y_0 = b\left(\frac{\mu v_r^2}{2\delta K}\right)^{1/\delta} \tag{7.30}$$

K is related to the Lennard-Jones 6–12 potential parameters and must include contributions from the intermolecular well-depth $\varepsilon_1\varepsilon_2^*$. Although this number has never been measured experimentally, Parmenter [2] has shown that the standard approximation

$$\varepsilon_1\varepsilon_2^* = (\varepsilon_1\varepsilon_2 * \varepsilon_1^*\varepsilon_2^*)^{1/2} \tag{7.31}$$

is valid in estimating several quantum state changing processes and can be deduced from such data already tabulated. This provides an order of magnitude estimate for the value of K, and since velocity changing collisions are expected only to sample the long-range part of the intermolecular potential, our expression for $\mathbf{u}(r)$ is expected to be qualitatively correct.

These expression for deflection angle χ and $\sigma(v_r)$ can be evaluated numerically. As expected for an inverse power repulsive potential, $\sigma(v_r)$ grows very quickly—up to hundreds of square Angstroms even for small relative velocities and weak potential values. Using the above relations and the potential parameters from the I_2 literature, these dynamic equations predict a small mean-square change, ≈ 2 m/sec, at 296°K. This first approximation implies that we can expect to observe velocity-changing collisions in I_2 with magnitudes of less than 10 m/sec.

7.4 PULSE SHAPE EFFECTS

A velocity change of ≈ 10 m/sec corresponds to a shift of ≈ 16 MHz in resonance frequency for an optical transition in the Rhodamine 6G wavelength. At room temperature I_2 has a Doppler spread of approximately $\Delta\nu = \pm 200$ MHz (± 0.006 cm^{-1}) associated with each transition. Since qualitatively $\Delta\nu\Delta t \approx 1$, a square-envelope radiation pulse with duration 100 ns will affect primarily molecules in the frequency range $v_0 \pm 5$ MHz (in I_2, velocities $v_0 \pm 3$ m/sec)—a very small section of the total velocity distribution. The exact excitation profile from such a pulse is neither particularly uniform or selective, mainly because of the high frequency components associated with the pulse edges (Figure 7.2).

However, this represents a vast oversimplification of the effects of a laser pulse on a molecular system. A major complicating factor comes from the laser system itself. The typical dye laser is not truly monochromatic; the dye jet has microphonics that can modulate the frequency of the laser on the timescale of tens to hundreds of ns. In addition, the molecular transition dipole moment in I_2 follows the projection of the molecular rotational angular momentum (the m_j value). Since one usually excites a high J state (J > 10), only a small number of molecules exactly on resonance will experience an electric field parallel to their transition dipole (Figure 7.3).

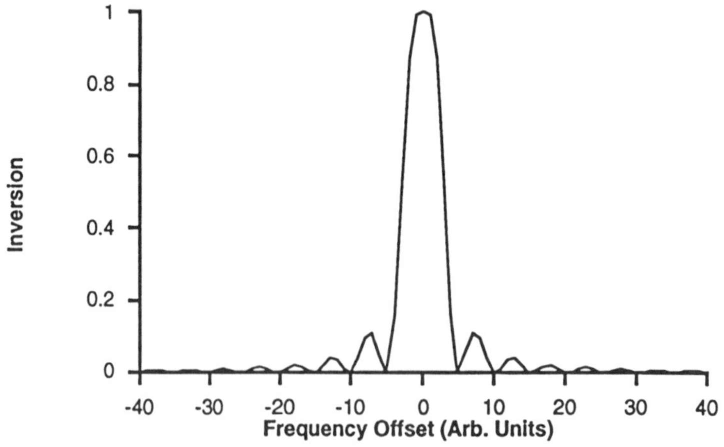

Figure 7.2 Inversion profile of a square enveloped pulse.

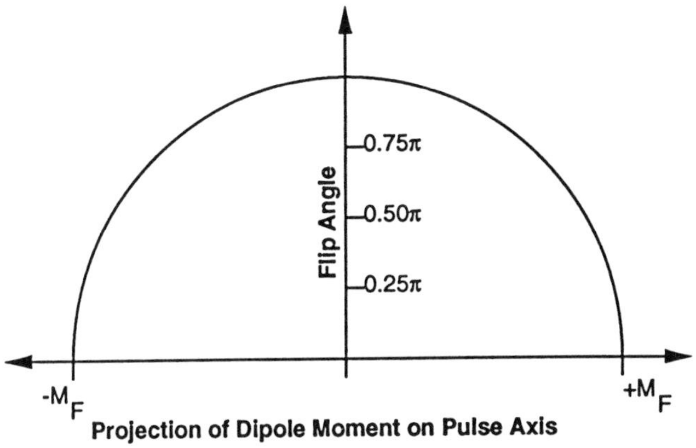

Figure 7.3 Variation of pulse flip angle with m_j.

All the others will experience a weakened pulse. All of these considerations make it difficult to design experiments that demand localized, selective and maximized excitation.

The most general and powerful approach to overcome these problems is to use shaped optical pulses. Here shaping can consist of modifying the pulse envelope as well as the pulse frequency. A square enveloped pulse has crudely a sinc(x) excitation profile (the exact profile is analytically solvable and changes as the flip angle $\theta = \omega_1 t$ is increased). This creates excitation far from resonance that can interfere with the velocity-selective experiments. Theoretically it should be possible to calculate the magnitude of this effect and then subtract it out, but with the actual small signal and experimental noise it is best not to deal with it at all. This can be eliminated by filtering the envelope to remove the high Fourier components. More elegant solutions replace the square envelope with a simple Gaussian envelope to produce a near-Gaussian excitation profile (Figure 7.4), while a Hermite envelope (a Gaussian curve multiplied by a Hermite polynomial) creates an even and localized inversion in a narrow bandwidth near resonance (Figure 7.5). The hyperbolic secant pulse envelope function $\mathrm{sech}(\alpha T)^{1+5i}$, based on an analytic solution to the Bloch equation, can make the profile insensitive to Rabi frequency inhomogeneity, but this pulse requires phase modulation as well.

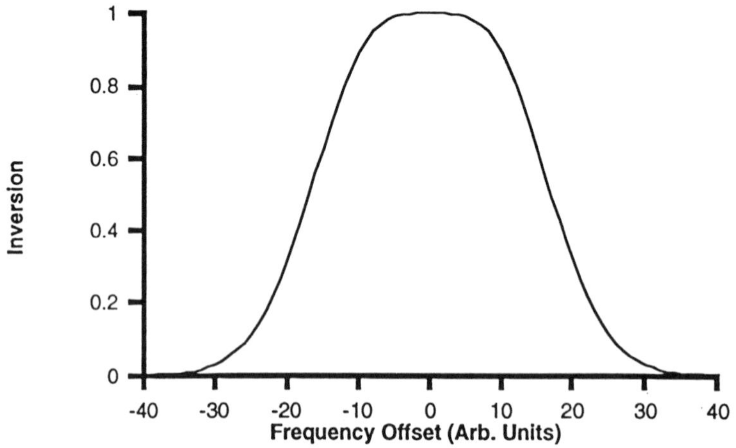

Figure 7.4 Inversion profile of a Gaussian enveloped pulse.

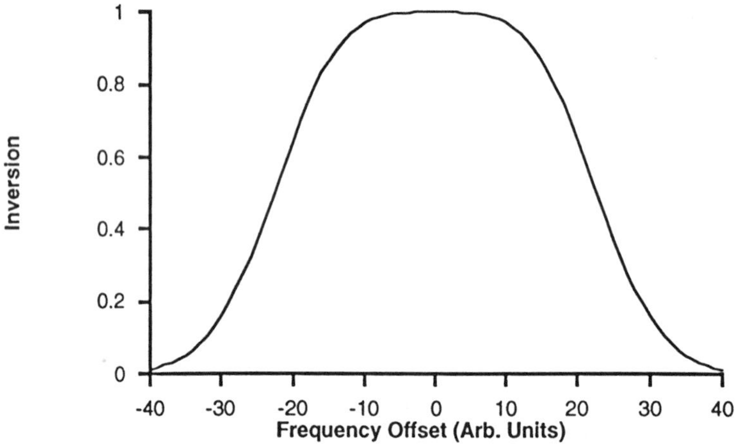

Figure 7.5 Inversion profile of a Hermite enveloped pulse.

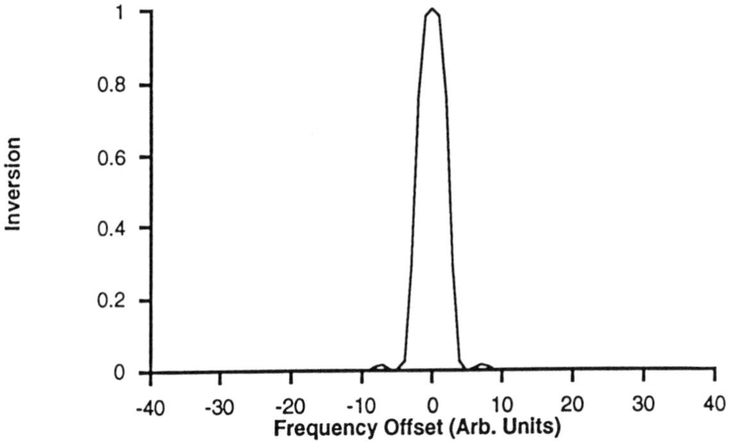

Figure 7.6 Autocorrelation function of a square enveloped pulse.

However, sophisticated pulse shaping is not always necessary. For example, the photon echo has been used successfully for over 25 years without much concern for the temporal profile (in fact, even incoherent pulses give echoes). It can readily be shown that, at least for gases, the magnitude of the echo is a strong function of the flip angles of the pulses, but the functional form of the decay and hence the measured value of T_2 does not. A large number of different pulse sequences will be presented in this work, but all share this insensitivity of the measured parameters to flip angle variation. For this reason, compensation for Rabi frequency inhomogeneity is unnecessary.

We will compare relaxation rates at different positions in the inhomogeneous line, so localization of the excitation is important. A useful way of determining localization is simply to determine the autocorrelation function of the excited population distribution. For square pulses, the autocorrelation function shows a long tail (Figure 7.6), making it a poor choice for experiments where good selectivity is desired. Gaussian and Hermite envelopes are much better at producing the desired resolution (Figures 7.7 and 7.8). Therefore, all experiments either used shaped pulses or square pulses that had their rise and fall times limited.

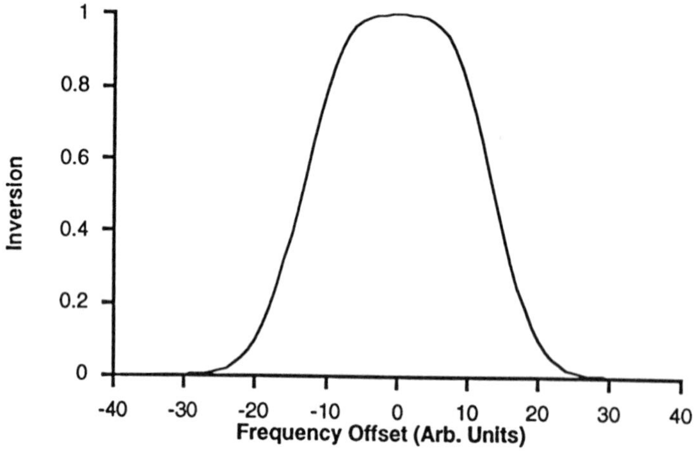

Figure 7.7 Autocorrelation function of a Gaussian enveloped pulse.

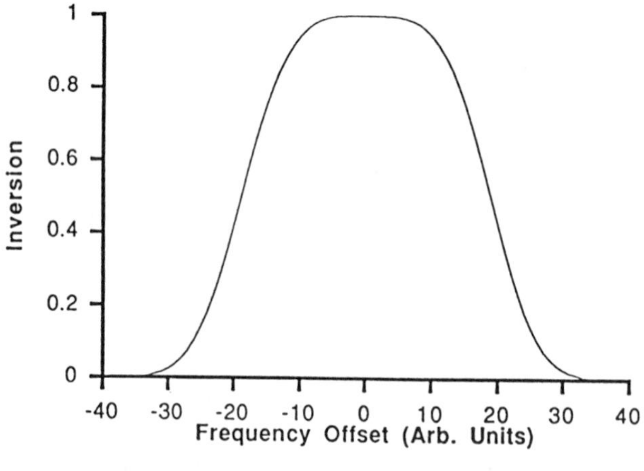

Figure 7.8 Autocorrelation function of a Hermite enveloped pulse.

7.5 FEATURES OF MOLECULAR IODINE SPECTRUM

An immense catalog of spectral information exists on I_2—the intermolecular potential diagram (Figure 7.9) shows the molecular states involved [9–12]. All of the transitions studied here involve $X^1\Sigma_g \rightarrow B^3\Pi_u$ transitions, but the $^1\Pi_u$ state is important because molecules in the B state predissociate spontaneously to it. This rate of predissociation is pressure independent; it varies with rovibrational level in the B state by the relationship (in the absence of a magnetic field) $\Gamma = k_v J(J+1)$, where k_v depends on the vibrational level in the B state. This constant has been catalogued for v′ = 7 to v′ = 24 [9, 13–15].

There are also hyperfine contributions to the zero-pressure lifetime. Since the atomic iodine nucleus has spin 5/2, the I_2 molecule will have a total nuclear angular momentum T. This couples with the total rotational angular momentum J in a state-symmetry restricted way to form F, the total angular momentum. The equation for the energy levels for a given level in either electronic state is [16]

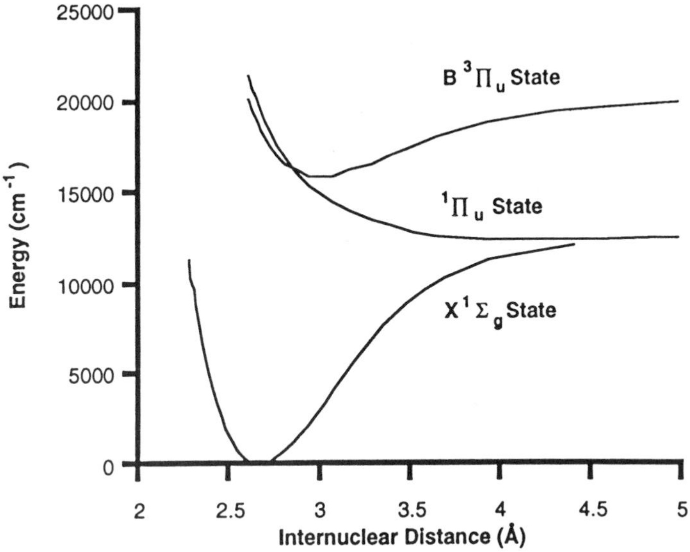

Figure 7.9 Intermolecular potentials for I_2 spectroscopic states involved in this study.

$$E(T,F,J) =$$

$$\left(\frac{-3eqQ}{8}\right)\frac{\{1-[T(T+1)+4I(I+1)]\}\,[G(G+1)-T(T-1)(2T+3)]}{[2(T-1)(2T+3)\,I\,(2I-1)(2J-1)(2J-3)]}$$

$$G = F(F+1) - T(T+1) - J(J+1) \qquad (7.32)$$

where $-eqQ = -2293$ MHz for the $X^1\Sigma_g$ state and -573 MHz for the $B^3\Pi_u$ state [17]. The selection rules are $\Delta T = 0$, $\Delta J = \pm 1$ and $\Delta F = \pm 1$ [18]. The pressure–independent predissociation rate as a function of hyperfine level is given by

$$\Gamma_{IJF} = \left[C_v^2 J(J+1) + \frac{a_v^2}{3}\left(\mathbf{I}^2 + \frac{3(\mathbf{I}\cdot\mathbf{J})^2 + 3\!/\!2\,\mathbf{I}\cdot\mathbf{J} - \mathbf{I}^2\mathbf{J}^2}{(2J-1)(2J-3)}\right) - a_c C_v \mathbf{I}\cdot\mathbf{J}\right] \times$$

$$[1 + p_v J(J+1) + q_v (J(J+1))^2] \qquad (7.33)$$

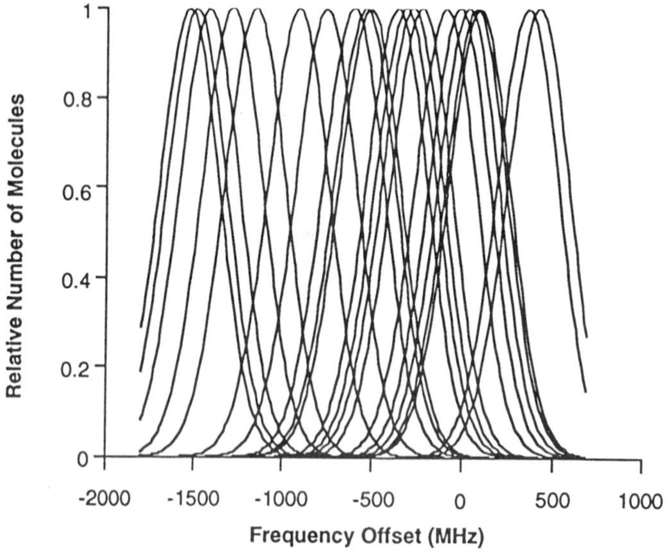

Figure 7.10 Frequency spacing and Doppler profile overlap for hyperfine components of the R61 (17-2) transition.

where the values of a_v, C_v, p_v, andq_v depend on vibrational level [19]. The complete zero-pressure lifetime is the sum of Γ_{IJF}, Γ, and τ_0^{-1} (the rate of purely spontaneous emission).

A radiation pulse will not pick out a single Doppler profile but instead a set of velocities from a distribution of profiles (see Figure 7.10 for an example). However, the following procedure can identify the exact transition, including the many hyperfine levels and molecular velocities involved. A monochromator or wavemeter can give an approximate reading for transition frequency. Exact position within a transition's bandwidth can be determined using a Fabry–Perot interferometer. Fluorescence decay spectra are taken at several pressures and a Sterm Vollmer plot is performed. The intercept of the plot is the described zero-pressure decay rate. This information provides a highly accurate assignment to the set of transition components that are involved.

7.6 EXPERIMENTAL DESIGN

Figure 7.11 shows the experimental schematic for generating nanosecond pulses with pulse shape and interpulse phase control. An IBM PC AT is specially modified with a digital/analog converter assembly consisting of an internally mounted interface card, an external patch panel and FORTRAN control software. The converter board panel is connected to a homemade counter (that sends out a fixed number of clocked pulses for a given trigger), a homemade interface board containing a digital number input/analog voltage output IC, and a parallel data bus into the pulse shape [20]. Upon receiving a trigger pulse from the computer, the homemade counter issues TTL pulses; external switches control their number and an external clock govern their frequency (typical frequencies are in the 5–10 KHz range). The pulses from this counter and the output from the digital number/voltage card connect to a 4100 series card from Evans Electronics. This card uses a voltage of 0 to +10 VDC to regulate the time delay between two TTL pulses that it issues upon receiving a trigger (that in this case comes from the external counter). These TTL pulses in turn serve as triggers for a high speed counter that drives the high speed D/A converter chip contained in the external pulse shaper construction. The parallel data bus from the computer connects to this shaper device and loads a digital approximation to the desired pulse envelope into the pulse shaper's fast ROM. This circuit approximates the continuous

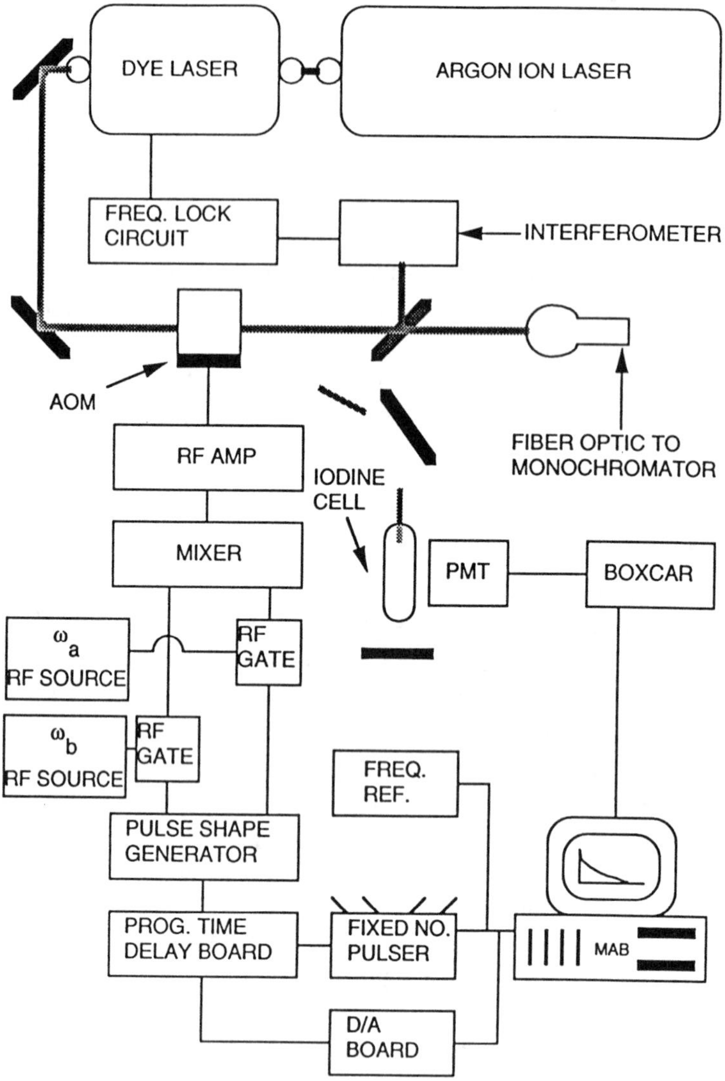

Figure 7.11 Experimental schematic.

envelope function with a 16 step digital representation where each step can be one of 256 possible amplitude gradations (the data bus loads 16 8-bit words). The trigger pulse from the Evans cards causes the ROM contents to clock into a fast D/A converter device at a rate determined by an on-board voltage controller oscillator. The D/A chip then produces the 16-step approximation with 8 nanoseconds between steps. Finally, passing this pulse shape through low-pass passive filters removes the high frequency Fourier components from the steps in the digital approximation to the desired pulse shape.

The pulse shape then feeds into an RF gate (Mini–Circuits models ZFM–2 and ZFM-4) where it imposes amplitude modulation on a CW RF frequency input. The output of this gate is a pulse with the envelope of the input shape but with the frequency of the RF source. This pulse is amplified and then sent to an acousto–optic modulator, where its wavevector adds to that of the input laser beam (Figure 7.12). Typical conversion efficiencies for an anti-reflection coated TeO_2 modulator are 60–75% of input laser power. The maximum conversion occurs for an RF input of 470 MHz with the modulator having a usable bandwidth of approximately 60 MHz. The AOM rise time is approximately 4 ns, so pulse sequences can consist of nanosecond scale pulse lengths with similar pulse delays. The single RF gate can also be replaced with two gates connected to different RF sources to give phase-incoherent pulses that differ in frequency by tens of MHz.

The laser system itself consists of an argon ion laser (Spectra–Physics model 171 or Coherent model Innova 100) that pumps a tunable dye laser (Spectra–Physics model 380A) to 500–800 mW single frequency output across the range of Rhodamine 590 dye (560–610 nm) or 200–400 mW single frequency output across the range of Rhodamine 610 dye (610–680 nm). This dye laser had its output frequency locked onto a particular transition through an external Fabry-Perot interferometer. The laser wavelength was determined by two methods. The earliest experiments used a beamsplitter to sample off part of the laser beam and then pass it into a three meter monochromator via a fiber optic cable (with a 10 μ slit, the monochromator was capable of determining wavelength to approximately 0.1 cm^{-1}). The second and better method replaced the monochromator with a homemade wavemeter that counted interference fringes between an input beam and an internal reference HeNe laser. This wavemeter (designed by E. Boyce of the National Research Council of Canada in Ottawa, Ontario) had a better accuracy than the monochromator (greater than 1 in 10^7). A photomultiplier tube (Thorn-

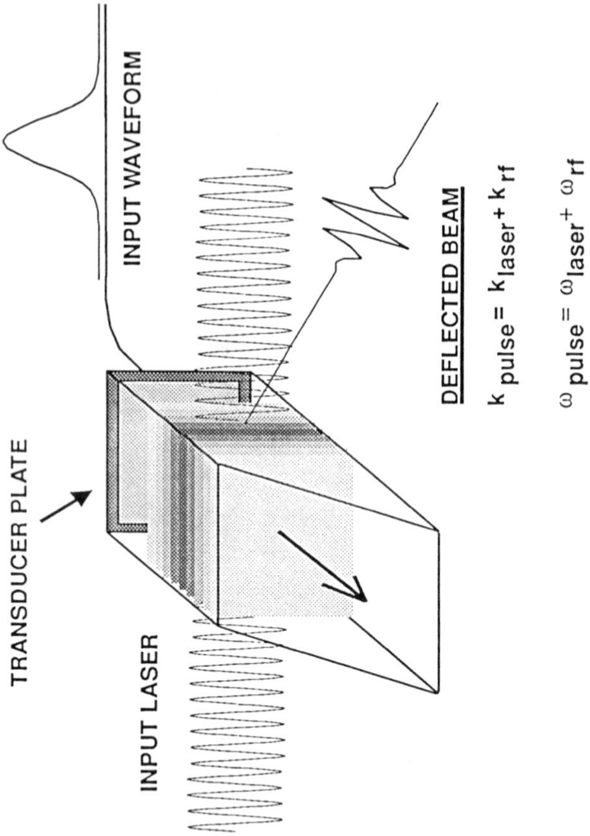

Figure 7.12 Acousto-optic modulator schematic.

EMI tube 9658B with maximum detection at 590 nm powered by a high voltage supply from Pacific Instruments), a fast amplifier (CLC Corporation), and a boxcar integrator (Stanford Research System model 250) in a NIMBIN crate (available from MechTronics Nuclear or Tennelec Electronics) processed sample fluorescence. For analysis, the integrator connected directly to a PC via a 16-bit D/A board. The computer displayed experimental results directly on its monitor using a homemade graphics program.

7.7 IODINE CELL PROCEDURE

These experiments used two different cell designs. Figure 7.13 shows the cell used for the earliest studies. Quartz Brewster's angle windows minimized scattered laser light. Iodine transferred into the cell finger from an attachable finger using a standard sublimation method. The vapor pressure of I_2 is well known over the range of temperatures [21] and by placing a temperature bath on the cell finger at a temperature equal to or lower than room temperature, the cell pressure could be controlled accordingly. For the later sets of experiments, Figure 7.14 shows an improved version that replaced the original cell. Though similar in operation, this stainless steel design possessed several advantages. First, a capacitance manometer measured actual gas pressure directly instead of relying on an estimated value. This manometer also provided constant monitoring for atmospheric contamination. Second, the stainless steel flanges held microtorr pressure consistently (the glass cell's Teflon stopcocks had leakage rates that varied wildly). Third, because the cell sat on an actual vacuum line, the cell could be evacuated as background pressure rose, which saved enormously in the cycling time of cell preparation.

To determine experimentally the purity of an iodine cell, the spectroscopic lifetime of a given transition was measured as a function of cell pressure. A Sterm-Vollmer plot (inverse lifetime versus cell pressure) gives a slope of $\sigma(v_r)$ where v_r is the mean relative velocity and s is the self–quenching cross–section. Since a contaminated cell will contain N_2 and O_2, each of which have fluorescence quenching cross-sections in the 10–20 sq. Å range [14, 22], a contaminated cell is quickly identified by a shortened lifetime. The intercept of this plot is the rate sum of the pressure independent decays that, as mentioned, give excellent information on line assignment. All spectroscopic quantities such as line assignments, exact hyperfine structure,

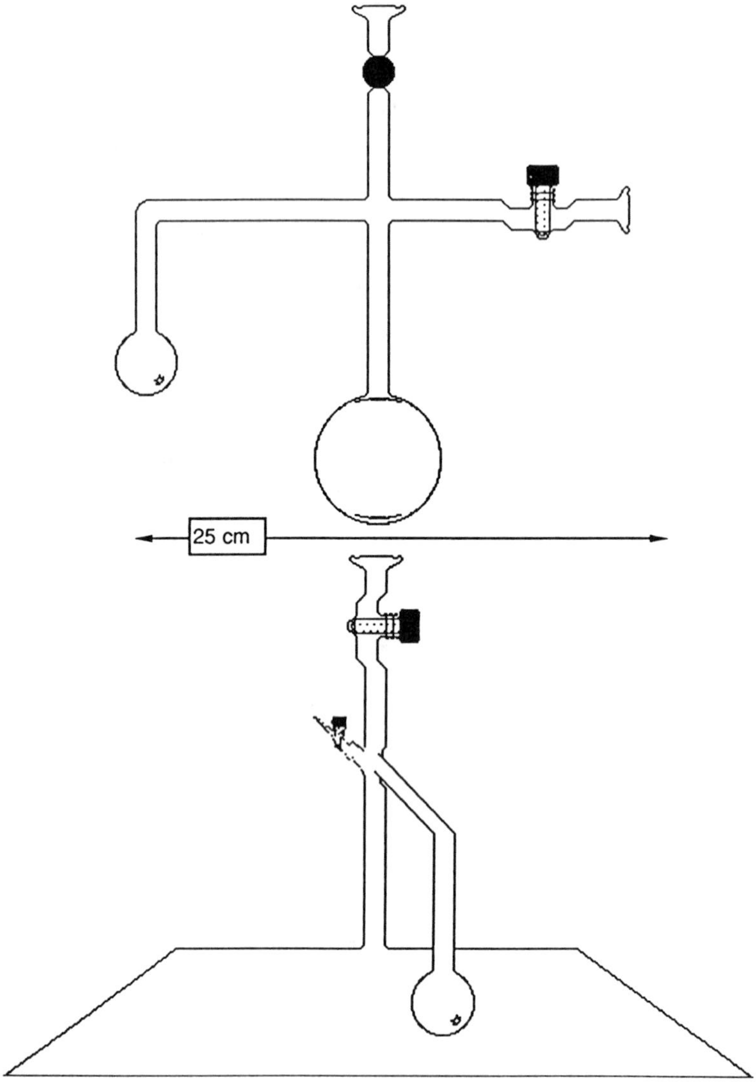

Figure 7.13 Original cell design.

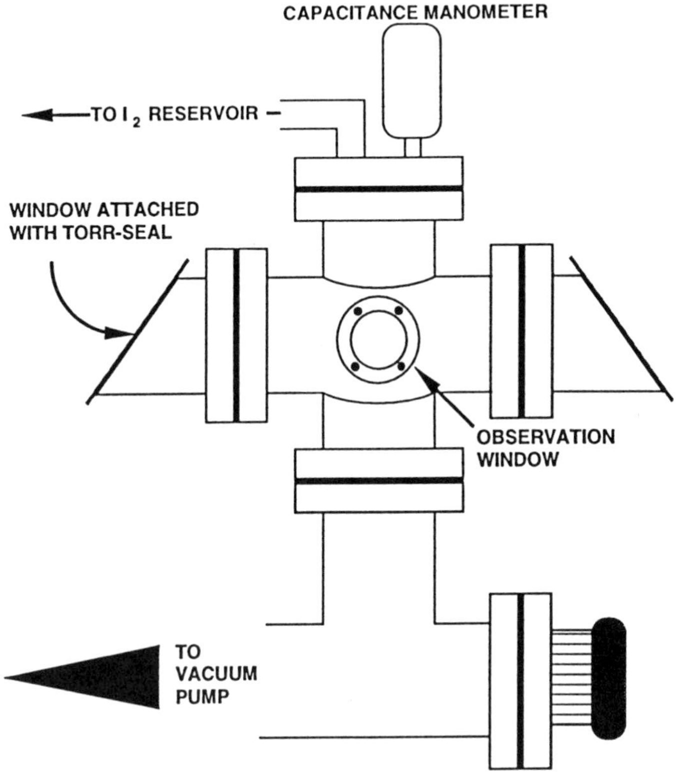

Figure 7.14 Stainless steel cell design.

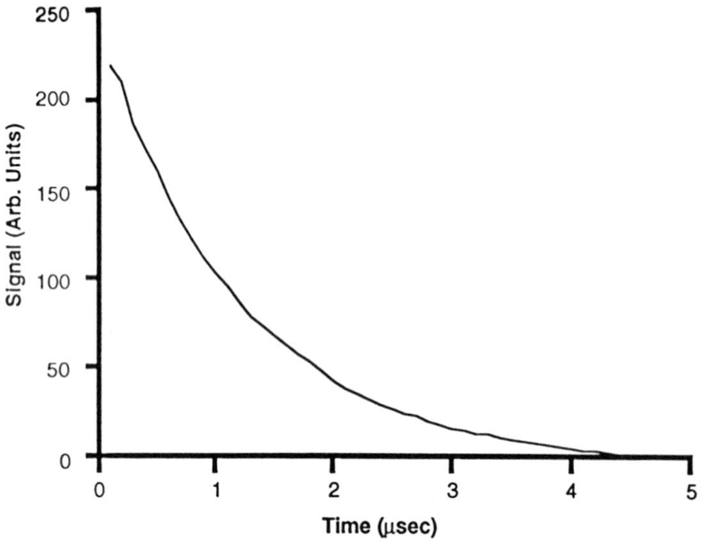

Figure 7.15 Typical fluorescence decay spectrum for 17192.47-cm^{-1} line.

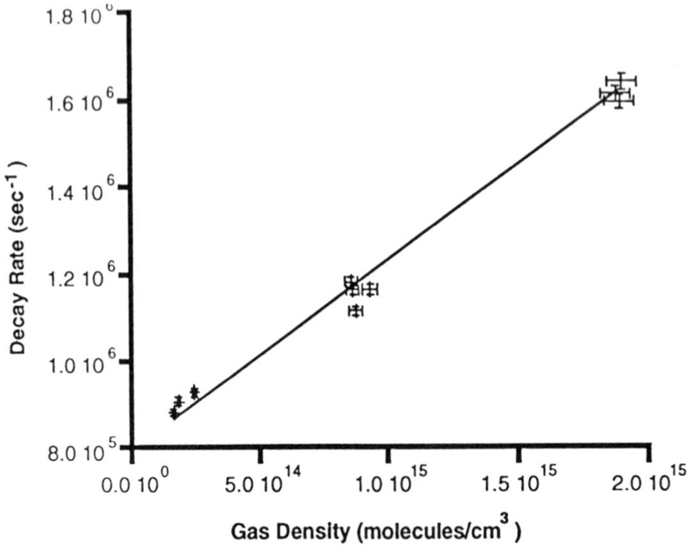

Figure 7.16 Sterm-Vollmer plot for 17503.75-cm^{-1} line.

and fluorescence decay fits were calculated on a microcomputer using established techniques [23, 24]. Figure 7.15 shows a typical fluorescence decay for a transition at 17192.47 cm^{-1} and Figure 7.16 shows the Sterm-Vollmer plot for the line at 17503.75 cm^{-1}.

7.8 PHOTON ECHO RESULTS

To generate photon echoes, these experiments used exclusively the sequence developed by Warren and Zewail [3] (Figure 7.17) because it allows rephasing polarization to be detected by monitoring fluorescence. As seen in Figure 7.18, the first half of the sequence transfers rephased polarization back into the ground state, while the second part transfers this polarization into the excited state. Simple subtraction of the fluorescence signal produced by one sequence from the other yields a difference spectrum that decays as a single exponential of time constant $T_2/2$ and that is free from the substantial background noise. The 1:2:1 ratio also produces slightly more signal when off-resonance effects are taken into account.

7.9 PHOTON ECHO ANALYSIS

Table 7.2 lists photon echo data and Figures 7.19 and 7.20 show the detailed results for the R61(17-2) transition at 17148.4 cm^{-1} (Figure 7.19 gives an echo spectra and Figure 7.20 shows the echo decay rate plots for each of the three frequency positions examined within the transition bandwidth). Table 7.2 shows that there is a wide range of values obtained, indicating that the phase-interrupting cross-section is not constant even for the same excited vibrational state. The photon echo numbers listed represent a bulk rate (i.e.,

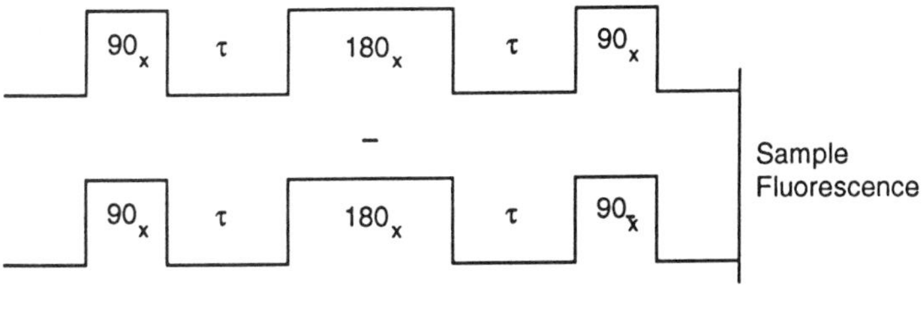

Figure 7.17 Three pulse photon echo sequence.

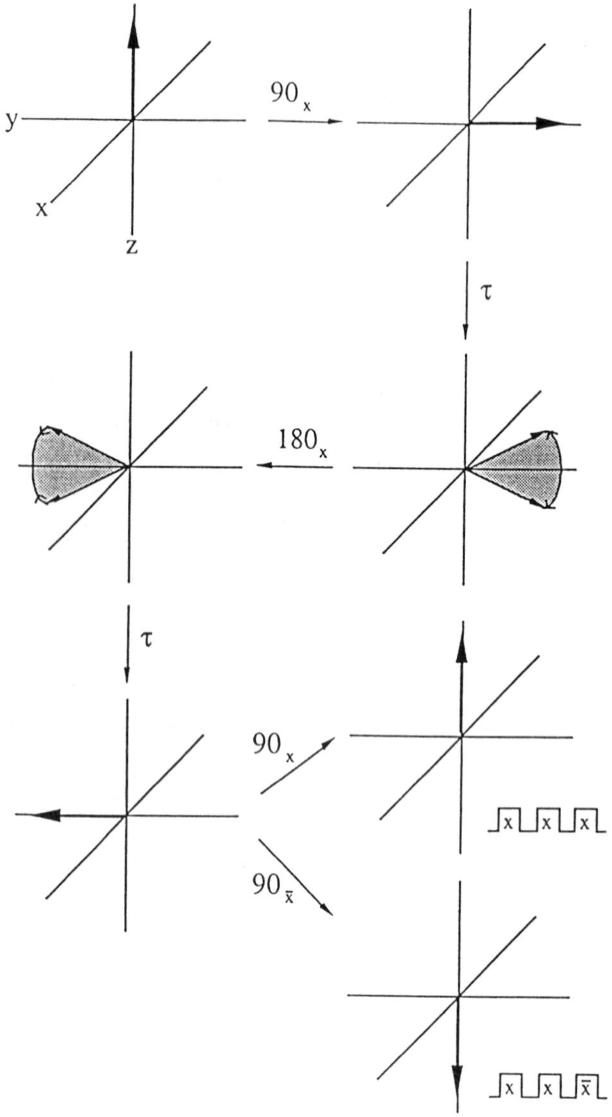

Figure 7.18 Bloch vector model describing photon echo formation and intensity detection.

Table 7.2 Measured Photon Echo Cross-Section in I_2.

Frequency	Assignment	$\sigma_{Dephasing}$ (sq. Å)	$\sigma_{Fluorescence}$ (sq. Å)
17503.75 cm^{-1}	R123 (20-1)	893 ± 37	362 ± 9
	P31 (16-0)		
17434.81 cm^{-1}	P70 (18-1)	834 ± 10	220 ± 7
17340.36 cm^{-1}	P70 (17-1)	776 ± 26	299 ± 32
17192.57 cm^{-1}	P29 (15-1)	934 ± 8	204 ± 6
17148.4 cm^{-1}	R61(17-2)		
360 MHz		471 ± 36	174 ± 13
720 MHz		351 ± 49	167 ± 6
1.1 GHz		612 ± 33	172 ± 7
17138.72 cm^{-1}	R79 (15-1)	711 ±66	206 ± 4
17104.57 cm^{-1}	R97 (15-1)		
300 MHz		841 ± 60	222 ± 8
800 MHz		560 ± 100	201 ± 2
1.3 GHz		713 ± 63	209 ± 5
17091.25 cm^{-1}	P90 (17-2)	479 ± 54	136 ± 9
16961.26 cm^{-1}	P50 (15-2)		
800 MHz		220 ± 10	
1.3 GHz		559 ± 75	233 ± 12
16943.3 cm^{-1}	R119 (14-1)	915 ± 57	181 ± 12

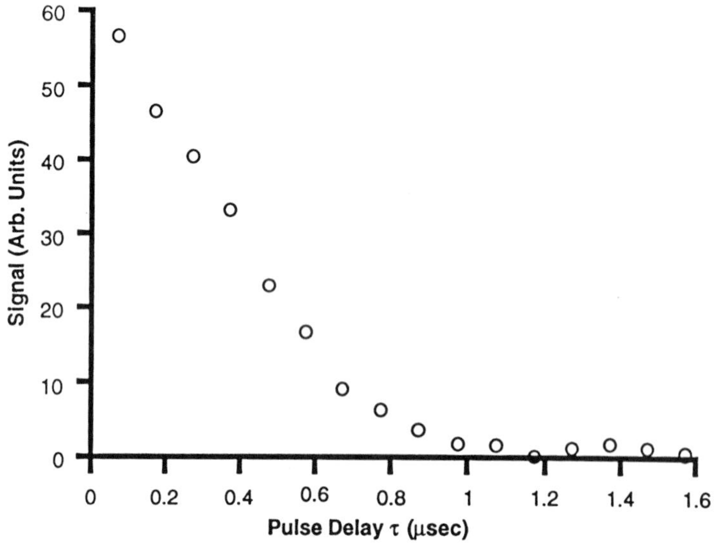

Figure 7.19 Typical photon echo decay spectrum observed with bandwidth of 17148.4-cm^{-1} line.

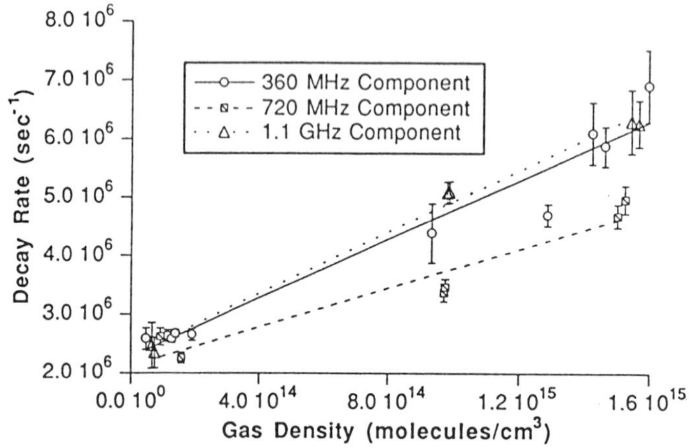

Figure 7.20 Sterm-Vollmer plot of dephasing rates versus gas density for different frequency positions within bandwidth of 17148.4-cm^{-1} line. *Source*: from [5].

Table 7.3a Contributions from Different Hyperfine Components for R61 (17-2) Line (17148.4 cm^{-1})

I	F	Relative Frequency (MHz)	360 MHz (308 m/sec)	720 MHz (325 m/sec)	1.1 GHz (304 m/sec)
1	61	−24.94	0.5556	0.5905	0.0050
1	62	−102.2	0.3065	0.8487	0.0198
1	63	−26.20	0.5511	0.5951	0.0051
3	59	86.19	0.9106	0.2447	0.0004
3	60	−287.8	0.0318	0.8759	0.2305
3	61	−523.3	0.0003	0.1659	0.9462
3	62	−613.0	≈ 0	0.0533	0.9814
3	63	−549.0	0.0002	0.1233	0.9836
3	64	−325.0	0.8629	0.7645	0.3271
3	65	−67.40	0.3356	0.2926	0.0007
5	57	412.0	0.0655	0.0016	≈ 0
5	58	−238.2	≈ 0	0.9751	0.1343
5	59	−764.7	≈ 0	0.0042	0.5564
5	60	−1160	≈ 0	≈ 0	0.0031
5	61	−1418	≈ 0	≈ 0	≈ 0
5	62	−1530	≈ 0	≈ 0	≈ 0
5	63	−1489	≈ 0	≈ 0	≈ 0
5	64	−1288	≈ 0	≈ 0	0.0002
5	65	−918.0	≈ 0	≈ 0	0.1404
5	66	−373.0	0.0075	0.5979	0.4788
5	67	355.0	0.5203	0.0050	≈ 0

the loss of coherence from a prepared velocity distribution in one excited state). This rate is made up of several components: collisions that scramble phase but retain molecular velocity, collisions that change velocity but retain phase, and collisions that do both. There is no distinction at this simple level and all of these processes remove molecules from the prepared distribution. Of particular interest are the 17104.4 and 17148.4-cm^{-1} lines where the echo cross-section changes at different locations within the profile (Tables 7.3a and 7.3b list the frequency positions and hyperfine contributions for all locations in both transitions). These previously unseen variations have several possible sources. First, there could be a rovibrational state dependence; this

Table 7.3b Contributions from Different Hyperfine Components for R97 (15–1) Line (17104.57 cm^{-1})

I	F	Relative Frequency (MHz)	300 MHz (286 m/sec)	800 MHz (319 m/sec)	1.3 GHz (311 m/sec)
1	97	−24.60	≈ 0	0.3858	0.3002
1	98	−100.4	0.0005	0.6455	0.1365
1	99	−25.49	≈ 0	0.3886	0.2978
3	95	81.15	≈ 0	0.1353	0.6481
3	96	−289.5	0.0222	0.9851	0.0081
3	97	−518.8	0.3679	0.3163	≈ 0
3	98	−602.0	0.6511	0.1339	≈ 0
3	99	−535.0	0.4189	0.2726	≈ 0
3	100	−312.0	0.0317	0.9545	0.0053
3	101	−69.00	≈ 0	0.1556	0.6050
5	93	394.0	≈ 0	0.0006	0.6653
5	94	−259.0	0.0133	0.9999	0.0138
5	95	−780.0	0.9931	0.0096	≈ 0
5	96	−1164	0.0605	≈ 0	≈ 0
5	97	−1407	0.0007	≈ 0	≈ 0
5	98	−1504	≈ 0	≈ 0	≈ 0
5	99	−1451	0.0002	≈ 0	≈ 0
5	100	−1243	0.0182	≈ 0	≈ 0
5	101	−875.0	0.7967	0.0015	≈ 0
5	102	−342.7	≈ 0.0500	0.8858	0.0029
5	103	359.0	≈ 0	0.0013	0.7839

possibility is illustrated in Figure 7.21. A plot of cross-sections versus the ground state vibrational level v″ (Figure 7.21(a)) shows there is a range of values associated with each ground state, but the center of that range increases as v″ does. Figures 7.21(b) and 7.21(c), that plot the dephasing cross-section versus v′ and J', respectively, do not show this effect. This suggests that the ground vibrational state does play a role in determine the

Figure 7.21 (opposite) (a) Plot of dephasing cross-sections versus ground vibrational state level v″. (b) Plot of dephasing cross-sections versus excited vibrational state level v′. (c) Plot of dephasing cross-sections versus excited rotational state level J'.

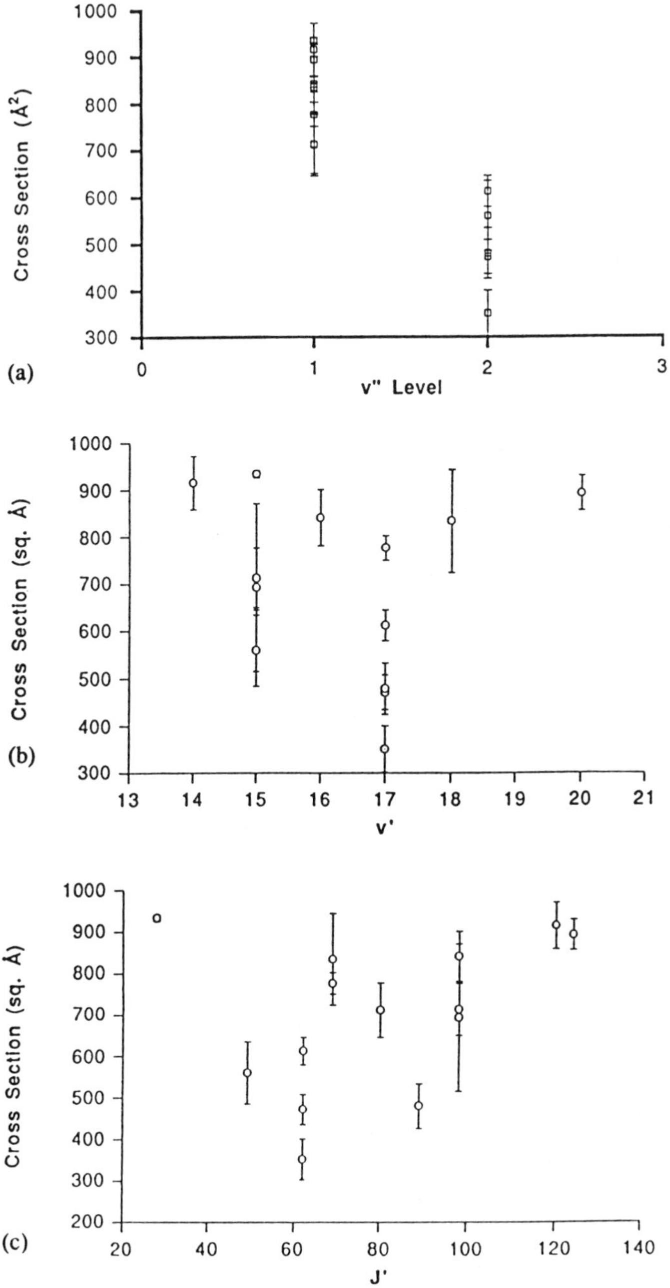
Cross Section (Å²)
v" Level
(a)
Cross Section (sq. Å)
v'
(b)
Cross Section (sq. Å)
J'
(c)

dephasing rate, but the limited number of v″ states sampled prevent further analysis.

A second dependence could be on hyperfine state. Although the combination of hyperfine components excited is a function of lineshape position, variations in this combination should not affect the size of the echo cross-section significantly. A simple geometric argument concerning the coupling of molecular angular momentum clarifies this; when *I* and *J* couple to form *F*, given the rotational states involved, the major contributor is *J*. The *F* vector therefore closely resembles *J* and they differ only slightly in precession angle. The nuclear momentum should therefore play only a small part in determining collisional effects.

The third dependence, that of molecular velocity, requires redefining the cross-section. The slope of the plot of echo decay rate versus gas density should equal σv_{rel}. For all values in Table 7.2, the cross-section listed is simply the slope from the respective data plots divided by $<v_{rel}>$. However, with pulses that can sample small sections of the transition the use of a completely averaged value for v_{rel} is wrong. There are two explanations for the differences in echo cross-section. The first is that the cross-section is constant and the variations are due to molecular velocity; however, this can be dismissed because one knows the position in the lineshape and therefore one can determine v_{rel} at that frequency by considering contributions from all hyperfine states by the following formula

$$<v_{rel}^2(v)> = \frac{5kT}{m} + \sum_{I,F} I_{I,F} v_{z,I,F}^2(v) \exp\left[\frac{-(v_{I,F} - v)^2}{2\sigma^2}\right] \tag{7.34}$$

where the intensities I_{IF} of the individual hyperfine transitions is given in [25]. This information is collected in Tables 7.3a and 7.3b for the 17148.4- and 17104.57-cm^{-1} lines, respectively. The variation in v_{rel} is only 10–15%, which is much too small to produce the larger variations in the dephasing cross-sections.

The second possibility is that the cross-section is a function of relative velocity ($\sigma = \sigma(v_{rel})$). For an intermolecular potential of the form $V_r = 1/r^s$, the cross-section has been shown to be proportional to the molecular speed by the relationship $\sigma \propto v^{-2/s-1}$ [26]. This can be explored by plotting the slope against the calculated relative speed, as shown in Figure 7.22(a) for

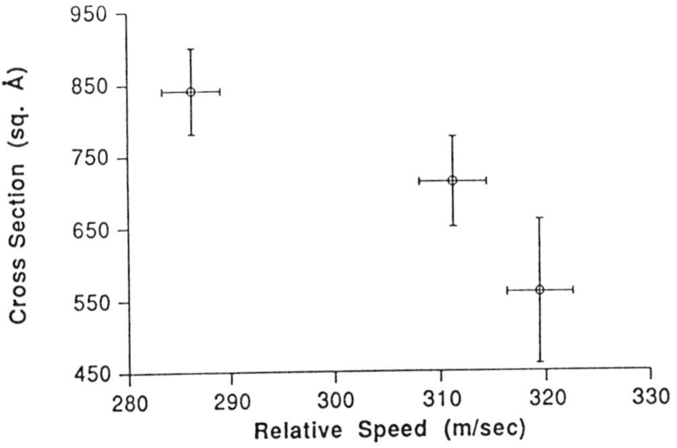

(a)

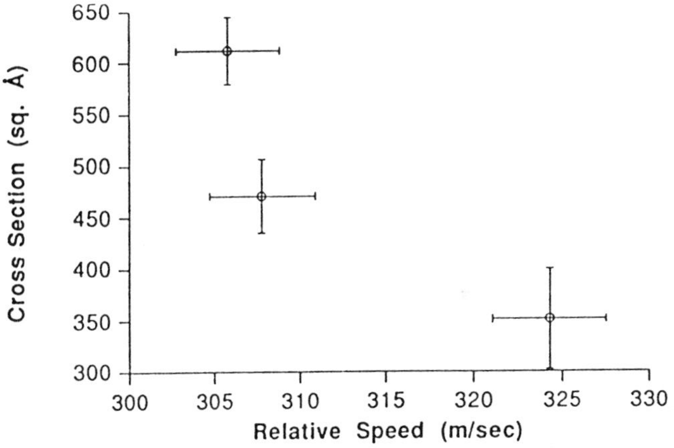

(b)

Figure 7.22 (a) Plot of dephasing cross-sections versus relative speed for 17104.57-cm^{-1} line. (b) Plot of dephasing cross-sections versus relative speed for 17148.4-cm^{-1} line.

the 17104.57-cm^{-1} line and in Figure 7.22(b) for the 17148.4-cm^{-1} line. However, the associated error with the cross-sections and the low number of points collected for each line makes extracting more definite conclusions impossible.

7.10 POPULATION TRANSFER STUDIES

Since large cross–sections were obtained for the dephasing rates and since elementary theory suggests that these cross-sections had a component dependent on velocity-changing collisions, an experiment that could measure the rate of velocity-changing collisions was the next one undertaken. A simple spectroscopic experiment that could extract the necessary information is a basic pump/dump experiment. Generally, such an experiment excites molecules at one frequency location and then allows a time before exciting molecules at a different location. Molecules that have shifted their position due to the appropriate dynamics will undergo stimulated emission, thereby decreasing the emitted signal. The magnitude of the decrease and the difference in frequency provides the quantitative information about population dynamics. Such an experiment requires the ability to prepare and observe a distribution of molecules with a well-known range of velocities; this range would have to be small compared to the total normal distribution of molecular velocities.

A. IDF Experiments

Such an experiment is elegantly simple hypothetically. First, a selective Gaussian pulse centered at frequency ω_a (Figure 7.23) excites a given population distribution. A time τ passes and this distribution broadens out into velocity space due to collisions. Now a second Gaussian pulse at ω_b excites a distribution centered at that frequency. If any of the excited molecules

---

Figure 7.23 (opposite) Concept behind IDF experiment. After first pulse the excited state population broadens out in frequency due to collisions. It also shrink in area due to T_1 processes and the hole in the ground state refills. After a delay τ a second pulse is given. If there were no molecules at frequency ω_b coming from ω_a then the second pulse would leave more molecules in the excited state (dashed line). However, the molecules present at ω_b at the time of the second pulse undergo stimulated emission, reducing signal.

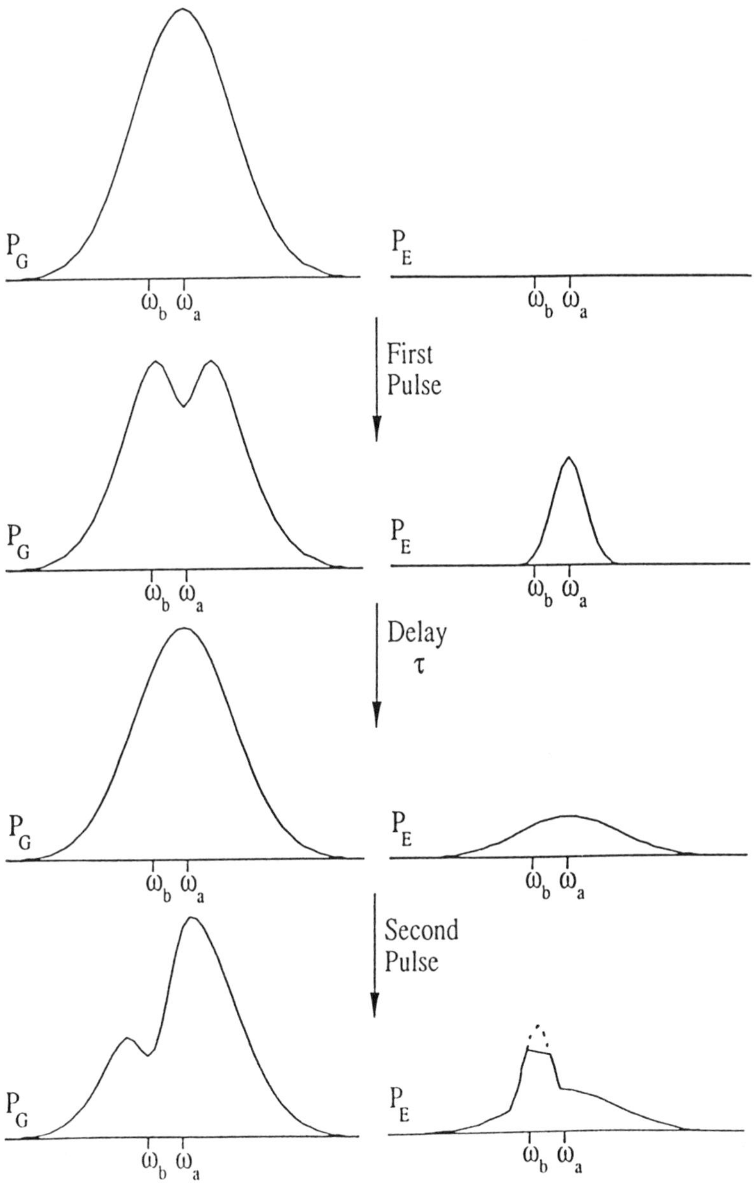

P_G
P_E
ω_b ω_a
First Pulse
Delay τ
Second Pulse

originally in (a) are now in (b) this second pulse will induce stimulated emission and return them to the ground state. This will produce a decrease in overall fluorescence when compared to the spectra gathered when both pulses are given individually. The magnitude of this effect should be small, since there are many different population depletion pathways in the gas and thus only relatively few molecules go from (a) to (b), so the background fluorescence created by the first pulse must be subtracted in order to observe this signal. Sweeping the delay τ performs a dynamic measurement and sweeping $\Delta\omega_{a-b}$ observes the magnitude of transfer between different velocity ranges. The lineshape obtained from this difference spectrum is thus directly indicative of the physical process involved. Since there are two frequencies used with no specific phase relationship between the pulse radiation, only the population will be sampled; these were called Incoherent Dual Frequency (IDF) experiments.

If the frequency difference $\Delta\omega_{a-b}$ is large compared to the pulse bandwidth, there is no initial overlap between the excited state populations created by the pulses. Velocity-changing collisions should create the overlap that will then decay due to the various T_1 processes present. Given the pressures involved, the difference spectrum minimum will occur at the point where most molecules have experienced a single collision that has transferred them from (a) to (b). However, in a survey of many different transitions in I_2, no

Table 7.4 Survey of Transitions Studied In IDF Experiments

Line Frequency	Assignment	$\sigma_{Fluorescence}$ (sq.Å)	Zero pressure lifetime
16956.7 cm^{-1}	R69 (11-0)	262 ± 5	1.22×10^{-6} sec^{-1}
16962.4 cm^{-1}	P49 (15-2)	307 ± 13	7.72×10^{-5} sec^{-1}
16965.5 cm^{-1}	P54 (15-2)	238 ± 6	1.18×10^{-6} sec^{-1}
	P57 (11-0)		
	R58 (11-0)		
17099.7 cm^{-1}	P85 (17-2)	233 ± 7	1.12×10^{-6} sec^{-1}
	P92 (17-2)		
17301.68 cm^{-1}	R131 (18-1)	402 ± 56	1.20×10^{-6} sec^{-1}
	R10 (16-1)		
17530.31 cm^{-1}	R72 (19-1)	242 ± 5	1.19×10^{-6} sec^{-1}

significant IDF difference spectrum was observed regularly (see Table 7.4 for the transitions studied). The IDF plots (Figure 7.24) show no appreciable signal from which we can measure quantitatively the number of molecules going between different frequency ranges. However, we can estimate a cross-section for a collisional event by noting a simultaneous minimum for all Δωs at a given pressure (Figure 7.25). If we plot the reciprocal of this time against gas density, we obtain the cross-section for a collision that transfers molecules from the original distribution excited by the first pulse and into all other ranges. From Figure 7.25, we measure a cross-section of approximately 50 sq. Å for the transition at 17099.7 cm^{-1}.

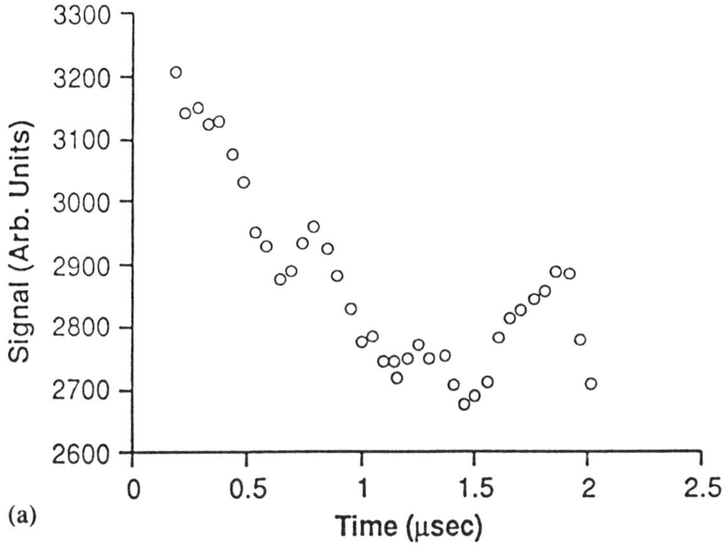

(a)

Figure 7.24 (a) IDF plot for 17099.7-cm^{-1} line for Δω = 10 MHz at 273°K. (b) IDF plot for 17099.7-cm^{-1} line for Δω = 20 MHz at 273°K. (c) IDF plot for 17099.7-cm^{-1} line for Δω = 30 MHz at 273°K. (d) IDF plot for 17099.7-cm^{-1} line for Δω = 40 MHz at 273°K. (e) IDF plot for 17099.7-cm^{-1} line for Δω = 10 MHz at 281°K. (f) IDF plot for 17099.7-cm^{-1} line for Δω = 20 MHz at 281°K. (g) IDF plot for 17099.7-cm^{-1} line for Δω = 30 MHz at 281°K. (h) IDF plot for 17099.7-cm^{-1} line for Δω = 40 MHz at 281°K. (i) IDF plot for 17099.7-cm^{-1} line for Δω = 10 MHz at 294°K. (j) IDF plot for 17099.7-cm^{-1} line for Δω = 30 MHz at 294°K. (k) IDF plot for 17099.7-cm^{-1} line for Δω = 40 MHz at 294°K.

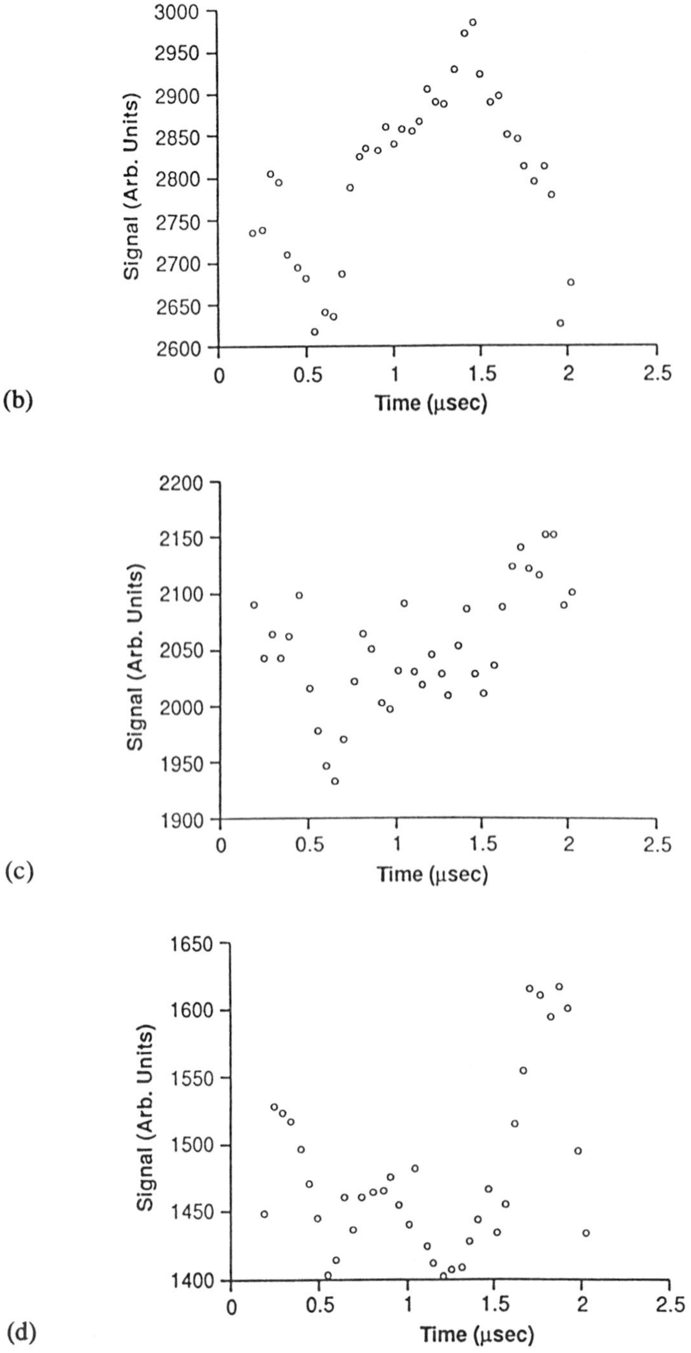

Figure 7.24 (cont.)

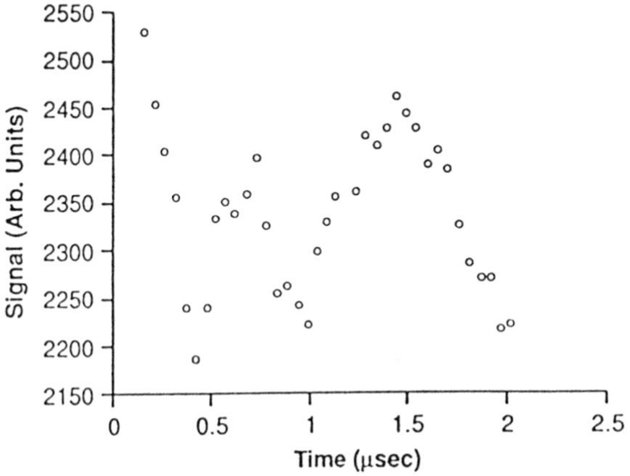
Signal (Arb. Units)
Time (µsec)

(e)

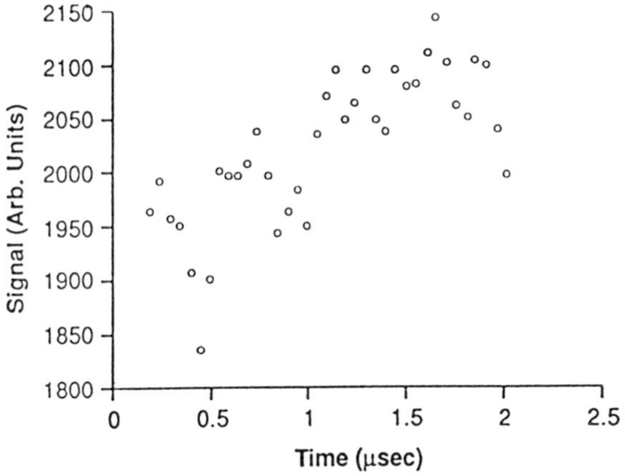
Signal (Arb. Units)
Time (µsec)

(f)

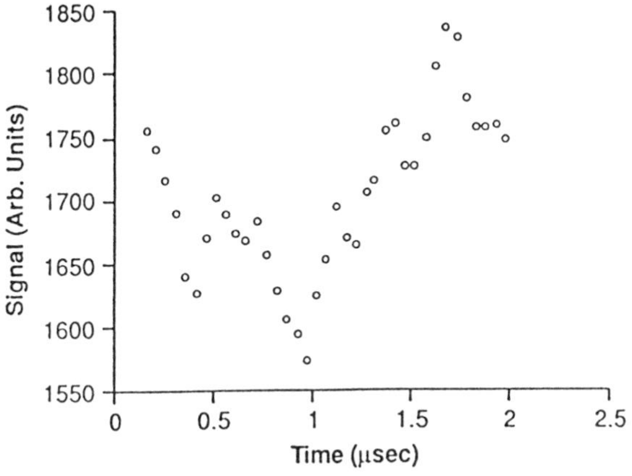
Signal (Arb. Units)
Time (µsec)

(g)

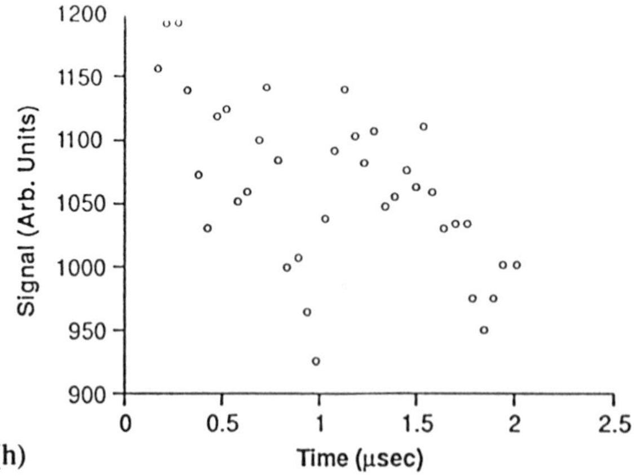

(h)

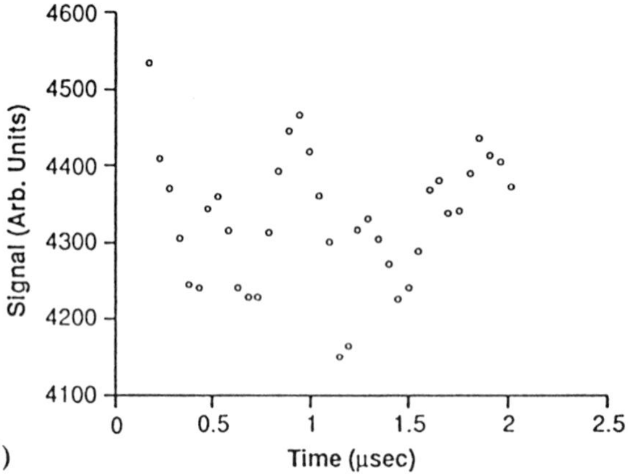

(i)

Figure 7.24 (cont.)

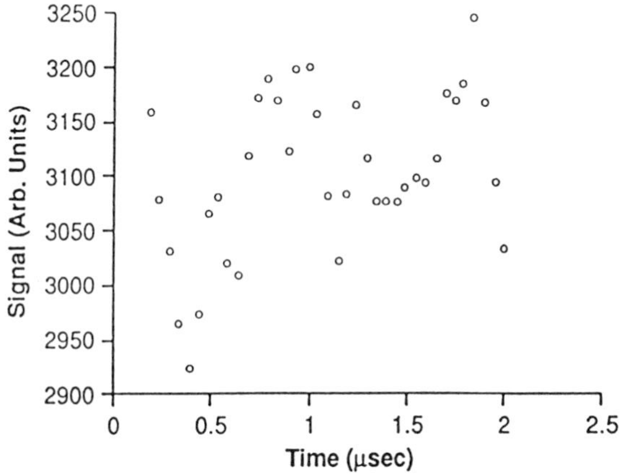
3250
3200
3150
3100
3050
3000
2950
2900
Signal (Arb. Units)
0
0.5
1
1.5
2
2.5
Time (μsec)

(j)

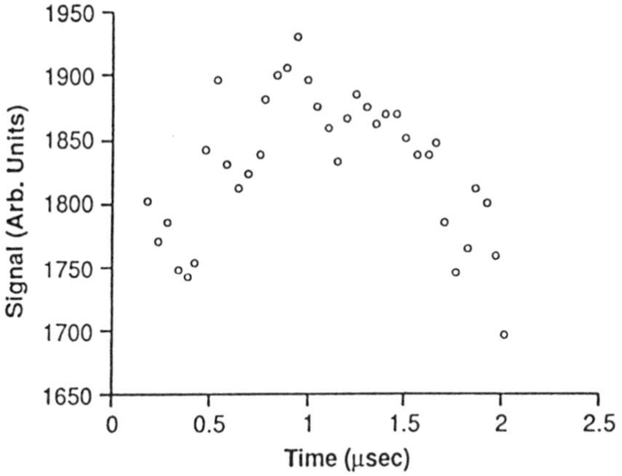
1950
1900
1850
1800
1750
1700
1650
Signal (Arb. Units)
0
0.5
1
1.5
2
2.5
Time (μsec)

(k)

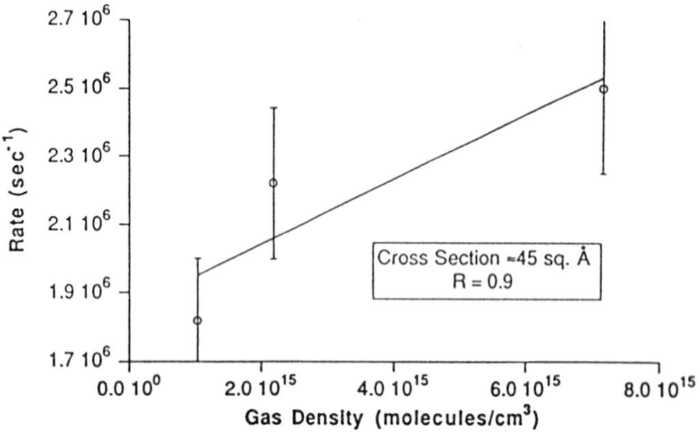

Figure 7.25 Plot of IDF signal minimum versus gas density for 17099.7-cm^{-1} line. *Source*: from [5] with permission.

B. Variable Bandwidth Experiments

These conclusions suggest the following single frequency experiment. A Gaussian laser pulse that excites a narrow region in velocity space is followed immediately by a second pulse that has a wider inversion profile (Figure 7.26). Only excited molecules that change their velocity by a small amount will be picked out by the wings of this second pulse. As pressure increases, so does the frequency of molecular collisions and therefore the effect should be increased. Therefore, if the collisions were small in magnitude of velocity change and fast in time and as pressure increased, the difference of the back-to-back fluorescence signal minus the sum of the signals created by the two pulses individually should be influenced proportionally. A plot of signal ratio versus gas density thus yields a lineshape unique to the physical process. Since these experiments sample different bandwidths in velocity space but have the same central frequency, they are called Variable Bandwidth Single Frequency (VBSF) experiments. They were conducted on a number of different transitions and at a number of different pressures. To ensure that the results were valid, the VBSF experiments consisted of two sets of experiments with a third companion study. The two VBSF were identical except for the inversion bandwidth of the narrow pulse, while the third had both pulses having the same inversion

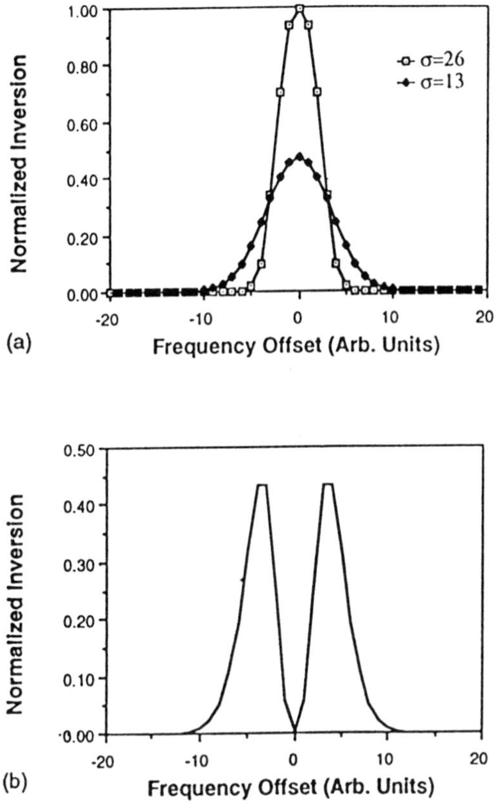

Figure 7.26 Concept behind VBSF experiment.

profile but having different base frequencies. This last run would detect if any molecules changed their velocity by a relatively large amount during the pulse interval. If the redistribution of translational energy is small then this last experiment should produce a pressure independent lineshape with the form of the cross-correlation function of the excited population distributions.

C. VBSF Results

VBSF results are given in Figures 7.27 and 7.28 for the case where the first pulse has a FWHM of 21 ns and the second pulse has a FWHM of 13 ns.

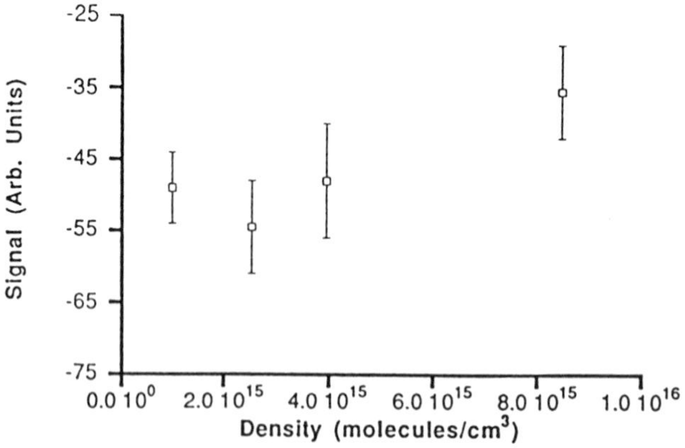

Figure 7.27 VBSF results for 17528.5-cm^{-1} line where first pulse FWHM = 21 nsec, second pulse FWHM = 13 nsec.

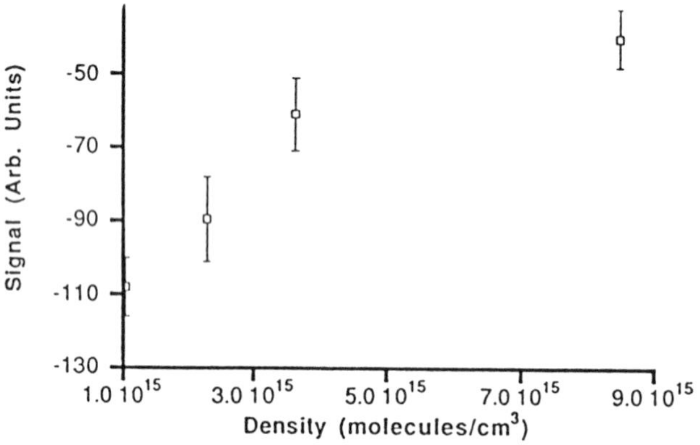

Figure 7.28 VBSF results for 17106.7-cm^{-1} line where first pulse FWHM = 21 nsec, second pulse FWHM = 13 nsec.

In Figures 7.29 and 7.30 spectra are given for the instance where the first pulse has FWHM of 26 ns and the second pulse has a FWHM of 13 ns. Of particular interest is the set involving both pulses having the same FWHM of 26 ns. Since they have identical bandwidth but different base frequencies, the pulse sequence creates an excited state distribution that is

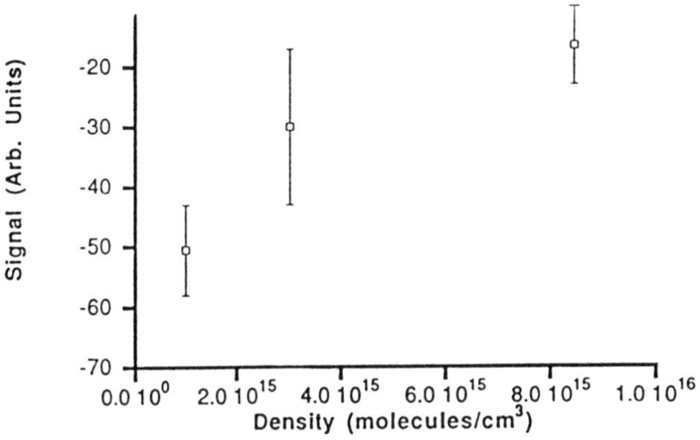

Figure 7.29 VBSF results for 17528.5-cm^{-1} line where first pulse FWHM = 26 nsec, second pulse FWHM = 13 nsec.

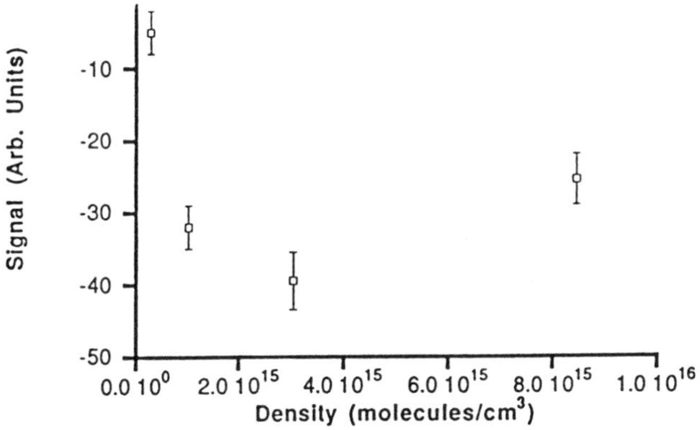

Figure 7.30 VBSF results for 17106.7-cm^{-1} line where first pulse FWHM = 26 nsec, second pulse FWHM = 13 nsec.

in frequency a cross-correlation function of the Gaussian pulse excitation profile, at least in the zero-pressure limit. Since this should also be nearly a Gaussian, this gives us the following method of detecting collisions that produce small velocity changes ($\Delta v \approx$ 1 m/sec). Molecules that relax back down to the ground state or are transferred out of the pulse bandwidth

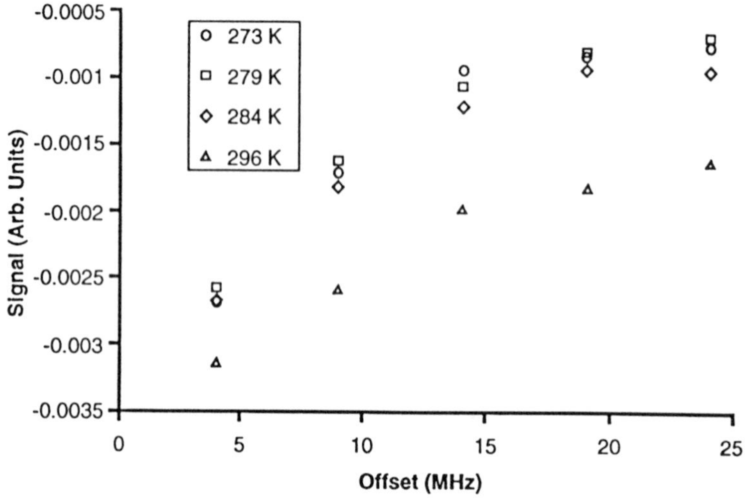

Figure 7.31 VBSF results for 17528.5-cm^{-1} line where both pulses FWHM = 26 nsec.

should not disturb the Gaussian distribution of molecules remaining in the excitation bandwidth. Only processes that rearrange the population within the pulse bandwidth will cause the function to be non-Gaussian. All that one needs to do is to increase the pressure in the gas cell and note any corresponding changes in the strength of the fluorescence signal as the frequency difference is swept.

To see this spectrum in practice, one subtracts the individual pulse spectra from the dual pulse spectrum (this difference in the spectrum's points are simply the integrals of the cross-correlation function). Representative spectra are shown in Figure 7.31 for the R114-(20-1) transition at 17528.5 cm^{-1} and Figure 7.32 for the R96 (15-1) transition at 17106.7 cm^{-1}. The salient features (distortion of the plot as gas pressure increases) indicate the aforementioned collisional redistribution of molecules within the pulse bandwidth. Since these pulses are phase-incoherent, population-changing collisions must cause this effect. However, while we cannot extract magnitudes, both this VBSF data and the IDF results suggest that velocity-changing collisions occur on the timescale of the pulse and in magnitude with predictions from classical mechanics.

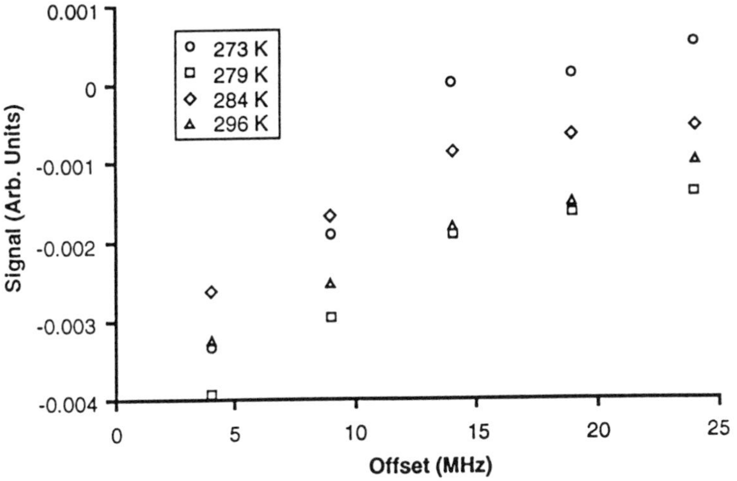

Figure 7.32 VBSF results for 17106.7-cm^{-1} line where both pulses FWHM = 26 nsec.

7.11 CONCLUSIONS

Polarization and population decays in molecular iodine were measured using variations on a probe theme of amplitude–crafted phase–specific frequency–shifted laser pulses. Two sets of experiments using phase–incoherent pulses examined the problem of population decay from a prepared velocity distribution in the excited state while a third set used a three-pulse photon echo technique to measure echo decays. The phase-incoherent sets each examined possible translational energy redistribution in fine detail; IDF experiments looked for relatively large transfers ($\Delta v > 6$ m/sec) while the VBSF set observed velocity changes in the range of $\Delta v \approx 1$–2 m/sec and on the time-scale of the laser pulses. Only in the VBSF experiments was population decay observed. The size of this population transfer is in agreement with forecasts by simple statistical mechanical arguments. Population decay in these experiments occurs at relatively high pressures (> 200 mtorr). The cross-sections determined for this process are similar in magnitude to those measured for the phase-interrupting cross-section; this suggests that velocity-changing collisions of this scale are a significant contributor to echo decay. This is in agreement with theory that states that the probability of coherence transfer is extremely small during collisions that transfer this amount of kinetic energy. The polarization decays show a correlation between their magnitude and the initial ground state vibrational level of the excited molecule. In two cases, the echo decay cross-sections vary at different frequency positions within the transition bandwidth, a previously unseen effect. In these instances the differences in echo cross-section magnitudes are greater than the changes in relative speed, indicating the cross-section is speed dependent. However, the error associated with the determined echo decay rates and the small number of measurements make an exact correlation between cross-section and relative speed impossible. The resulting conclusions are that the rate and distribution of population transfer is in line with that expected from thermodynamics and that the population decay rate, relative speed and quantum state specifications of the molecule affect the rate of polarization decay. This last point deserves further investigation.

ACKNOWLEDGMENTS

This work was supported by the National Science Foundation under grant CHE-9101544.

REFERENCES

1. J. O. Hirschfelder, C. S. Curtis and R. B. Bird, *Molecular Theory of Gases and Liquids*, John Wiley, New York, 1954, p. 148.
2. C. S. Parmenter and M. Seaver, *J. Chem. Phys.*, **70(12)**, 5458 (1981).
3. W. S. Warren and A. H. Zewail, *J. Chem. Phys.*, **78(5)**, 2279 (1982).
4. F. Spano and W. S. Warren, *J. Chem. Phys.*, **89**, 5492 (1988).
5. M. A. Banash and W. S. Warren, *J. Chem. Phys.*, (submitted).
6. P. R. Berman, T. W. Mossberg and S. R. Hartmann, *Phys. Rev. A.*, **25(5)**, 2550 (1982); P. R. Berman, *Phys. Rep.*, **43**, 101 (1978).
7. L. Schiff, *Quantum Mechanics*, McGraw-Hill, New York, 1968, p. 117.
8. D. A. McQuarrie, *Statistical Mechanics*, Harper and Row, New York, 1976, p. 358.
9. J. Tellinghuisen, *J. Chem. Phys.*, **57(6)**, 2397 (1972).
10. J. Tellinghuisen, *J. Chem. Phys.*, **82(9)**, 4012 (1985).
11. R. J. LeRoy, *J. Chem. Phys.*, **52(5)**, 2683 (1970).
12. S. Gernsenkorn and P. Luc, *J. Physique*, **46**, 355 (1985).
13. M. Broyer, J. Vigue and J. C. Lehmann, *J. Chem. Phys.*, **64(11)**, 4793 (1976).
14. G. A. Capelle and H. P. Broida, *J. Chem. Phys.*, **58(10)**, 4212 (1973).
15. A. Chutjian, *J. Chem. Phys.*, **51(12)**, 5414 (1969).
16. S. R. Jeyes, A. J. McCaffery and M. D. Rowe, *Mol. Phys.*, **36(3)**, 845 (1978).

17. R. Bacis, M. Broyer, S. Churassy, J. Vergès and J. Vigué, *J. Chem. Phys.*, **73(6)**, 2641 (1980).
18. C. H. Townes, and A. L. Schawlow, *Microwave Spectroscopy*, McGraw-Hill, New York, 1955, p. 221.
19. J. Vigué, M. Boyer and J. C. Lehmann, *J. Physique*, **42**, 949 (1981).
20. J. L. Bates, Doctoral Thesis, Princeton University (1991).
21. A. N. Nesmeyanov, *Vapor Pressure of the Elements*, Academic Press, NY, 1963, p. 370.
22. J. I. Steinfeld, *J. Phys. Chem. Ref. Data*, **13(2)**, 445 (1984).
23. K. J. Johnson, *Numerical Methods in Chemistry*, Marcel Dekker, New York, 1980, p. 279.
24. P. R. Bevington, in *Data Reduction and Error Analysis in the Physical Sciences, McGraw-Hill, New York, 1969, p.104.*
25. C. H. Townes, and A. L. Schawlow, in *Data Reduction and Error Analysis in the Physical Sciences*, McGraw-Hill, New York, 1969, p. 152.
26. S. L. Coy, *J. Chem. Phys.*, **73(11)**, 5531 (1980).
27. R. P. Runyon and A. Haber, *Fundamentals of Behavioral Statistics*, Random House, New York, 1988, p. 359.

8

Design of Femtosecond Optical Pulse Sequences to Control Photochemical Products

David J. Tannor

University of Notre Dame, Notre Dame, Indiana

8.1 INTRODUCTION

There have been great strides in the generation of ultrashort (femtosecond) laser pulses and their use to both probe and control molecular dynamics [1]. Zewail et al. have used pairs of femtosecond pulses to probe molecular dissociation and predissociation in real time [2]. By controlling the time delay between pulses they have been able to monitor the evolution of the wave-packet on the excited electronic state potential energy surface or surfaces. In addition, Zewail et al. have demonstrated, to a limited extent, control of the fragment species by changing the time delay between pulses [3]. Scherer et al. have succeeded in developing a method to control the relative phase between a pair of femtosecond pulses, in addition to the time delay, thereby fully exploiting the coherence properties of laser light. Their initial application was to I_2, where the phase of the second pulse relative to the first pulse determines whether absorption or stimulated emission is induced [4, 5]. These workers have aptly called their technique *wavepacket interferometry*, and have shown how the second pulse serves as a probe of the recurrence time and the phase of the wavepacket which returns to the Franck-Condon region. Great progress has also been made in the amplitude and phase shaping of individual pulses. Heritage [6] and Nelson [7] have developed pulse shaping techniques based on gratings and spatial filters. Complementary pulse shaping techniques, using electrooptical modulation on the picosecond time scale, have been developed by Warren [8]. Shank has generated pulses of 6 fs duration and these pulses have been applied by Mathies et al. to probing photoisomerization in bacteriorhodopsin [9]. Finally, the advent of the titanium-sapphire laser promises to make the generation of femtosecond pulses much simpler than has previously been possible, thereby opening the door to sophisticated programs of molecular probing and control [10].

Given the great experimental progress the inexorable question is: How is it possible to take advantage of the unique properties of lasers—amplitude pulse shaping and sequencing, high power, frequency tunability and phase coherence—to bring about selective and energetically efficient photochemical reactions? Many intuitive approaches to laser selective chemistry have been tried in the past twenty years. Most of these approaches have focused on depositing energy in a sustained manner, using monochromatic radiation, into a particular state or mode of the molecule. Virtually all such schemes have failed due to rapid intramolecular energy redistribution.

Several new approaches have emerged in which laser pulse shapes and pulse sequences, designed to achieve a desired chemical objective, are calculated systematically. Brumer and Shapiro developed a scheme that use interference between cw lasers, although they have proposed some picosecond pulse pair experiments as well [11]. Tannor et al. developed the idea of breaking a desired chemical bond with a sequence of two ultrashort laser pulses [12, 13]. They assumed a triatomic molecule ABC which interacts with two laser pulses where the control parameter is the delay time between those two pulses. By changing the delay time, they could break the AB bond, getting A + BC final product, or the BC bond, leading to AB + C final product.

In 1985, Tannor and Rice [12] (see also [13] and [14]) formulated the problem of the search for *optimal* optical pulses, subject to suitable constraints, as a problem in the calculus of variations. Their formulation was based on a perturbation theory expression for the time-dependent wavepacket amplitude and led to a nonlinear integral equation for the optimal pulse(s). A formal and computational breakthrough was achieved in 1988 by Rabitz et al. [15–18] who recognized the utility of optimal control techniques, used widely in engineering, for the calculation of optimal pulses. A variety of extensions of this approach have since been carried out by Manz et al. [19, 20]. Optimal control theory (OCT) is nothing more and nothing less than the extension of the classical calculus of variations to problems with differential equations constraints. In the context of molecular control, this differential equation is the time-dependent Schrödinger equation (TDSE), which is introduced via a Lagrange multiplier into the variational equations. The OCT methodology, in addition to providing a simplified algorithm for obtaining optimal fields numerically, has no difficulty in handling fields of arbitrary strength. This results in a significant generalization of the perturbative formalism not only because greater chemical yields can be obtained with strong fields but also because many new mechanisms are available to strong fields that are not accessible in the weak-field regime. Many of the components of the TDSE/OCT formalism developed by Rabitz et al. were discovered independently by Kosloff et al. [21] in the context of control using two electronic states. The approach of Kosloff et al. was developed intuitively and as such is extremely valuable pedagogically as it provides a heuristic explanation for the optimal control methods. A fairly complete discussion of the method of Kosloff et al. will be provided here because of the physical insight it provides.

It is interesting to note that after the initial explosion of interest in being able to perform optimizations using strong fields, there has been renewed interest in the optimal solutions in the perturbative regime. Some interesting analytical results have been obtained in this regime [23–26]. Using a perturbative approach, Demiralp and Rabitz have proven the existence of multiple solutions of the optimization problem. Yan et al. have developed a method to find the globally optimal solution, within first-order perturbation theory. In the laboratory, the yield from a weak-field will obviously be reduced but the action of the weak-field pulse sequences should be free of many of the unwanted side effects of strong fields, including ionization and nonlinear dependence on the orientation of the molecule.

The first part of this chapter focuses on population transfer between electronic states proceeding via transition dipole coupling. This process is the precursor to using different electronic states to control the nature of chemical products. The central physical quantity which governs population transfer is the instantaneous transition dipole moment. This object represents a generalization of the Einstein *B* coefficient to include time dependence and, in particular, phase. Several illustrations are given of the use of the instantaneous transition dipole moment to achieve desired objectives, including monotonic population transfer to the excited electronic state and vibrational heating or cooling without population transfer to the excited electronic state.

This second part of this chapter introduces a model triatomic ABC (masses H,H,D), which has been previously studied [12, 13, 21]. This system, with two degrees of freedom, is the simplest paradigm for control of chemical products. The Tannor-Rice *pump-dump* scheme is reviewed briefly to illustrate that some measure of control is indeed possible and to set the stage for the type of mechanisms that may emerge from the optimization procedure. This same example is then used to illustrate the formulation of the variational problem. The "seat of the pants" optimization procedure, developed by Kosloff et al. is then presented [21]. Finally, the optimal control formalism is applied to the variational problem, leading to an expression for the optimal field which requires the solution to a set of coupled partial differential equations. The numerical solution of these coupled equations is then shown to involve many of the components from the "seat of the pants" procedure! The pulses that emerge from the optimization are analyzed using the Husimi transformation technique to learn about their mechanism. To paraphrase the Talmud: a pulse which is not interpreted is like a letter which is not opened.

Interspersed throughout this chapter is some discussion of the mathematical properties of the optimal pulses, including the existence of an optimization "Hamiltonian", a scalar quantity which is nominally time dependent, but which is independent of time for evolution under the influence of the optimal control. Some speculation is offered regarding the number of locally optimal pulses and the properties of the set of such pulses (e.g., orthogonality). Finally, the question of the limits on quantum controllability is touched on briefly.

8.2 THE INSTANTANEOUS TRANSITION DIPOLE MOMENT

The time-dependent Schrödinger equation for a two electronic state system with transition dipole moment μ can be written as:

$$i\hbar \frac{\partial}{\partial t}\begin{pmatrix} \psi_g(t) \\ \psi_e(t) \end{pmatrix} = \begin{pmatrix} H_g & \mu\varepsilon^*(t) \\ \mu\varepsilon(t) & H_e \end{pmatrix}\begin{pmatrix} \psi_g(t) \\ \psi_e(t) \end{pmatrix} \tag{8.1}$$

The subscripts g and e refer to the ground and excited electronic state indices, respectively. $H_{g/e}$ refers to the Born-Oppenheimer Hamiltonian for the ground/ excited electronic state, respectively. The two electronic states are coupled by the transition dipole moment, μ, which interacts with the electric field $\varepsilon(t)$ associated with the laser pulse. Here and throughout the rest of the chapter I assume that μ is real. Complex values of the field are considered admissible, in keeping with the spirit of the rotating wave approximation in which the complex conjugate of the active part of the field is neglected on the grounds that it is off resonant. $\psi_{g/e}$ is the wavefunction (wavepacket) associated with the ground/excited Born-Oppenheimer potential energy surface. The importance of the instantaneous transition dipole moment becomes immediately apparent when considering the change in excited electronic state population [21, 22]:

$$\frac{dP_e}{dt} = \frac{d\langle\psi_e|\psi_e\rangle}{dt} = \frac{-2}{\hbar} Im[\langle\psi_e|\mu|\psi_g\rangle \cdot \varepsilon(t) \tag{8.2}$$

where $\psi_{g/e}$ is the wavefunction of the ground/excited electronic state, μ is the dipole moment operator, and $\varepsilon(t)$ is the instantaneous electric field. I will refer to the quantity $\mu_{eg} = \langle\psi_e|\mu|\psi_g\rangle$ as the instantaneous (complex) transition dipole moment; in other contexts it is referred to as (one component of) the induced polarization. I shall often call it simply the instantaneous dipole moment for short. It is clear that by changing the phase of $\varepsilon(t)$ it is possible to adjust the sign of dP_e/dt (i.e., to control whether population is transferred from the ground to the excited state, the excited to the ground state, or is 'locked' with no population transfer). Of course, all this presupposes experimental control of the phase of the light on a vibrational time scale, something that has only just become possible. The instantaneous dipole moment can be viewed as a generalization of the Einstein B coefficient; I elaborate on this perspective below. First I provide some illustrations of the use of the phase of $\varepsilon(t)$ to achieve specific objectives.

A. Electronic State Population Transfer Using a Series of Phase Locked Pulses

In studies of optimal control of chemical reactions using shaped laser pulses, several groups have obtained intriguing periodic pulse sequences [26, 27]. In a number of cases, the effect of these periodic pulse sequences has been to give a striking *staircase* pattern for the objective as a function of time (Figure 8.1). The staircase is characterized by the following features:

1. There are discontinuous "steps" in the excited state population (or more generally, in the objective function) for every period of the oscillator, with flat plateau regions in between; and
2. The height of the steps grows quadratically with time.

On one level, this behavior can be understood very easily. The periodic steps come during the interval when the field is on. Although the intensity of each of the pulses in the field is constant the steps grow quadratically because of an interference effect. The first pulse transfers a certain amount of amplitude, A_1, to the excited electronic state, giving an excited state population of $P = |A_1|^2$. The second pulse transfers amplitude $A_2 = A_1$ giving total amplitude $A = 2A_1$ and $P = 4|A_1|^2$. The third pulse transfers amplitude $A_3 = A_1$ giving total amplitude $A = 3A_1$ and $P = 9|A_1|^2$. While the amplitude A grows linearly in time the population P grows quadratically. This behavior continues, within first-order perturbation theory, for all time. However, the exact solution to the two surface time-dependent Schrödinger equation reveals that as the excited state population begins to become significant (saturation) P will begin to rise less quickly and will eventually turn over. In other words, the global pattern of population change is a combination of the staircase behavior in Figure 8.1(b) superimposed on a sinusoidal rise and fall of the excited state population, characteristic of the Rabi solution for two level systems.

A different perspective on the staircase pattern is obtained by considering the instantaneous dipole moment. The steps in Figure 8.1(b) come once per period because that is when the ground and excited state wavepackets overlap; these are the only windows of time for which μ_{eg} is nonzero. During the first few periods, when there is little or no amplitude upstairs $\psi_e \approx 0$; hence $\mu_{eg} \approx 0$ and the steps are smaller. After a quarter of a Rabi cycle ($\pi/2$ pulse) 50% of the population is on the excited electronic state; it is during this interval that μ_{eg} reaches its maximum value. ($1 = .5 + .5$ is

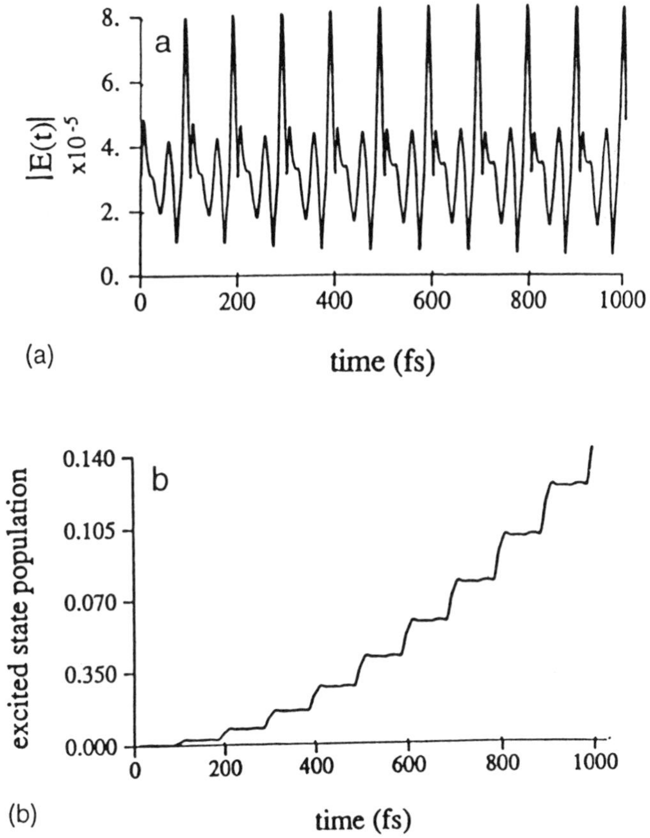

Figure 8.1 a) Electric field versus time, the result of an optimization designed to achieve maximum population inversion in a system consisting of two displaced harmonic oscillators in one dimension. b) Excited state population versus time using the pulse from a). *Source*: from [27] with permission.

the best way of partitioning the number 1 to maximize the product of its terms). This is the most favorable time for absorption, consistent with the fact that in this interval the slope in the plot of P versus t is maximum. After one half of a Rabi cycle (π pulse), with 100% of the population upstairs, μ_{eg} is again close to zero and the rate of absorption goes to zero. An implicit assumption throughout this discussion is that a classical treatment of the radiation is valid and thus there is a sea of photons (i.e., more available than can possibly be absorbed). Thus, as the dipole is increased, the rate of absorption of photons increases commensurately.

A second example, which highlights the use of the phase of the electric field to control the *sign* of dP_e/dt is provided by the experiments of Scherer et al. [4, 5]. These workers have developed an elegant experimental method to prepare a sequence of two pulses, each with about 30 fs half-width half-max, with a precise phase relationship between the pulses at a given locking frequency. They have applied this pulse sequence to the $B \leftarrow X$ electronic transition in molecular iodine and collected the total fluorescence from the excited state. What was found was an interference contribution to the total fluorescence if the pulses are spaced by some multiple of the excited state vibrational period (Figure 8.2(a)). Moreover, the phase of the second pulse can be used to control whether the interference is constructive or destructive. The simplest explanation of these results is as follows. The first pulse prepares a wavepacket on the excited electronic state. After a variable time delay, the second pulse prepares a second wavepacket on the excited state. If the time delay is a multiple of the excited state vibrational period the wavepackets interfere (Figure 8.2(b)). The phase of the second pulse relative to the first controls the phase of the second wavepacket relative to the first and hence whether there is constructive or destructive interference between the packets.

An alternative perspective on these experiments is obtained using the formula for dP_e/dt. The first pulse prepares a moving wavepacket, and hence a μ_{eg} which is time-dependent with the period of the excited state iodine. If the second pulse is equal to $-i\langle\psi_g|\mu|\psi_e\rangle \cdot C(t)$ then absorption takes place; if the phase is given by $i\langle\psi_g|\mu|\psi_e\rangle \cdot C(t)$ stimulated emission takes place and excited state population decreases (C(t) is any real function of time). In other words, through the phase of the light one may control absorption versus stimulated emission because the generalized $B_{ge} \neq B_{eg}$! Thus one may

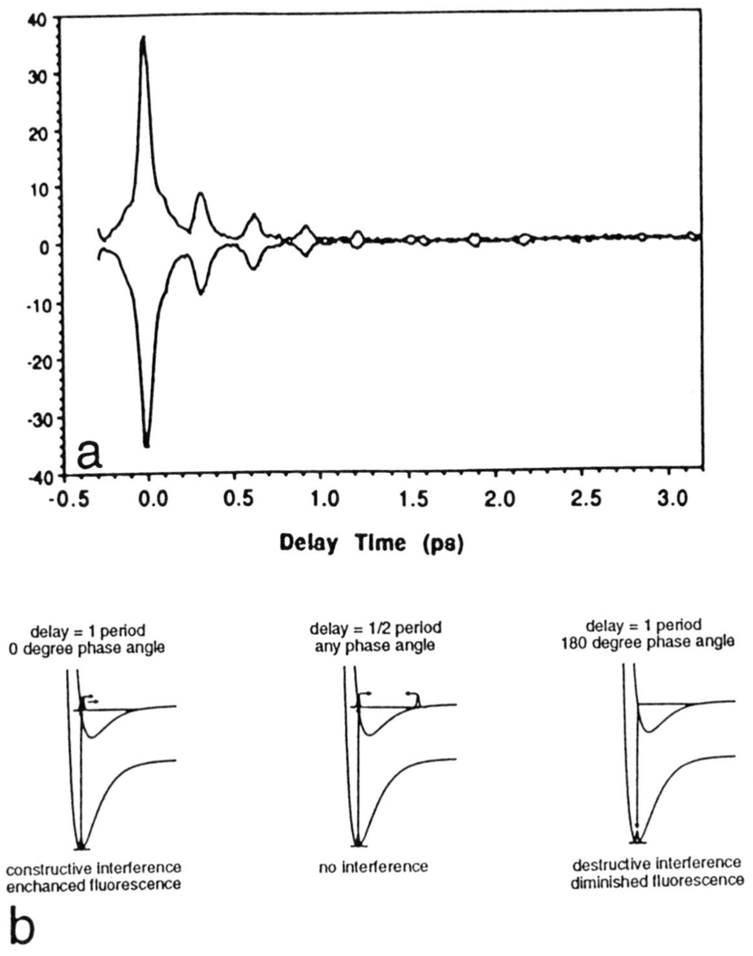

Figure 8.2 a) Interference contribution to the total fluorescence from I_2 as a function of time delay between a pair of phase locked pulses. The upper curve is for in phase pulses, producing constructive interference of amplitude on the excited state, and the lower curve is for out of phase pulses, producing destructive interference of amplitude on the excited state. b) Schematic diagram of the excited state wavepacket motion, showing why an interference signal can be generated only at multiples of an excited state vibrational period. An alternative perspective on this experiment is that the second pulse produces either absorption or stimulated emission, depending on the relative phase between the second pulse and the instantaneous dipole. *Source*: from [5] with permission.

conceive of lasing without population inversion and population inversion without lasing through the action of phase locked laser pulse sequences.

B. Heating and Cooling of Internal Degrees of Freedom Without Electronic Population Transfer

It is possible to view Equation 8.2 as a generalization of the third of the optical Bloch equations; this perspective allows for the use of the Feynman-Vernon-Hellwarth (FVH) geometrical picture for the optical Bloch equations. As a result, it is natural to consider (geometrically) a variety of generalizations of coherence phenomena in two-level systems to the case where there is a wavepacket on each of two electronic states potential energy surfaces.

One example is given by photon echoes which will be discussed in the next section. Another very interesting example is that of photon locking which provides an elegant solution to the following problem [28], originally posed by Nelson et al. [7]: Is it possible to design a pulse sequence, using an Impulsive Stimulated Raman Mechanism, to give large amplitude vibrational motion on the ground electronic state without creating significant amounts of excited electronic state population? The problem with excited state population is two-fold: it complicates the interpretation of the experimental results and it is often a precursor to dissociation or ionization. By applying the FVH picture to two electronic states, each with its own evolving wavepacket, it was realized that there must be a generalization of *photon locking*: the design of a field with specific phase characteristics to keep the population on each surface fixed while the field in on. For a simple, isolated, two-level system this *population locking* is hardly more than a curiosity: if the probability for being in each electronic state is fixed there is little else that matters. However, if the two electronic states have structure (i.e., coordinate dependence) the phase can be chosen to keep the populations locked and the sign and amplitude of the field can be used to vary some other property of the system. This leads to the following solution to Nelson's problem:

1. With a first pulse prepare some small (e.g., 1%) of population in the excited electronic state; and
2. Subsequently, adjust the phase to keep the excited state population locked, but use the sign and amplitude of the field to monotonically increase the vibrational energy on the ground state.

In a sense, the excited electronic state is being used as an optical lever, or catalyst, without itself being populated. Numerical experiments show that this works beautifully [28], both for weak-fields and strong fields (Figure 8.3); for strong fields the heating is simply faster.

This work has been extended to achieve *cooling* of a thermal vibrational ensemble, again using an approach based on photon locking [29]. The idea is to cool a thermal vibrational distribution down to 0°K using the excited electronic state as an optical lever but without transferring population to the excited state. Here, the density matrix formalism must be used to describe correctly the absence of coherence among the states in the initial distribution.

Fourier analysis of the heating and cooling pulse sequences show a fascinating result: the peaks in frequency fall at positions *in between* the excited electronic state vibrational resonances (Figure 8.3(d)). This is consistent with an underlying Stokes (anti-Stokes) resonance Raman process for heating (cooling), but by tuning between excited state vibrational resonances, excited state population is avoided. This suggests that it may be profitable to consider continuous wave (CW) analogs of such a scheme. However, it should be recalled that the phase of the pulse sequence (actually the relative phase of the latter portion of the pulse sequence to the first pulse) is critical in obtaining zero population transfer; furthermore, the sign of the phase determines whether heating or cooling is achieved. Thus any CW analogs may require sensitive phase control.

With the experimental capability of phase-locked pulse sequences now possible, it is hoped that this scheme will be tested in the laboratory before long.

C. Generalization of the Einstein Coefficients

The instantaneous transition dipole moment, μ_{eg} can be viewed as a generalization of the Einstein B coefficient. As is known from the Einstein theory for blackbody radiation, B_{eg} determines the rate of absorption, B_{ge} determines the rate of emission, and A_{ge} determines the rate of spontaneous emission. However, in the Einstein treatment, which deals with two-level systems, ψ_g and ψ_e are single complex numbers with no spatial dependence and the phase is taken as random; as a result, the magnitude of the instantaneous dipole is time-independent. In reality, ψ_g and ψ_e are moving wavepackets on the ground and excited state potential energy surfaces and as a result both

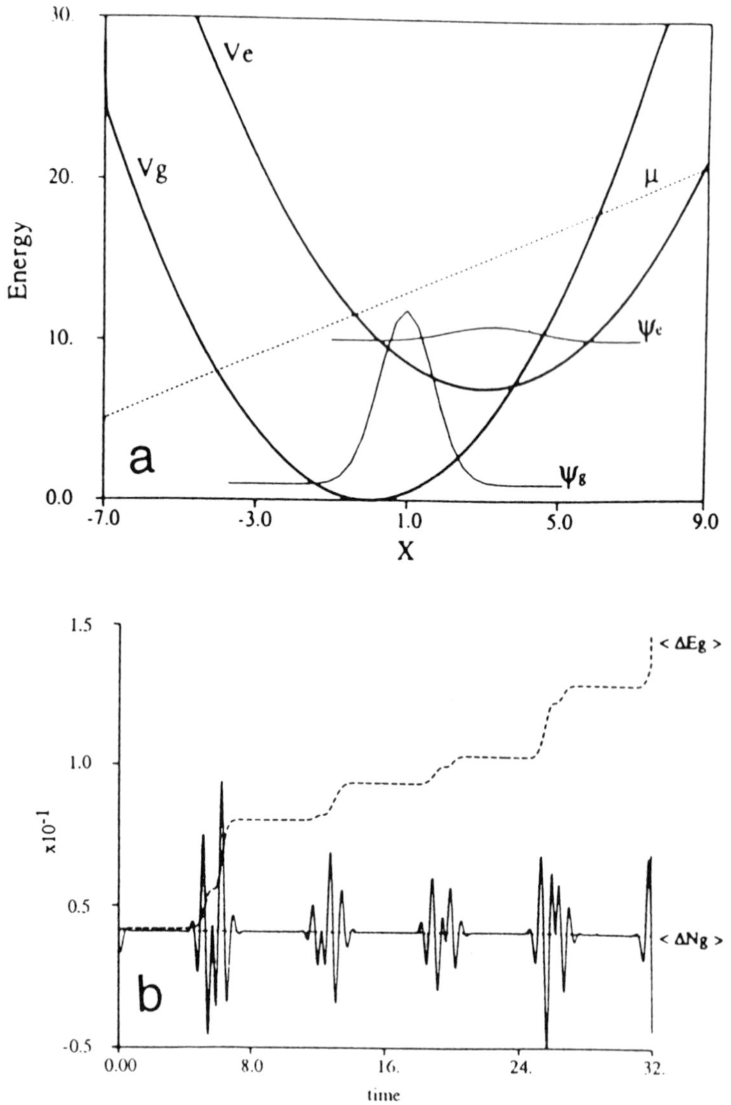

Figure 8.3 a) Potential energy curves used in the study of photon locking vibrational heating. The frequency of the ground and excited surfaces is 1.0 and 0.8, respectively. The excited state curve is shifted by 7.0 units of energy and 3.0 units of distance which leads to a vertical energy distance of 10.6 units. The dipole operator has a slope of 1. The ground and excited absolute values of the wavefunction after the first exciting pulse are shown (not to scale).

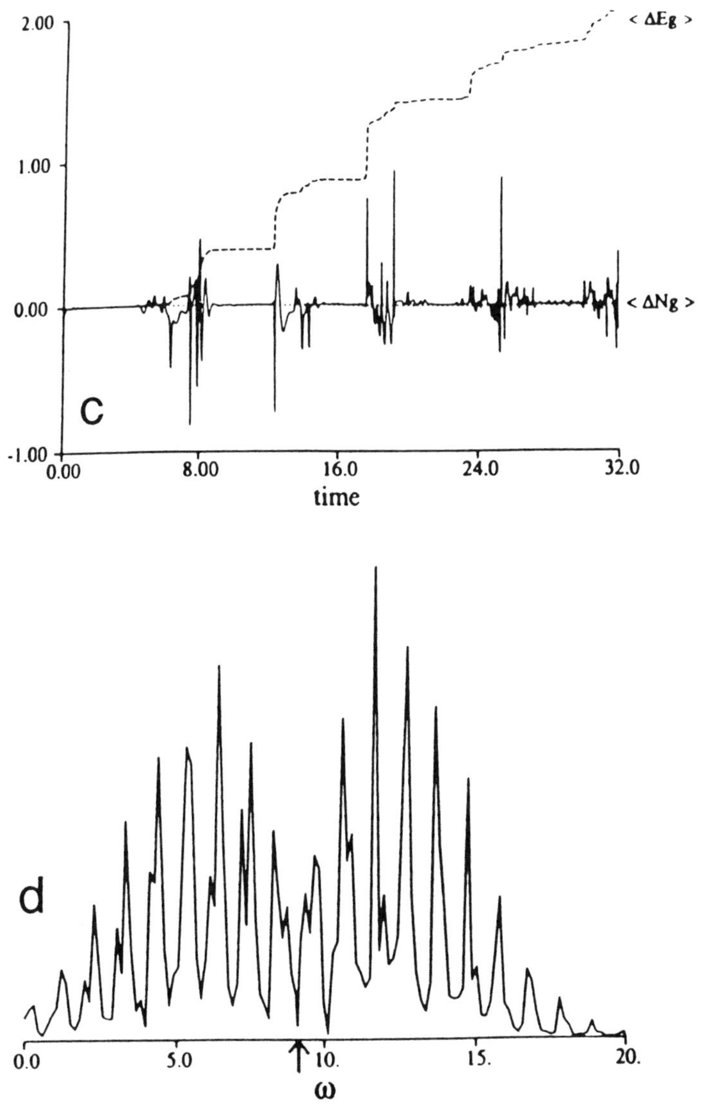

Figure 8.3 (cont.) b) Change in population on the ground state <δN_g> (dotted line), change in ground-state energy <δE_g> (dashed line), and the real part of the field ε (solid line) as a function of time for a weak-field excitation. c) Same as b) but for a strong field excitation. The pulse sequence increases the ground-state energy by a factor of 4 in approximately five vibrational periods. d) The power spectrum of the pulse in b). The arrow indicates the frequency of the vertical transition. *Source*: from [28] with permission.

the magnitude and phase of the instantaneous dipole are functions of time. The generalized μ_{eg} and μ_{ge} are not equal but are complex conjugates. As I have shown above, if the phase of the light is controlled it can interact completely differently with μ_{eg} and μ_{ge} (e.g., resonantly with the former and nonresonantly with the latter to produce absorption, or vice versa to produce stimulated emission).

Until now, I have focused on the analog of the Einstein *B* coefficient, with μ_{eg} and μ_{ge} playing the role of the *B* coefficient for absorption and stimulated emission, respectively. I now comment briefly on the generalization of the Einstein *A* coefficient. A good example of the importance of the generalized *A* coefficient is the photon echo. A photon echo is a form of spontaneous emission in the sense that it occurs in the system at a time when there are no external fields. Tannor and Rice [30] developed a generalized picture of photon echoes, valid for two wavepackets on two electronic states, in which they were able to relate the echo with the time of a favorable overlap of ground and excited state wavepackets (Figure 8.4). Yet this raises the following question: if the emission is truly spontaneous, why should one instant in time be more favorable for emission than any other? How does the excited state amplitude know that there is ground state amplitude directly underneath, and the time is right to emit?

Equation 8.2 makes it clear why spontaneous emission is enhanced when the ground and excited state wavepackets overlap: the rate of spontaneous emission is governed by the same quantity μ_{ge}. What is usually emphasized in the echo phenomenon is the collective, macroscopic aspect—that an ensemble of systems rephases. However, without significant overlap between the ground and the excited electronic state wavepackets in the individual molecules there can be no macroscopic transition dipole moment and no echo can be observed.

This older concept of the Einstein *A* and *B* coefficients as constants can give rise to paradoxes and misconceptions about optical coherence experiments. For example, the usual thinking is that the condition for lasing is when there is a population inversion (i.e., more than 50% population in the excited electronic state). This criterion comes from a random phase assumption: if the phase of the excited and ground state amplitude is random the probability of absorption versus emission will be statistical; thus if excited state population exceeds that in the ground state, emission will exceed absorption. The random phase assumption has been valid for the vast

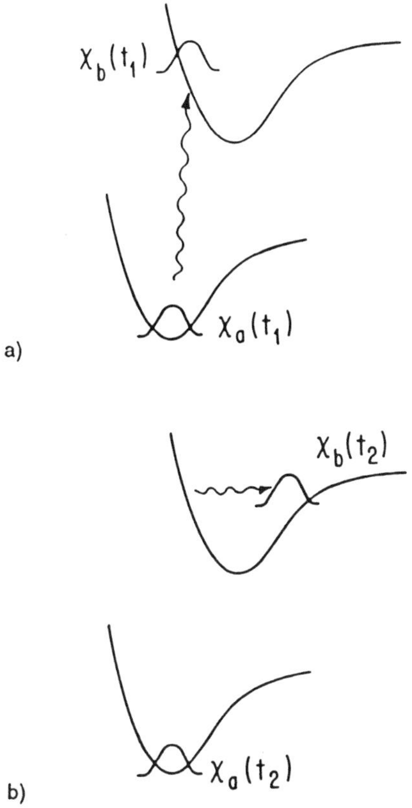

Figure 8.4 a) Schematic diagram of the ground and excited electronic state Born-Oppenheimer curves for photon echoes. The $\pi/2$ pulse moves half of the initial amplitude, $\chi_a(0)$ from surface a to surface b. After the pulse the nuclear wavefunctions on the ground state and excited state curves are denoted $\chi_a(t_1)$ and $\chi_b(t_1)$. b) Wave-packet evolution of χ_b. Motion of the wavepacket causes the overlap of the ground state wavefunction to decay, resulting in "free induction decay". χ_a remains in place in coordinate space. c) The π-pulse exchanges the amplitude of surface a with surface b. d) Wavepacket evolution proceeds on both surface a and surface b. When the two wavepackets overlap as some later time a photon echo results. *Source*: from [30] with permission.

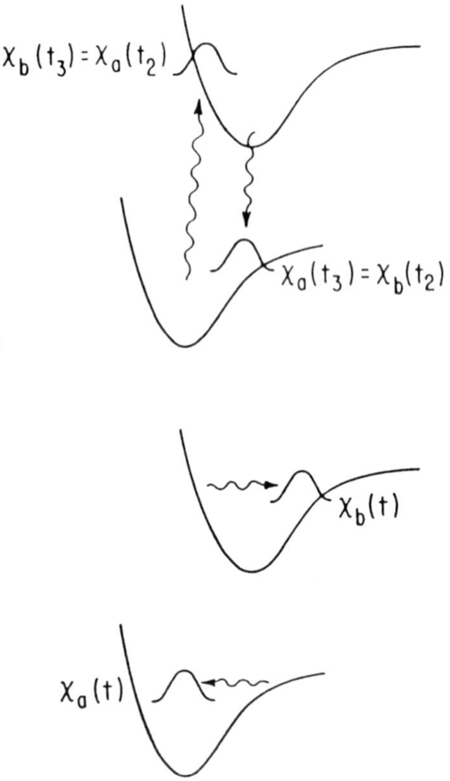

c)

d)

Figure 8.4 (cont.)

majority of optical experiments, giving rise to the implicit belief that complete population inversion (100% population in the excited electronic state) was impossible. With the advent of phase control on the femtosecond time scale the random phase assumption should no longer be taken for granted. In view of the considerable experimental progress in femtosecond technology, it seems timely to adopt a new view of coherent optical processes which at once generalizes the older frameworks while freeing us from the constrained concepts of now archaic experimental methods.

8.3 Control of Photochemical Reactions

In this section, I turn to controlling chemical product formation. This involves moving from bound systems to continuum systems and from one degree of freedom to at least two so that there is a choice of exit channels which will define the challenge of control. The concepts introduced in the previous section, particularly the instantaneous dipole moment, will play a significant role in this section as well.

A. Variational Formulation of Control of Product Formation

Consider the ground electronic state potential energy surface in Figure 8.5. This potential energy surface, corresponding to collinear ABC, has a region of stable ABC and two exit channels, one corresponding to A + BC and one to AB + C. This system is the simplest paradigm for control of chemical product formation: a system with two degrees of freedom is the minimum that can display two distinct chemical products. The objective is, starting out in a well-defined initial state ($v = 0$ of the ABC molecule) to design an electric field as a function of time which will steer the wavepacket out of channel 1, with no amplitude going out of channel 2, and vice versa.

I introduce a single excited electronic state surface at this point. The motivation is several-fold:

1. Transition dipole moments are generally much stronger than the coordinate dependence of permanent dipole moments, which governs infrared transitions.
2. The difference in functional form of the excited and ground potential energy surface will be our dynamical kernel. The use of excited elec-

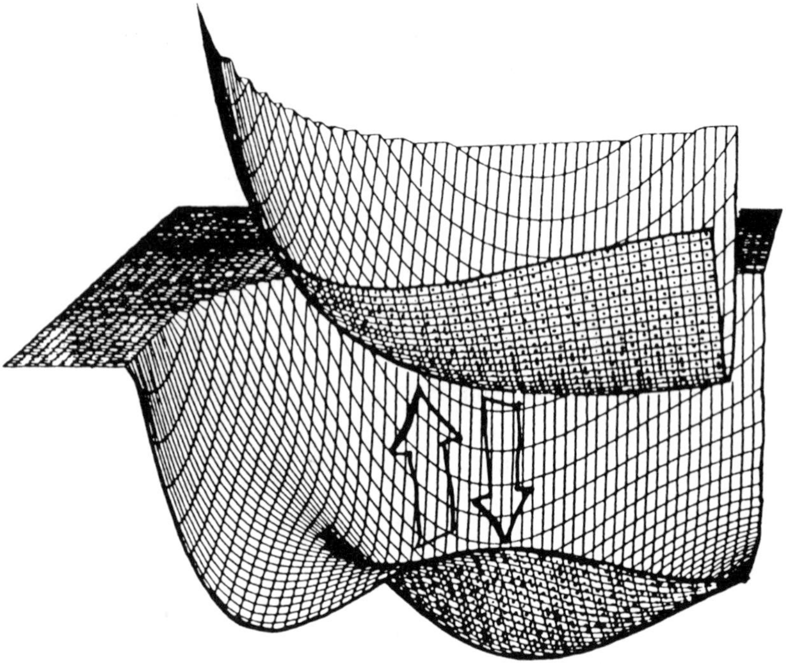

Figure 8.5 Stereoscopic view of the ground and excited state potential energy surfaces for a model collinear ABC system with the masses of HHD. The ground state surface has a minimum, corresponding to the stable ABC molecule. This minimum is separated by saddle points from two distinct exit channels, one leading to AB + C the other to A + BC. The object is to use optical excitation and stimulated emission between the two surfaces to 'steer' the wavepacket selectively out one of the exit channels. An intuitive method for steering the wavepacket, put forward by Tannor and Rice, involves adjusting the time delay between a pair of femtosecond pulses. The first pulse excites a wavepacket to the excited electronic state and a second pulse stimulates emission at some later time—a pump-dump scheme. By controlling the time delay between the pump and the dump pulses one is controlling the propagation time on the excited state potential energy surface; this in turn may profoundly affect the selectivity of product formation (AB + C versus A + BC). Figure 6 shows this classically and Figures 8.7–8.9 quantum mechanically. *Source*: from [21] with permission.

tronic states facilitates large changes in force on the molecule, effectively instantaneously, without necessarily using strong fields.

3. The technology for amplitude and phase control of optical pulses is significantly ahead of the corresponding technology in the infrared.

The object now will be to steer the wavefunction out of a specific exit channel on the ground electronic state, using the excited electronic state as an intermediate. In so far as the control is achieved by transferring amplitude between two electronic states, all the concepts regarding the central quantity μ_{eg} introduced above will now come into play.

Posing the problem mathematically, one seeks to maximize

$$J \equiv \lim_{T \to \infty} \langle \psi(T) | P_\alpha | \psi(T) \rangle \tag{8.3}$$

where P_α is a projection operator for chemical channel α (here, α takes on two values, referring to arrangement channels A + BC and AB + C; in general, in a triatomic molecule ABC, α takes on three values—1, 2, 3—referring to arrangement channels A + BC, AB + C, and AC + B). The quantity J is a *functional* of $\varepsilon(t)$, and the problem of maximizing J with respect to $\varepsilon(t)$ falls into the class of problems belonging to the calculus of variations.

I will return to the variational formulation and its solution in Section 8.3.C. Before that, however, I will anticipate on physical grounds the types of mechanisms that may be expected to come into play and introduce the elements of an intuitive procedure for iterative optimization.

B. Pump-Dump Scheme

Tannor and Rice developed an intuitive scheme, based on the close correspondence between the center of a wavepacket in time and that of a classical trajectory (Ehrenfest's theorem), to control product formation within this model [12]. Their approach was to use the timing between a pair of pulses—the first pulse produces an excited electronic state wavepacket and the second pulse stimulates emission. The time delay between the pulses controls the time that the wavepacket evolves on the excited electronic state. By the Franck-Condon principle, the second step instantaneously prepares a wavepacket on the ground electronic state with the same position and momentum as the excited state wavepacket. By controlling the position and momentum

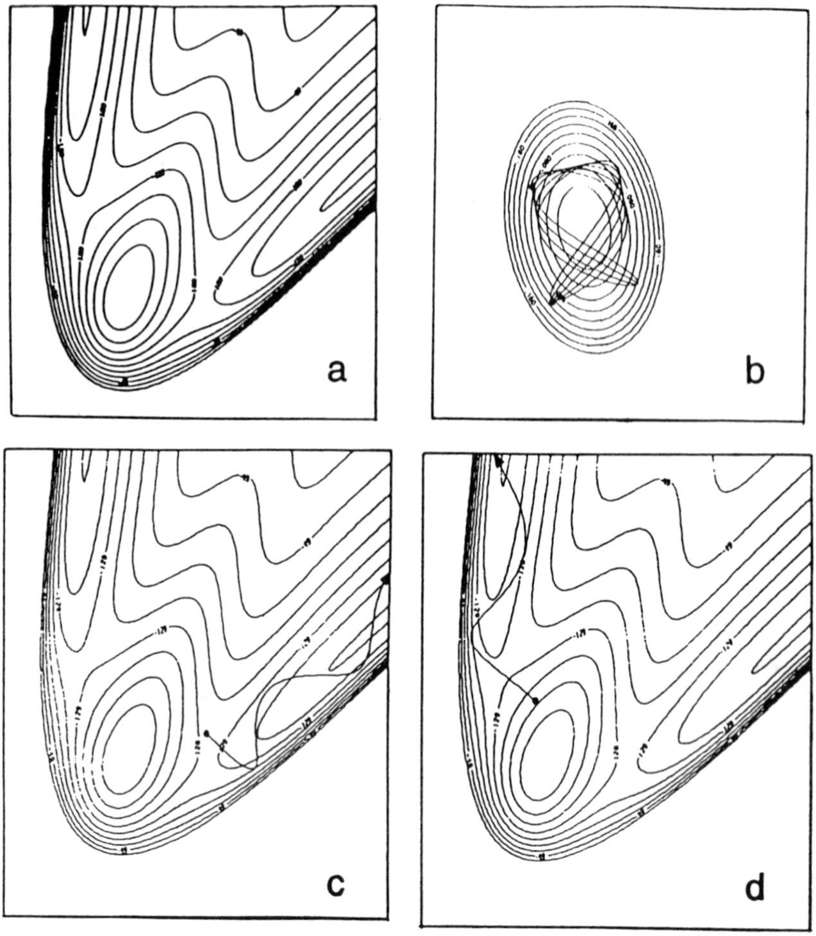

Figure 8.6 a) Equipotential contour plots of a) the ground and b) excited state potential energy surfaces (here a harmonic excited state is used because that is the way the first calculations were done). The classical trajectory that originates from rest on the ground state surface makes a vertical transition to the excited state, and subsequently undergoes Lissajous motion, which is shown superimposed. c) Assuming a vertical transition down at time t_1 (position and momentum conserved) the trajectory continues to evolve on the ground state surface and exits from channel 1. d) If the transition down is at time t_2 the classical trajectory exits from channel 2. *Source*: from [13] with permission.

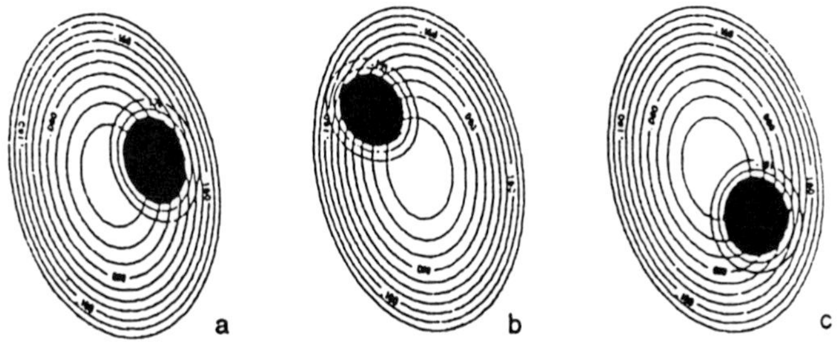

Figure 8.7 Magnitude of the excited state wavefunction for a pulse sequence of two Gaussians with time delay of 610 a.u. = 15 fs. a) t = 200 a.u., b) t = 400 a.u., c) t = 600 a.u. Note the close correspondence with the results obtained for the classical trajectory (Figure 8.6b). *Source*: from [13] with permission.

of the wavepacket produced on the ground state through the second step, one can gain some measure of control over product formation on the ground state. The Tannor-Rice pump-dump scheme is illustrated classically in Figure 8.6 and quantum mechanically in Figures 8.7–8.9.

The procedure developed by Tannor and Rice is significant for three reasons: it shows that control is possible, it gives a starting point for the design of optimal pulse shapes, and it gives a framework for interpreting the action of two pulse and more complicated pulse sequences. Nevertheless, the approach is limited: in general with the best choice of time delay and central frequency of the pulses one may achieve only partial selectivity. Perhaps most importantly, the Tannor-Rice scheme does not exploit the phase of the light. Intuition breaks down for more complicated processes and classical pictures cannot adequately describe the role of the phase of the light and the wavefunction. The next step is therefore to address the question: How is it possible to take advantage of the many additional available parameters: pulse shaping, multiple pulse sequences, etc., (i.e., in general an $E(t)$ with arbitrary complexity)—to maximize and perhaps obtain perfect selectivity?

Hence attempts were made to develop a systematic procedure for improving an initial pulse sequence. The first attempts, by Kosloff et al. [21], were based entirely on intuition and proceeded as follows:

1. Propagate the initial state forward in time until some final time, using the initial guess for the field (Figure 8.10(a–d)).

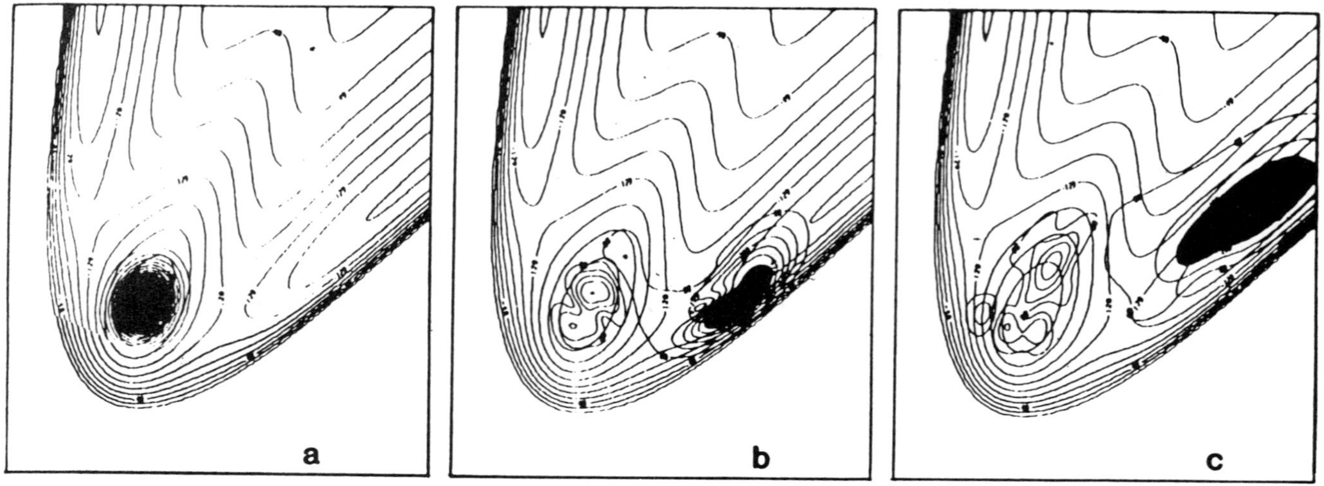

Figure 8.8 Magnitude of the ground state wavefunction for the pulse sequence in Figure 8.6. a) $t = 0$, b) $t = 800$ a.u., c) $t = 1000$ a.u. Note the close correspondence with the classical trajectory of Figure 8.6(c). Although some of the amplitude remains in the bound region, that which does exit does so exclusively from channel 1. *Source*: from [13] with permission.

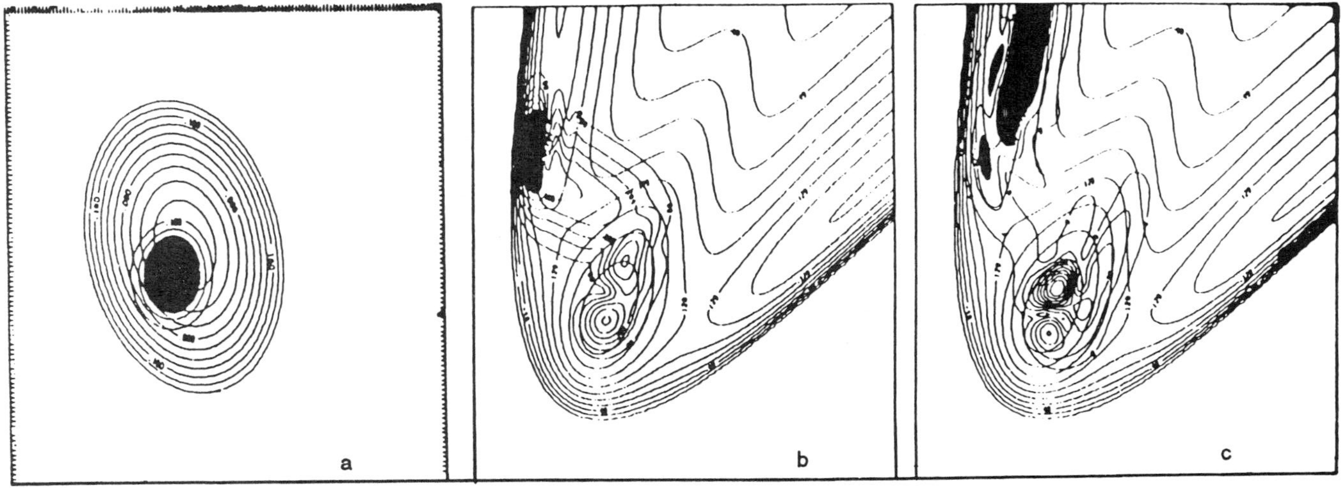

Figure 8.9 Magnitude of the ground and excited state wavefunctions for a sequence of two Gaussian pulses with time delay of 810 a.u. a) excited state wavefunction at 800 a.u., before the second pulse. b) ground state wavefunction at 1000 a.u. c) ground state wavefunction at 1200 a.u. That amplitude which does exit does so exclusively from channel 2. Note the close correspondence with the classical trajectory of Figure 8.6(d). *Source*: from [13] with permission.

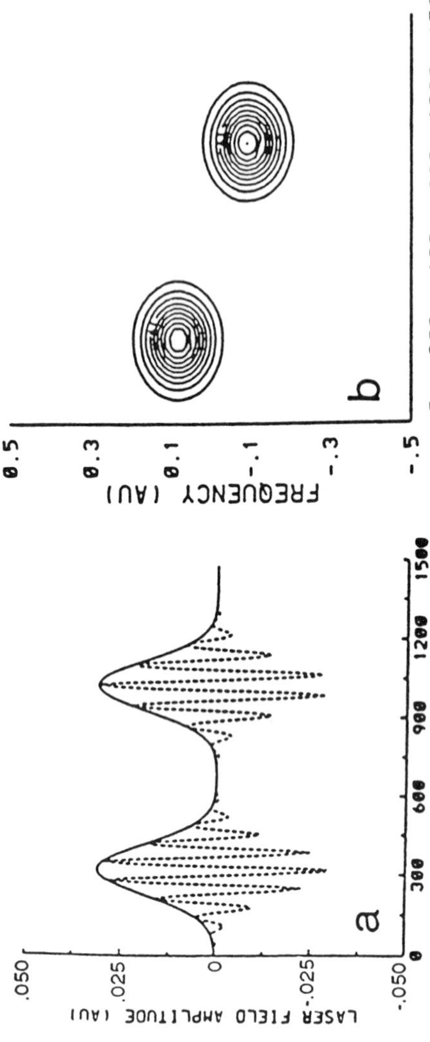

Figure 8.10 Pictorial explanation of an iterative method for optimizing the amount of wavepacket amplitude which exits out the desired chemical channel. First a guess at a good pulse sequence is used to propagate the wavefunction forward to some final time T. Here the initial guess consists of two gaussians, shown in a), in the spirit of the Tannor-Rice scheme described above. b) The Husimi distribution corresponding to the pulse sequence in a). The Husimi distribution allows one to visualize the pulse in time and frequency simultaneously.

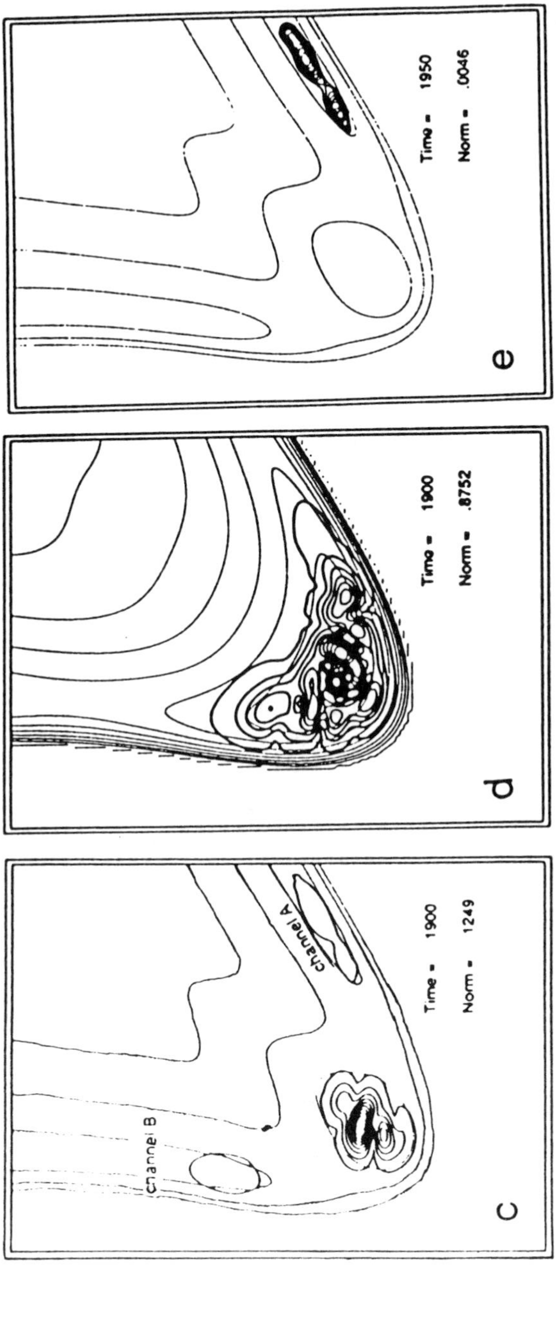

Figure 8.10 (cont.) c) The ground state wavefunction at the final time *T*. Note that there is amplitude in *both* exit channels, which is the generic result obtained with an intuitively designed pulse sequence. d) The excited state wavefunction at the final time, $\psi_e(T)$. e) The effect of applying a projection operator to the ground state wavefunction at the final time, which sets to zero all amplitude which is not in the desired channel. This new wavefunction is called χ and the figure therefore illustrates $\chi(T)$. In the optimization procedure, $\chi(T)$ is propagated backwards to $t = 0$; χ and ψ are then propagated forward in parallel, χ under the old field, ψ under a new field which is determined by the instantaneous overlap between $\psi(t)$ and $\chi(t)$. At the final time *T*, the projection operator is applied to the new $\psi(T)$ to construct a new $\chi(T)$. These steps are repeated until convergence is obtained. The procedure has the following interpretation: at every instant the field is chosen to make ψ at the current iteration resemble χ at the previous iteration as closely as possible. *Source:* from [21] with permission.

2. At the final time, remove all the amplitude in the undesired channels (Figure 8.10(e)).
3. Form a new wavefunction consisting of the filtered wavefunction on the ground state (Figure 8.10(e)) and the unfiltered wavefunction on the excited state (Figure 8.10(d)).
4. Propagate this new wavefunction backwards in time to t=0, using a field given by $-i$ times the overlap between the instantaneous excited state wavefunction and the instantaneous ground state wavefunction. In other words, the new field is calculated "on the fly" at each time step and fed back immediately into the propagation for the next time step. This procedure is continued until $t = 0$. Store the field generated in this fashion, and use that field for the forward propagation of the initial state at the next iteration.
5. Go to step (2) and repeat steps (2–5) until convergence is obtained.

Note the central role being played in step (4) of the instantaneous dipole moment, not of the entire wavefunction but of a filtered ground state wavefunction and an unfiltered excited state wavefunction. The rationale for this choice of the field is that it has precisely the right phase to transfer amplitude which is in the desired channel from the ground state to the excited state going backwards in time; it is therefore argued that by microscopic reversibility this field should transfer additional amplitude from the excited to the ground state, in the desired channel when propagating forward in time.

This procedure, while it does not generally lead to monotonic increase of the objective at each iteration, is of considerable pedagogical value. It contains many of the components of the procedure which will be derived from optimal control theory in the next section. The reader is encouraged to come back to this algorithm after reading Section 8.3.C to compare and contrast the details of the two procedures.

C. Optimal Control Methodology

I now return to the variational formulation of the problem of control of photochemical products. As described above, one seeks to maximize

$$J \equiv \lim_{T \to \infty} \langle \psi(T) | P_\alpha | \psi(T) \rangle \tag{8.4}$$

where P_α is a projection operator for chemical channel α. In the original formulation of [12], which was in the perturbative regime, it was necessary to introduce a constraint on the norm of the field; otherwise, the larger the field the larger the yield. This constraint took the form:

$$\|E\| \equiv \int_0^T dt|\varepsilon(t)|^2 = E \tag{8.5}$$

where E is a constant. This leads to a maximization problem for the modified objective functional,

$$\bar{J} \equiv \lim_{T\to\infty} <\psi(T)|P_\alpha|\psi(T)> - \lambda \int_0^T dt|\varepsilon(t)|^2 \tag{8.6}$$

where λ is a Lagrange multiplier. Alternatively, it is possible to let the norm of the field vary but put a penalty on the size of the norm. In this case, one treats λ as fixed and the higher the value of λ, the higher the penalty. In the presence of strong fields it is not clear that such a constraint or penalty is necessary, although it does serve to keep the field intensity within physically reasonable limits. However, if this constraint is abandoned the resulting optimal control problem is singular and somewhat more sophisticated numerical methods are required. All the work discussed below uses the penalty factor formulation.

In the OCT formulation, the time-dependent Schrödinger equation written as a 2 × 2 matrix in a Born-Oppenheimer basis set, Equation 8.1 is introduced into the objective functional with a Lagrange multiplier, $\chi(x,t)$ [15, 21]. The modified objective functional may now be written as:

$$\bar{J} \equiv \lim_{T\to\infty} <\psi(T)|P_\alpha|\psi(T)> + 2Re \int_0^T dt<\chi(t)| - \frac{\partial}{\partial t} - \frac{iH}{\hbar}|\psi(t)> - \lambda \int_0^T dt|\varepsilon(t)|^2 \tag{8.7}$$

where

$$H \equiv \begin{pmatrix} H_g & \mu\varepsilon(t)^* \\ \mu\varepsilon(t) & H_e \end{pmatrix} \tag{8.8}$$

It is clear that as long as ψ satisfies the time-dependent Schrödinger equation, the new term in $\bar{J}$ will vanish for any $\chi(x,t)$. The function of the new term is to make the variations of $\bar{J}$ with respect to ε and with respect to ψ independent, to first-order in $\delta\varepsilon$ (i.e., to deconstrain ψ and ε).

The requirement that $\delta\bar{J}/\delta\psi = 0$ leads to the equations:

$$i\hbar\frac{\partial\chi}{\partial t} = H\chi \tag{8.9}$$

$$\chi(x,T) = P_\alpha\psi(x,T) \tag{8.10}$$

that is, the Lagrange multiplier must obey the time-dependent Schrödinger equation, subject to the boundary condition at the *final* time T that χ be equal to the projection operator operating on the Schrödinger wavefunction. These conditions conspire, so that a change in ε, which would ordinarily change $\bar{J}$ through the dependence of $\psi(T)$ on ε, does not do so to first-order in the field.

For a physically meaningful solution it is *required* that

$$i\hbar\frac{\partial\psi}{\partial t} = H\psi \tag{8.11}$$

$$\psi(x,0) = \psi_0(x) \tag{8.12}$$

Some workers prefer to derive this latter equation from the condition that $\delta\bar{J}/\delta\chi = 0$ The physical interpretation is that if one removes the constraint that ψ must satisfy the time-dependent Schrödinger equation, one can only attain a maximum value of J higher than or equal to the physical maximum. If one then proceeds to minimize this "fictitious" $\bar{J}$ with respect to χ one can hope to recover the maximum value of $\bar{J}$ corresponding to the physical problem. This forms the basis for a powerful but unconventional approach to functional maximization [36].

Equations 8.9–8.12 form the basis for a double ended boundary value problem. ψ is known at $t = 0$, while χ is known at $t = T$. One can propagate

ψ forward in time to obtain $\psi(t)$, and propagate χ backwards in time to obtain $\chi(t)$, once one has a form for $\varepsilon(t)$. Note the emergence from the OCT formalism of the same dynamical ingredients that appeared in the intuitive optimization scheme of Kosloff et al., discussed previously. Equation 8.10 has the physical interpretation that at the final time T all amplitude that is not in the desired chemical channel be discarded. Equation 8.9 is then solved by propagating this amplitude backwards in time!

The optimal $\varepsilon(t)$ is given by the condition that $\delta\bar{J}/\delta\varepsilon = 0$ which leads to the equation:

$$\varepsilon(t) = \frac{-i}{\hbar\lambda}\,[\langle\chi_g|\mu|\psi_e\rangle - \langle\psi_g|\mu|\chi_e\rangle] \tag{8.13}$$

This equation also contains components of the procedure of Kosloff et al. The first term involves the overlap (i.e., the instantaneous dipole moment) of the filtered wavefunction on the ground state and the unfiltered wavefunction on the excited state; even the prefactor $-i$ matches the prefactor in the Kosloff procedure. Moreover, note that knowledge of the optimal $\varepsilon(t)$ involves knowledge of $\psi(t)$ and $\chi(t)$, while knowledge of $\psi(t)$ and $\chi(t)$ requires knowledge of $\varepsilon(t)$. Therefore, in general, the equations must be solved iteratively, another parallel to the procedure of Kosloff et al. Note that Equation 8.13 differs by a minus sign from Equation 2.15 in [21]; this comes from a difference in the definition of the sign of λ. Equation 8.13 differs from Equation 11 in [27] by complex conjugation, since the ε defined there is the complex conjugate of the ε here.

I will now discuss two different iterative procedures for solving the full set of coupled equations, Equations 8.9–8.12, both methods having the property that the objective increases at every iteration. The first is the more conventional gradient and gradient-type methods [15–20, 27, 30], which require a line search at every iteration. The second, which is based on novel methods of Krotov et al. [32–35], does not require any line search. We have adopted the latter in our own work [36, 37] and found it to be 4–5 times faster than gradient type methods.

The gradient of the objective with respect to the field is given by (gradient is used here and below to refer to the direction of maximum *positive* slope):

$$\frac{\delta \bar{J}}{\delta \varepsilon} = -2\lambda \left\{ \varepsilon(t) + \frac{i}{\hbar\lambda} [<\chi_g|\mu|\psi_e> - <\psi_g|\mu|\chi_e>] \right\} \tag{8.14}$$

Therefore, given the field at the k iteration, the field at the $k + 1$ iteration may be written as:

$$\varepsilon^{k+1}(t) = \varepsilon^k + \alpha \frac{\delta \bar{J}}{\delta \varepsilon} \bigg|_{\varepsilon = \varepsilon^k}$$

If α is small (α a positive real parameter) the objective functional must increase or stay the same at each new iteration. However, the second-order dependence of the objective on the field is not completely known a priori (particularly because of the nonlinear dependence of $\psi(T)$ on the field). This implies that there is no a priori way of knowing the value of α which will maximize the objective in the direction of the gradient. One approach to the selection of α is to perform a line search along the direction of the gradient (i.e., the best value of α is found by empirical comparison of many different values). Once the best value of α is found the new gradient is calculated and a new search performed [27]. Some workers have reported rapid convergence to the optimal pulse using a conjugate gradient (CG) method [16, 19, 20]; however, it is not clear that the conjugate gradient method should be better than the gradient method unless the initial guess is near the extreme.

A fundamentally different numerical procedure has been derived by Krotov which does not require a line search at each iteration [34, 36, 37]. This method is unique in that, since an assumption of small variations is not used in the derivation, the procedure leads to macro changes in the electric field at each iteration without a line search. In [36] an explicit proof that the field at the $k + 1$ iteration gives an objective greater than or equal to the field at the k iteration is given for the two electronic state problem at hand. The numerical procedure is as follows:

1. Propagate the initial state forward in time until some final time, using the initial guess for the field (Figure 8.10(a–d)).
2. At the final time remove all the amplitude in the undesired channels (Figure 8.10(e)).
3. Propagate the filtered wavefunction backwards in time to $t = 0$.
4. Propagate the filtered wavefunction forward in time using the old field; propagate the initial wavefunction forward in time using a new field,

which is calculated as follows: calculate the overlap between the instantaneous filtered wavefunction and the instantaneous full wavefunction times $-i$,

$$\varepsilon(t) = \frac{-i}{\hbar\lambda}\,[<\chi_g^k|\mu|\psi_e^{k+1}> - <\psi_g^{k+1}|\mu|\chi_e^k>] \tag{8.15}$$

In other words, the new field in calculated "on the fly" at each time step and fed back immediately into the propagation for the next time step. This procedure is continued until the final time T, always propagating the filtered wavefunction using the field from the previous iteration.

5. Go to step (2) and repeat steps (2–5) until convergence is obtained. In tests on a variety of applications, I have found this method to be 4–5 times faster than gradient type methods [37].

The method of Krotov has an interesting feature in common with the procedure of Kosloff et al. in that it has immediate feedback of the instantaneous overlap in determining the field at the next time step. This allows us to give a precise interpretation of the optimization procedure which retains some of the flavor of the interpretation of Kosloff et al. At every instant in time, the optimal field at the $k + 1$ iteration is the one which makes ψ^{k+1} as close as possible to χ^k. Since we know that χ^k makes a complete exit, the closer ψ^{k+1} is to that state the more complete its exit will be.

D. Digression on the Hamiltonian Structure of the Necessary Conditions

There is an alternative formulation of the set of Equations 8.9–8.12 that bears the same relationship to the original formulation as Hamiltonian mechanics does to Lagrangian mechanics [38–40]. A control Hamiltonian, **H**, is constructed and the optimal pulse is that for which the Hamiltonian is an extreme (i.e., $\partial\mathbf{H}/\partial\varepsilon = 0$). Algorithmically, the search for the optimal pulse can then be performed as a maximization of **H**. One should not confuse the optimization Hamiltonian **H**, with the physical Hamiltonian, H.

The optimization Hamiltonian is defined as

$$\mathbf{H} \equiv 2Re<\chi\,|\frac{H}{i\hbar}|\,\psi> - \lambda|\varepsilon(t)|^2$$

This is the Legendre transform of the original objective, neglecting the boundary term $<\psi|P|\psi>$. Hamilton's equations take the form

$$\dot{\chi} = -\frac{\partial \mathbf{H}}{\partial \psi} = \frac{H}{i\hbar}\chi \tag{8.16}$$

$$\dot{\psi} = \frac{\partial \mathbf{H}}{\partial \chi} = \frac{H}{i\hbar}\psi \tag{8.17}$$

in agreement with Equations 8.9 and 8.11. The condition for the optimal field is

$$\tilde{\varepsilon} = arg \max_{\varepsilon} \mathbf{H}(t,\varepsilon,\psi(\varepsilon)) = \frac{-i}{\hbar\lambda}[<\chi_g | \mu | \psi_e> - <\psi_g | \mu | \chi_e>] \tag{8.18}$$

in agreement with Equation 8.13.

It is straightforward to show that

$$\frac{d\mathbf{H}}{dt} = 0 \text{ for } \varepsilon = \tilde{\varepsilon}(t)$$

that is, the control Hamiltonian is time-independent on the optimal trajectory. This relationship can be used as a diagnostic of how close a particular control is to one which satisfies the necessary conditions for optimality. Note the analogy to the relationship in classical mechanics, that if $\partial H/\partial t = 0$ then $dH/dt = 0$ (i.e., energy is conserved) on a trajectory for which the integral of the Lagrangian is an extreme (i.e., any physical trajectory) [41].

8.4 APPLICATIONS

A. Model ABC System

Figures 8.11–8.12 show the effect of pulse design on chemical branching for the model ABC system described earlier. Parameters for the ground and excited state potential energy surfaces are given by Kosloff et al. [21]. Note the two exit channels differ only by an isotopic substitution (the channel to the right corresponds to D + HH while the channel to the left corresponds

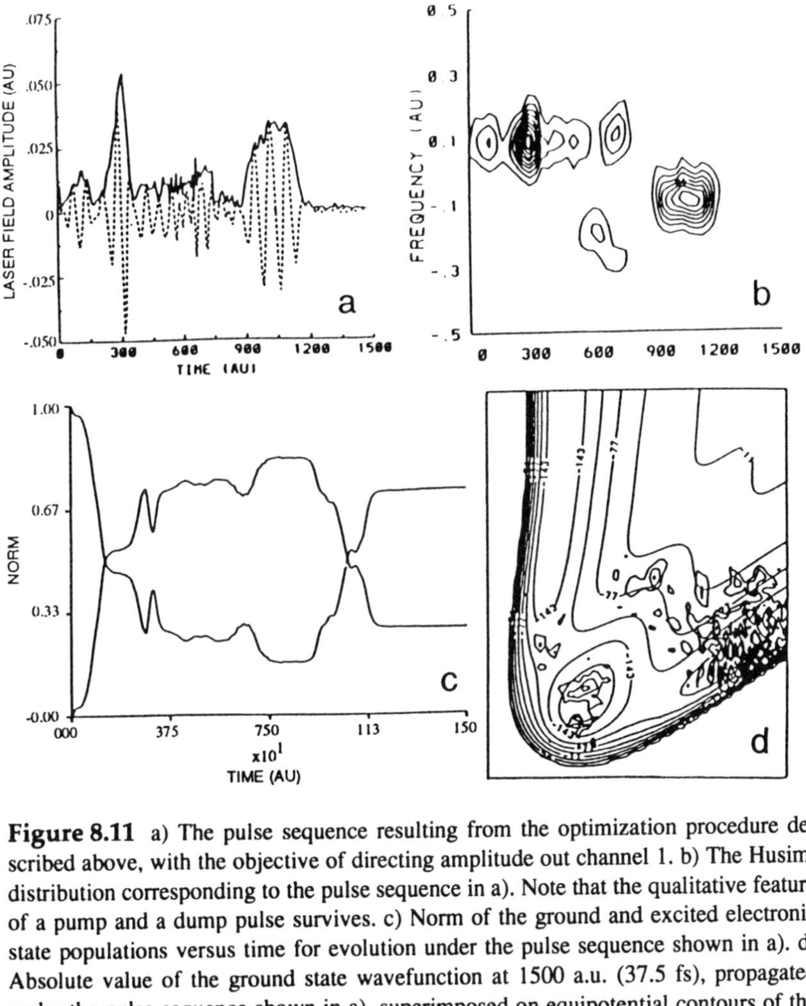

Figure 8.11 a) The pulse sequence resulting from the optimization procedure described above, with the objective of directing amplitude out channel 1. b) The Husimi distribution corresponding to the pulse sequence in a). Note that the qualitative feature of a pump and a dump pulse survives. c) Norm of the ground and excited electronic state populations versus time for evolution under the pulse sequence shown in a). d) Absolute value of the ground state wavefunction at 1500 a.u. (37.5 fs), propagated under the pulse sequence shown in a), superimposed on equipotential contours of the ground state surface. The exit out channel 1 is extremely selective. *Source*: from [27] with permission.

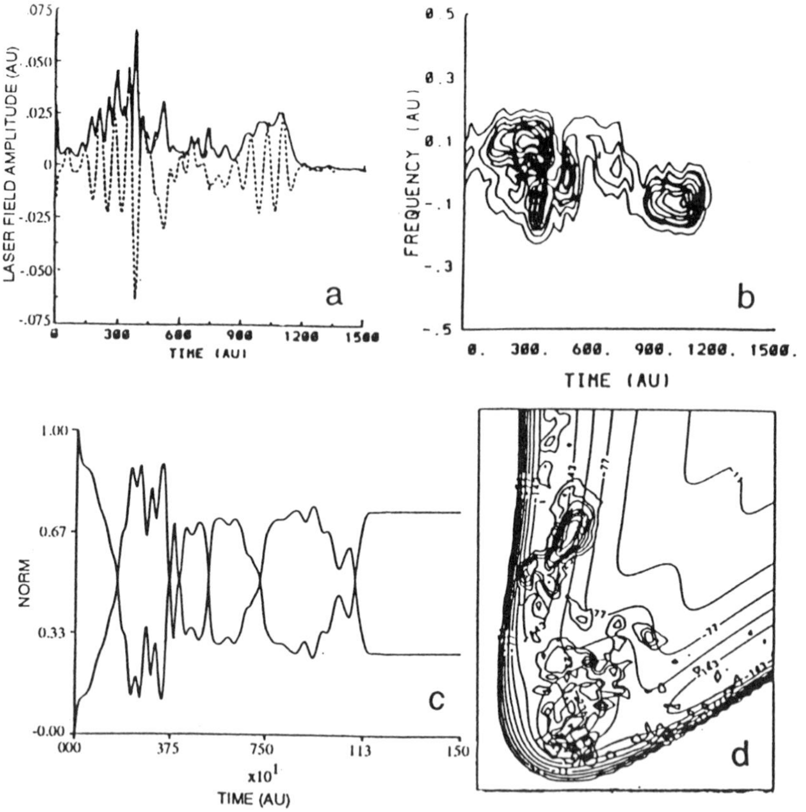

Figure 8.12 a) The pulse sequence resulting from the same iterative optimization procedure used in Figure 8.10, but with the objective of directing amplitude out channel 2 (i.e., the projection operator for channel 2, rather than channel 1 is used). b) The Husimi distribution corresponding to the pulse sequence in a). Note that some of the character of a pump and a dump pulse is still present, but there is a significant amount of pulse amplitude at times in between. This pulse sequence, viewed in its entirety, is actually closer to a periodic frequency sweep. c) Norm of the ground and excited electronic state populations versus time for evolution under the pulse sequence shown in a). Note the multiple Rabi cycling of amplitude between the two surfaces, related to the sustained frequency swept character of the pulse sequence. d) Absolute value of the ground state wavefunction at 1500 a.u. (37.5 fs), propagated under the pulse sequence shown in a), superimposed on equipotential contours of the ground state surface. The exit out channel 2 is now extremely selective. This pulse has to work harder than the pulse in Figure 8.10. Inspection of the wavefunction at intermediate times (not shown) indicates that the Rabi cycling is amplifying the motion in the ABC symmetric stretch. Once sufficiently large amplitude motion is achieved a little push steers the amplitude out channel 2. *Source*: from [27] with permission.

to H + HD). Figure 8.10(a) shows the initial guess for the pulse (two Gaussians), and Figure 8.10(b) shows its Husimi transform (Gaussians in both time and frequency). Figure 8.11(a) shows the pulse designed to yield high probability in channel 1 and Figure 8.11(b) shows its Husimi transform. Figure 8.11(c) shows the norm on the ground and excited state surfaces as a function of time and under the influence of the pulse in Figure 8.11(a). Figure 8.11(d) shows the wavepacket on the ground state potential surface after 1500 a.u. (37.5 fs) and under the influence of the pulse in Figure 8.11(a). Note the high degree of selectivity for exit out of channel 1 (D + HH). Figure 8.12(a) shows the pulse designed to yield high probability in channel 2, and Figure 8.12(b) shows its Husimi transform. Figure 8.12(c) shows the norm on the two surfaces as a function of time and under the influence of the pulse in Figure 8.12(a). Figure 8.12(d) shows the wavepacket on the ground state potential surface after 1500 a.u. (37.5 fs) and under the influence of the pulse in Figure 8.12(a).

It is clear that these pulses are remarkably effective in directing wavepacket amplitude out one channel or the other. Equally intriguing is the superficial similarity between the two pulses, and between both pulses and the initial guess. The pulses in Figures 8.11(a) and 8.12(a) occupy identical windows in time, in contrast with the simple pump-dump control scheme of Tannor and Rice [12] which makes use of different windows in time to achieve different products. The Husimi plots show that the two pulses are even similar in their joint time-frequency distributions. There are significant differences, however. The pulse in Figure 8.11(a) is still primarily a two-pulse, pump-dump sequence. The pulse in Figure 8.12(a), however, has substantial amplitude between the pump and dump and is much closer to a being a continuous excitation/de-excitation. This is confirmed by inspection of the norm in Figures 8.11(b) and 8.12(b). The norm in Figure 8.11(c) changes once from the pump pulse and once again from the dump pulse. The norm in Figure 8.12(c), corresponding to the pulse in Figure 8.12(a), undergoes three full cycles. It should be noted that there may be significant phase differences between the two pulses in Figures 8.11(a) and 8.12(a) which are not evident in the Husimi plots, the latter showing only the absolute value of the distribution.

The pulses in Figures 8.11(a) and 8.12(a) provide examples of how simple mechanisms may underlie the pulses calculated using the TDSE/OCT machinery. The pulse sequence in Figure 8.11(a) is qualitatively a simple pump-dump form (i.e., the time delay between the two pulses is such that the wavepacket on the excited state surface has the correct position and momentum to exit out of channel 1 on the ground state surface) [12, 13]. The pulse sequence in Figure 8.12(a) reflects a different mechanism, closer to a cyclically repeating pump and dump. The cyclical pump-dump pulse sequence provides a natural mechanism for preparing highly excited symmetric stretch vibration, since this is the coordinate with greatest difference in forces on the two potential surfaces and thus this motion is amplified on each cycle. This can be seen in pictures of the wavepacket motion (not shown). If amplified sufficiently this motion will lead to selective three body fragmentation (A + B + C). In the present study, however, with the objective of A + BC fragmentation, the symmetric stretch motion is amplified only in the first stage of the pulse sequence. In the second stage of the pulse sequence the energy in the symmetric stretch is channeled into the A + BC (H + HD) channel.

I will now touch briefly on two issues concerning the optimal pulses of a more fundamental nature. The optimization performed here was terminated arbitrarily when a sufficient degree of product formation was achieved because of the CPU intensive nature of the calculations. However, the numerical results shown here are not inconsistent with the possibility that 100% product yield can be achieved in any channel desired. Is it possible to achieve 100% in any channel desired and what are the conditions that a system must satisfy in order for this to be the case? This is related to the general problem of quantum controllability, a subject on which we have only incomplete answers at present.

The second issue is whether the optimal pulses leading out the different channels are orthogonal, either in a straightforward or in some generalized sense. Although the Husimi plots suggest that there is in fact significant overlap between the pulses, note that the Husimi plots show only the absolute value of the time-frequency distributions, while the phase can be extremely important in determining orthogonality. Moreover, it is conceivable that pulses that are locally optimal in the function space are mutually orthogonal. The possibility of such orthogonality is consistent with the formulation of the optimal control problem in the weak-field regime, where the optimal pulses can be viewed as eigenvectors of an integral operator and, in some cases, a linear Hermitian operator.

B. A System With Realistic Parameters: I_2

The I_2 system, which has been the subject of a substantial amount of femtosecond work, is studied theoretically as a prospect for experimental control. The I_2 system lacks the complexity of a polyatomic system with the prospect for selective bond breaking. Nevertheless, its optical properties make it well suited for study with available femtosecond technology, and the vibrational time scale is sufficiently long for currently achievable pulse sequences to effect the dynamics on the time scale of a fraction of a period. Furthermore, because its spectroscopy and potential energy curves are well known it is an ideal system for initial comparisons between theory and experiment.

Several theoretical studies have been undertaken to explore the use of femtosecond pulse sequences to achieve dissociation of I_2 [37, 42, 43]. Dissociation on the *B* state is taken as the objective because of its experimental accessibility, although dissociation on any other electronic state (including the *X* state) could have been chosen as well. The scientific objective of the study is to find one or several *interesting* pulse sequences (i.e., whose mechanisms for achieving the objective is neither overly trivial nor so complex so as to defy a simple interpretation). If the pulse sequences are sufficiently interesting it is anticipated that they would stimulate experiments, despite the absence of issues of chemical branching in this simple system.

Somloi et al. [37] performed an unconstrained optimization of the electric field designed to achieve maximal photodissociation yield on the *B* state. This was to give the field the maximum freedom so that unexpected mechanisms might be discovered. Analysis of the optimal pulse sequence indicates that it is exploiting a quasi-CW mechanism: the frequency of the excitation is resonant with the *B* state continuum and hence produces efficient dissociation, albeit in an uninteresting way.

To find a different mechanism from the quasi-continuous-wave excitation the electric field was restricted to the frequency interval from 0 cm^{-1} to 19752 cm^{-1} using a method developed by Rabitz et al. [44]. The optimal pulse sequence under these conditions takes the form of a *pump-dump-pump* sequence (Figure 8.13). (Although at first glance the pulse sequence looks like a four-pulse sequence, the third and fourth pulse are both performing part of the same function. This point will be expanded below.) This pulse sequence is a natural extension of the Tannor-Rice *pump-dump* scheme [12,

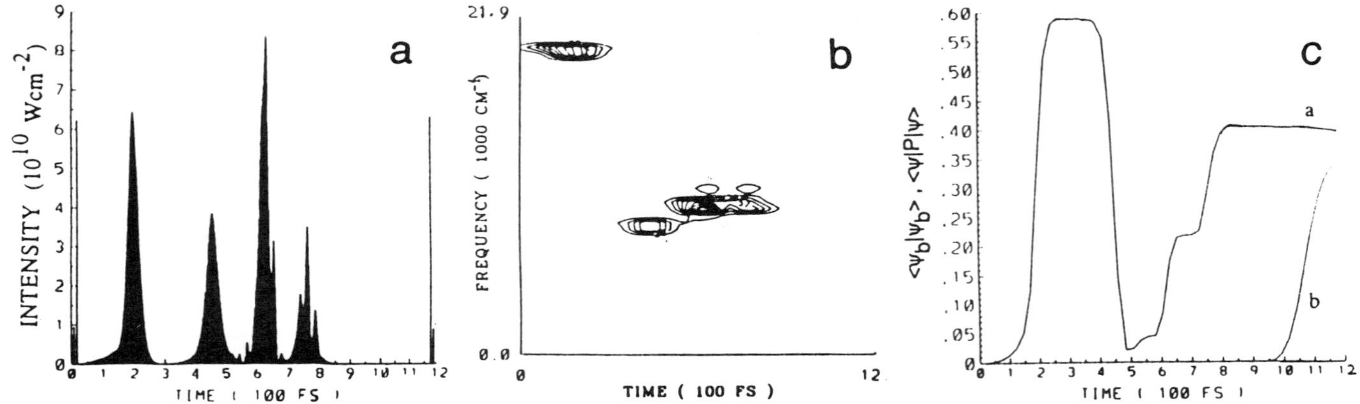

Figure 8.13 a) Optimal pulse for dissociation on the *B* electronic state of I_2 with frequencies corresponding to direct excitation from the ground vibrational state into the continuum filtered. b) Husimi distribution for the pulse sequence in a). c) *B* state population (upper) and probability for exit on the *B* state (lower) as a function of time for the pulse sequence in a). Although the pulse sequence in a) appears at first sight to be a four pulse sequence, the Husimi distribution and the *B* state population indicate that it works by a three 'photon' pump-dump-pump sequence (i.e., the third and fourth pulse are both pumps). *Source*: from [37] with permission.

13]. The first photon excites the molecule, which then evolves on the excited electronic state potential surface until the second photon dumps it by stimulated emission. Since the ground state potential energy surface is steeper than the excited, the kinetic energy gained by the wave function on the ground state potential energy surface is enough to generate dissociation on the excited state potential energy surface after the molecule interacts with the third photon.

Although the pulse sequence consists of four pulses, the third and fourth pulses are performing the same function, namely the final pump. This statement is supported by the Husimi distribution, Figure 8.13(b), which indicates that the third and fourth pulse are at the same frequency. Moreover, the plot of the excited state population versus time, Figure 8.13(c), shows clearly that both the third and fourth pulse are serving to pump amplitude to the excited state. Extensive tests removing the third and fourth pulses separately indicate that their effect is simply additive; they are not using interference to increase the excited state population. With the help of the Husimi transforms of the optimal electric field a sequence of four Gaussian-shape pulses was constructed as a possible candidate for experimental application. This approximating field gave 11% for dissociation probability and the wavefunction is well localized at the final time *T*. With higher values of the amplitudes the dissociation probability can be increased further since no final optimization was done on the Gaussians with respect to their intensity. Both with respect to intensity and pulse duration these Gaussian pulses should be producible in the laboratory with current technology.

The remaining question is: How good is the physical model for I_2? A similar model was used by Scherer et. al. [5] and they were able to reproduce the qualitative features of the experimental interferogram. That fact is encouraging, and suggests that the qualitative effects described here should persist in the real system. Nevertheless, a more sophisticated model—including a more accurate potential and the rotational degree of freedom (in particular, averaging over orientations)—is required to give completely accurate experimental predictions.

8.5 CONCLUSIONS

This chapter has been written from the author's very personal vantage point. Many applications and alternate points of view have not even been touched on. A notable omission is a discussion of optimal control in the collisional regime [45, 46] and in the condensed phase [47]. The aim of the article has been to present a coherent and reasonably comprehensive pedagogical treatment. I have tried to emphasize the physical concepts that underlie my present understanding of laser control of chemical reactions and the associated optimal control mathematical apparatus. In this Section I will make a few final comments on the relationship between phase and amplitude and between physical intuition and optimization.

The pulses obtained in the study of electronic population transfer and heating and cooling of vibrational motion involve a phase sensitive mechanism. By using a periodic pulse sequence whose period is equal to the period of the excited state oscillator and the right choice of phase between pulses, it is possible to ensure that the amplitude that is transferred at later times adds constructively with the amplitude at earlier times. With a different choice of phase it is possible to ensure that the population stay locked (i.e., that there is no transfer of amplitude between electronic states).

The Tannor-Rice pump-dump scheme for control of chemical branching, is primarily amplitude control (i.e., timing) and, in its simplest realization, is not phase sensitive. Within a perturbative treatment the second pulse will transfer some amplitude back to the ground state and it is not important whether that amplitude is in phase or out of phase with the preexisting amplitude.

The optimal pulses computed here make use, to some extent, of both the amplitude (timing) and phase of the light. The phase of the light is primarily being used to maximize population transfers and ultimately the chemical yield. One mechanism which was anticipated but has not yet been explicitly observed is a first pulse which leads to amplitude which, if allowed to evolve undisturbed would go out both product channels, then a second pulse creates amplitude in the two channels with just the right phase to enhance the amplitude in one channel and cancel in the other.

Let me now turn to some general comments concerning the relationship between physical intuition and optimal pulse sequences. It is unlikely that computer optimization of pulses can ever replace physical intuition completely. Even with the apparently "black box" methodology of optimal control, physical intuition enters it every turn: a) in the definition of the Hamiltonian and therefore the restricted sets of mechanisms that are possible; b) the choice of constraints, which can profoundly affect the solution c) the initial guess for the pulse sequence; a well-motivated initial guess is more likely to lead to a physically meaningful and experimentally realizable optimal pulse sequence. Moreover, physical intuition is required in interpreting the optimal pulses and using them to develop experimentally successful analogs. This brings us to our next point, which concerns experimental realization.

To what extent are the pulses that come out of the optimization going to be useful in the laboratory? It is fairly certain that the pulses as generated on the computer will not be immediately effective in the lab. The most important reason for this is the high degree in uncertainty in the Hamiltonian of the molecule (the Born-Oppenheimer potential energy surfaces), particularly for polyatomics. Moreover, if the pulse is overly detailed it will be difficult to replicate exactly in the laboratory and the results may be overly sensitive to minute features of the pulse sequence. This has led to a great deal of methodological discussion of how to bridge the gap between theory and experiment. Several interesting approaches have been explored by Rabitz et al. First, those workers have performed systematic studies of the effect of Hamiltonian uncertainty on the effect of their pulses. The have gone on to extend methods from the engineering control literature to deal with optimal control in the presence of system uncertainties [47, 48]. One such approach is based on a minimax principle: to find the control such that the uncertainty in the Hamiltonian does the least damage in the worst case scenario [48]. Another, potentially very useful approach being explored by Rabitz et al. falls under the general heading of *learning algorithms*, of which the genetic algorithm is one example [49]. In the genetic algorithm, an analogy is exploited between the search for good pulses and survival of the fittest in Darwinian evolution. In one formulation the algorithm would proceed as follows. First, a large family of pulse sequences is generated by any method desired, for example random number generation. These pulse sequences are then applied to the system. Only the two pulse sequences which gave the highest value of the objective are retained; the remainder are discarded. From these two "parents" a large new family of pulse sequences is generated, by creating many new combinations of the parents' "genetic material" (i.e., time segments of the two parent pulse sequences are interspersed in many different ways). Again, the new family of pulse sequences is tested and all but the two best sequences are discarded. This process is then repeated for many generations. The approach has been successful on some simple test cases [49]. What is particularly appealing about this approach is that is can be carried out directly in the laboratory (i.e., the original family of pulses synthesized and the time segments of the two best interspersed in the laboratory without the need for intermediate computations).

This author's bias is that a necessary part of optimization methodology is the "post mortem" on the optimal pulse—performing a series of additional tests to isolate the mechanism by which the pulse works. Even if a true experimental in situ optimization procedure becomes practical it will be important to analyze the physical mechanism by which the pulses work in order to optimize them further or extend them to new systems. It is highly likely that an interplay—a feedback loop, if you will—between intuitive guesswork and optimization will always be profitable: the intuitive guess can be improved by optimization and the optimal pulse, once analyzed, can improve our intuition. In summary, given the present prospectus for obtaining sufficiently accurate potential energy surfaces for polyatomics into the next century, it may be that the most enduring legacy of the calculation of optimal pulses will be the new mechanisms and qualitative insight which they can provide, with the understanding that particular experimental realizations will require close collaboration between theory (modeling) and experiment.

ACKNOWLEDGEMENTS

It is a pleasure to acknowledge the many significant contributions of my collaborators, Ronnie Kosloff, Stuart Rice, Pierre Gaspard, József Somlói, Yijian Jin, Jeff Kantor, Daniela Kohen, András Lorincz, Audrey Dell Hammerich, Vladimir Kazakov, and Vladimir Orlov to this research. I am grateful to the Departments of Chemistry at UC San Diego and Columbia University, where the bulk of this chapter was written. The Telluride Summer Workshop on Control of Molecular Systems, in 1991, generated many vigorous discussions in this area, and this chapter has been profoundly colored by the reflections of the participants in that Workshop. Some of the computer calculations were performed on the Cray Y-MP machine at the National Center for Supercomputing Applications (University of Illinois at Urbana-Champaign). This work was supported by a grant from the US Office of Naval Research. The author is an Alfred P. Sloan Foundation Fellow.

REFERENCES

1. *Science*, **255**, 1643 (1992).

2. R. M. Bowman, M. Dantus, A. H. Zewail, *Chem. Phys. Lett.*, **156**, 131 (1989).
3. E. Potter, J. L. Herek, S. Pedersen, Q. Liu, A. H. Zewail, *Nature*, **355**, 66 (1992).
4. N. F. Scherer, A. J. Ruggiero, M. Du, G. R. Fleming, *J. Chem. Phys.*, **93**, 856 (1990).
5. N. F. Scherer, R. J. Carlson, A. Matro, M. Du, A. J. Ruggiero, V. Romero-Rochin, J. A. Cina, G. R. Fleming, S. A. Rice, *J. Chem. Phys.*, **95**, 1487 (1991).
6. A. M. Weiner, J. P. Heritage, R. N. Thurston, *Opt. Lett.*, **11**, 153 (1986).
7. A. M. Weiner, D. E. Leaird, G. P. Wiederecht, K. A. Nelson, *Science*, **247**, 1317 (1990).
8. F. Spano, M. Haner, W. S. Warren, *Chem. Phys. Lett.*, **135**, 97 (1987).
9. R. A. Mathies, C. H. Brito Cruz, W. T. Pollard, C. V. Schank, *Science*, **240**, 777 (1988).
10. Y. Yan, B. E. Kohler, R. E. Gillilan, R. M. Whitnell, K. R. Wilson, S. Mukamel, in *Ultrafast Phenomena VIII*, Springer-Verlag, Berlin (1992).
11. M. Shapiro, P. Brumer, *J. Chem. Phys.*, **84**, 4103 (1986); P. Brumer, M. Shapiro, *Chem. Phys. Lett.*, **126**, 54 (1986); P. Brumer, M. Shapiro, *Annu. Rev. Phys. Chem.*, **43**, 257 (1992).
12. D. J. Tannor and S. A. Rice, *J. Chem. Phys.* **83**, 5013 (1985).
13. D. J. Tannor, R. Kosloff and S. A. Rice, *J. Chem. Phys.* **85**, 5805 (1986).
14. D. J. Tannor and S. A. Rice, *Adv. Chem. Phys.*, **70**, 441 (1988).
15. A. P. Pierce, M. A. Dahleh and H. Rabitz, *Phys. Rev. A.*, **37**, 4950 (1988).
16. S. Shi, A. Woody and H. Rabitz, *J. Chem. Phys.*, **88**, 6870 (1988).
17. S. Shi and H. Rabitz, *J. Chem. Phys.*, **92**, 364 (1990).
18. S. Shi and H. Rabitz, *Comp. Phys. Comm.*, **63**, 71 (1991).
19. W. Jakubetz, B. Just, J. Manz and H-J. Schreier, *J. Phys. Chem.*, **94**, 294 (1990).
20. J. E. Combariza, B. Just, J. Manz and G. K. Paramonov, *J. Phys. Chem.* **95**, 10351 (1991).
21. R. Kosloff, S. A. Rice, P. Gaspard, S. Tersigni and D. J. Tannor, *Chem. Phys.*, **139**, 201 (1989).
22. A. D. Hammerich, R. Kosloff and M. A. Ratner, *J. Chem. Phys.* **97**, 6410 (1992).
23. D. J. Tannor and P. Salamon, *Rep. Math. Phys.*, **30**, 233 (1991).
24. M. Demiralp and H. Rabitz, *Int. J. Eng.*, **31**, 307 (1993).
25. M. Demiralp and H. Rabitz. *Int. J. Eng.*, **31**, 333 (1993).
26. Y. Yan, R. E. Gillilan, R. M. Whitnell, K. R. Wilson, S. Mukamel, *J. Phys. Chem.*, **97**, 2320 (1993)

27a. Y. Yin, D. J. Tanner, J. Somloi, and A. Lorintz, (unpublished).

27b. D. J. Tannor and Y. Jin, in *Mode Selective Chemistry*, (B. Pullman, J. Jortner, and R. D. Levine, Eds.) Kluwer, Dordrecht, 1991.

28. R. Kosloff, A. Hammerich and D. J. Tannor, *Phys. Rev. Lett.*, **69**, 2172 (1992).
29. A. Bartana, R. Kosloff and D. J. Tannor, *J. Chem. Phys.* **99**, 196 (1993).
30. D. J. Tannor and S. A. Rice, in *Understanding Molecular Properties*, (J. Avery et al., Eds.), Reidel, 1987.

31. S. H. Tersigni, P. Gaspard, S. A. Rice, *J. Chem. Phys.*, **93**, 1670 (1990).
32. V. F. Krotov, *Control and Cybernetics*, **17**, 15 (1988).
33. V. F. Krotov, *Automation and Remote Control*, **34**, 1863 (1973); *Automation and Remote Control*, **35**, 1 (1974); *Automation and Remote Control*, **35**, 345 (1974).
34. V. F. Krotov and I. N. Fel'dman, *Engrg. Cybernetics*, **21**(1983), 123 (1984).
35. V. A. Kazakov and V. F. Krotov, *Automation and Remote Control*, **5**, 430 (1987).
36. D. J. Tannor, V. A. Kazakov, and V. Orlov, in *Time-Dependent Quantum Molecular Dynamics*, (J. Broeckhove and L. Lathouwers, Eds.), Plenum Press, New York, (1992).
37. J. Somloi, V. A. Kazakov and D. J. Tannor, *Chem. Phys.*, **172**, 85 (1993).
38. A. E. Bryson, Jr. and Y-C. Ho, *Applied Optimal Control*, Hemisphere,Washington, DC, (1975).
39. L. I. Rozonoér, *Automation and Remote Control* **20**, 288 (1959).
40. L. I. Rozonoér, *Automation and Remote Control* **20**, 405 (1959).
41. H. Goldstein, *Classical Mechanics*, Addison-Wesley, Reading, Massachusetts, 1950.
42. B. Amstrup, R. J. Carlson, A. Matro, S. A. Rice, *J. Phys. Chem.* **95**, 019 (1991).
43. K. Wilson, private communication.
44. P. Gross, D. Neuhauser and H. Rabitz, *J. Chem. Phys.*, **96**, 834 (1992).
45. M. Shapiro and P. Brumer, *J. Chem. Phys.*, **90**, 6179 (1989).
46. P. Gross, D. Neuhauser and H. Rabitz, *J. Chem. Phys.*, **94**, 158 (1991).
47. M. Dahleh, A. Peirce and H. Rabitz, *Phys. Rev. A*, **42**, 1065 (1990).
48. J. G. B. Beumee and H. Rabitz, *J. Chem. Phys.*, **97**, 1353 (1992).
49. R. S. Judson and H. Rabitz, *Phys. Rev. Lett.*, **68**, 1500 (1992).

Index